Developments in Hydrobiology 1

DR. W. JUNK BV PUBLISHERS   THE HAGUE – BOSTON – LONDON 1980

Rotatoria

Rotatoria

Proceedings of the 2nd International Rotifer Symposium
held at Gent, September 17-21, 1979

Edited by
H. J. DUMONT and J. GREEN

DR. W. JUNK BV PUBLISHERS   THE HAGUE – BOSTON – LONDON 1980

Distributions:

for the United States and Canada

Kluwer Boston, Inc.
160 Old Derby Street
Hingham, MA 02043
USA

for all other countries

Kluwer Academic Publishers Group
Distribution Center
P.O. Box 322
3300 AH Dordrecht
The Netherlands

ISBN-13: 978-94-009-9211-5   e-ISBN-13: 978-94-009-9209-2
DOI: 10.1007/978-94-009-9209-2

Reprinted from *Hydrobiologia*, Volume 73, Nos. 1-3, 1980

CONTENTS

PREFACE

At the end of the first international Symposium on Rotifers in Lunz, Austria, September 1976, entousiastic pleas were made for a second gathering of possibly the same format and spirit that had made the first one such a great success. One of us (HJD), 'supported' herein by his friends Charles E. King and Jens Petter Nilssen in particular, tentatively suggested that Gent University might host a meeting of this kind, having in mind that since the XIth congress of SIL in 1950, no other SIL activity had take places in Belgium. In view of the relatively large number of Belgian rotifer workers, many among which are or were active in Gent, this proposal sounded acceptable to the attendance of the Lunz meeting, who then gave each other rendez vous in three years time.

In 1977, at the SIL conference in Kopenhagen, Agnes Ruttner-Kolisko had the second international rotifer conference officially endorsed by SIL, and from then on, made sure that the idea did not go dormant again, by reminding the organisers of their responsibilities in many a letter.

In 1978, at the ASLO special symposium on the ecology of zooplankton, held at Dartmouth College, New Hampshire, USA, all rotifer workers from the Lunz family present, and in addition some American collegues, were called together by John J. Gilbert. During this mini-meeting, the main topics of the present symposium were defined. Five reviews were invited, and three workshops were planned. Care was taken to avoid overlap with the topics already covered during the first symposium, except in those fields where major breakthroughs had occurred since. Importantly, it was decided to keep the number of attendants down to about 50, a number which–already at Lunz–had been recognized by most as the upper limit for good interaction. The dates were fixed at 17-21 sept. 1979.

Response from rotiferologists to the first circular was overwhelming. At one time, it looked as if the attendance was going to be of the order of 70, but various cancellations, due to a great variety of sad circumstances, finally reduced the number of registrants to 51, representing 16 different countries.

Grants-in-aid were obtained from the Belgian Ministry of National Education and from the National Fund for Scientific Research.

As for the first Symposium, it was suggested at an early date that the proceedings be published. Junk B.V. immediately accepted to do this, and kindly made an entire special volume of its Journal 'Hydrobiologia' available for this purpose. Also, arrangements were made to have the proceedings out as quickly as possible.

The proceedings include all papers that were presented in Gent, and two communications by collegues who could not attend in person. They thus amount to 23 original contributions, including the summaries of the workshops. All have been reviewed by the editors, and many, in addition by one or more referees. Manuscripts have been adjusted, wherever needed, for grammar, clarity of meaning, and length. However, where possible, the flavor of some personal styles was left unaltered.

A word needs to be said about the choice of the conference site. Rather than having the meetings take place in the University buildings, and disperse participants over various hotels in town, it was decided to take the symposium to the outskirts of Gent, and let it take place in the isolation of a former abbey, now transformed into a conference center. Isolation and increased interaction between participants were expressedly aimed at. We understand that the symposium's private 'pub' and all facilities it offered greatly contributed to this goal.

Isolation was ruptured on two occasions: a Mozart concert, in the framework of the festival of Flanders, and a visit to Lake Donk at Berlare, the cradle of limnological research in this country. It was the site of a former limnological institute, known for having produced the 'Annales de Biologie lacustre' between 1906 and 1926. The municipal council of the township of Berlare also hosted the Symposium Banquet, for which a most generous grant was made available, while transportation was provided by the University of Gent. The post-symposium excursion, in turn, was devoted to a visit of the ancient cities of Bruges and Gent.

Finally, it is a refreshing thought that among the participants 18 (i.e. exactly 50% of the total attendance at Lunz) were present again–and quite a few expressed regret for not being able to make it to Gent–while on the other hand more than 30 'new faces' turned up. The former is proof of that rotiferologists of the world are truly a 'scientific family'; the latter is a most reassuring indication of the growth and viability of the study of rotifers.

H. J. Dumont & J. Green

LIST OF PARTICIPANTS

Amsellem, J. Miss. Laboratoire Histologie, Université Lyon I, 43, Boulevard du 11 Nov., 69621 Villeurbanne, France.

Angeli, N., Dr. Université des Sciences et Techniques. Biologie animale 59655 Villeneuve d'Ascq Cedex France.

Clément, P., Dr. Laboratoire Histologie, Université Lyon I, 43 Boulevard du 11 Nov., 69621 Villeurbanne, France.

Coomans, A., Prof. Dr. Instituut voor Dierkunde, R.U. Gent, K. Ledeganckstr. 35, 9000 Gent, Belgium.

Cornillac, A., Miss. Laboratoire Histologie, Université Lyon I, 43, Boulevard du 11 Nov., 69621 Villeurbanne, France.

Coussement, M., Mr. Instituut voor Dierkunde, R.U. Gent, K. Ledeganckstr. 35, 9000 Gent, Belgium.

De Maeseneer, J., Prof. Dr. Faculteit der Landbouwwetenschappen, R.U. Gent, Coupure Links, 533, 9000 Gent, Belgium.

De Ridder, M., Mrs. Kloosterstraat 13, 1710 Dilbeek, Belgium.

Donner, J., Dr. A/2801 Katzelsdorf, Austria.

Dumont, H. J., Dr. Instituut voor Dierkunde, R.U. Gent, K. Ledeganckstr. 35, 9000 Gent, Belgium.

Epp, R. W., Dr. Hale Scientific Building, University of Colorado Boulder, Colorado 80309, U.S.A.

Fankhauser, H., Miss. Waldheimstrasse, 43, 3012 Bern, Switzerland.

Fiers, F., Mr. Instituut voor Dierkunde, R.U. Gent, K. Ledeganckstr. 35, 9000 Gent, Belgium.

Gilbert, J. J., Prof. Dr. Department of Biological Sciences, Dartmouth College, Hanover, New Hampshire 03755, U.S.A.

Gillard, A., Prof. Dr. Faculteit der Landbouwwetenschappen, R.U. Gent Coupure Links 533, 9000 Gent, Belgium.

Green, J., Prof. Dr. Zoology Dept., Westfield College, Hampstead London NW3 7ST, England.

Grundström, R., Dr. Institute of Limnology, Box 557, 75122 Uppsala, Sweden.

Herzig, A., Dr. Limnologisches Institut, Ost. Akademie d. Wisschenschaften Berggasse 18/19, 1090 Wien, Austria.

Hofmann, W., Dr. Max-Planck-Institut für Limnologie, Postfach 165, 2320 Plön, Germany.

Hussey, C., Dr. British Museum (Natural History), Cromwell Road London SW7 5BD Department of Zoology, England.

King, Ch. E., Prof. Dr. Department of Zoology, Oregon State University, Corvallis, OR 97331, U.S.A.

Koste, W., Mr. Ludwig-Brill-str., 5, 4570 Quakenbrück, Germany.

Kunicki-Goldfinger, W., Prof. Dr. Institute of Microbiology, Warsaw University, Nowy Swiat 67, 00-046 Warszawa, Poland.

Kutikova, L. A., Dr. Zoological Institute, Academy of Sciences of the USSR, Leningrad, B 164, USSR.

Lair, N., Dr. Université de Clermont-Ferrand II, Equipe d'Hydrobiologie regionale, 24 Av. des Landais, B.P. 45, 63170 Aubière, France.

Leentvaar, P., Mr. Rijksinstituut voor Natuurbeheer, Leersum, Kasteel Broekhuizen, Nederland.

Leimeroth, N., Mr. Fachbereich Biologie der J. W. Goethe-Universität Zoologie, AK Okologie, Siesmayerstrasse 70,6000 Frankfurt/Main, Germany.

Lubzens, E., Dr. Israel Oceanographic & Limnological Research Institute Tel-Shikmona, P.O. Box 8030, Haifa, Israel.

Martens, K., Mr. Instituut voor Dierkunde, R.U. Gent, K. Ledeganckstr. 35 9000 Gent, Belgium.

May, L., Dr. Institute of Terrestrial Ecology, Wetlands Research Group 12, Hope Terrace, Edinburg, EH9 2AS, Scotland.

Nilssen, J. P., Dr. Zoological Institute, University of Oslo, Blindern Oslo 3, Norway.

Nogrady, T., Prof. Dr. Chemistry Department, Concordia University, Loyola Campus 7141 Sherbrooke Street West, Montreal Quebec, Canada H4B 1R6.

Pejler, B., Dr. Limnological Institute, Box 557, S-751, 22 Uppsala, Sweden.

Pensaert , J., Miss. Instituut voor Dierkunde, R.U. Gent, K. Ledeganckstr. 35, 9000 Gent, Belgium.

Pizay, M.-D., Mrs. Université des Sciences et Techniques. Biologie animale. 59655 Villeneuve d'Ascq Cedex Lille, France.

Pourriot, R., Prof. Dr. E.N.S. Laboratoire de Zoologie, 46, rue d'Ulm 75230 Paris Cedex 05, France.

Preissler, K., Dr. Universität München. Fachbereich Biologie am Stadtpark 20, D. 8000 München 60, BRD.

Ricci, C., Dr. Universita di Milano, Istituto di Zoologia, Via Celoria, 10, Milano, Italy.

Rostan, J. C., Dr. Laboratoire de Biologie animale/zoologie, Universite Lyon I, 43 Boulevard du 11 Nov. 69621 Villeurbanne, France.

Ruttner-Kolisko, A., Prof. Dr. Biolog. Station Lunz, A-3293 Lunz a. See, Austria.

Schlüter, M., Dr. Abteilung für Algenforschung und Algentechnologie der Gesellschaft für Strahlen- und Umweltforschung mbH Munchen Bunsen-Kirchhoffstr. 13, 46 Dortmund/West-Germany.

Shiel, R. J., Dr. Zoology Department, University of Adelaide, Box 498 G.P.O. Adelaide, South Ausralia.

Sinkeldam, J., Mr. Rijksinstituut voor Natuurbeheer, Leersum, Kasteel Broekhuizen, Nederland.

Snell, T. W., Dr. Division of Science and Mathematics, University of Tampa, Tampa, Florida 33606, U.S.A.

Sorgeloos, P., Dr. Centrum voor Studie van Waterverontreiniging R.U. Gent, Rozier, 9000 Gent, Belgium.

Starkweather, P. L., Dr. Department of Biological Sciences, University of Nevada, Las Vegas, Nevada 89154, U.S.A.

Turner, P. N., Dr. 809 Woods RD[25], Newport News, Virginia 23602, U.S.A.

Wallace, R. L., Dr. Biology Department, Ripon College, Ripon WI 54791 U.S.A.

Walz, N., Dr. Zoologisches Institut der Universität München, Seidlstrasse 25, 8000 München 2, Germany.

Wattiez, C., Miss. Université de l'Etat à Mons, Fac. de Medecine, Service de Biologie generale et d'Ecologie, 7000 Mons, Belgium.

Wurdak, E., Dr. Department of Biological Sciences, Dartmouth College, Hanover, N.H. 03755, U.S.A.

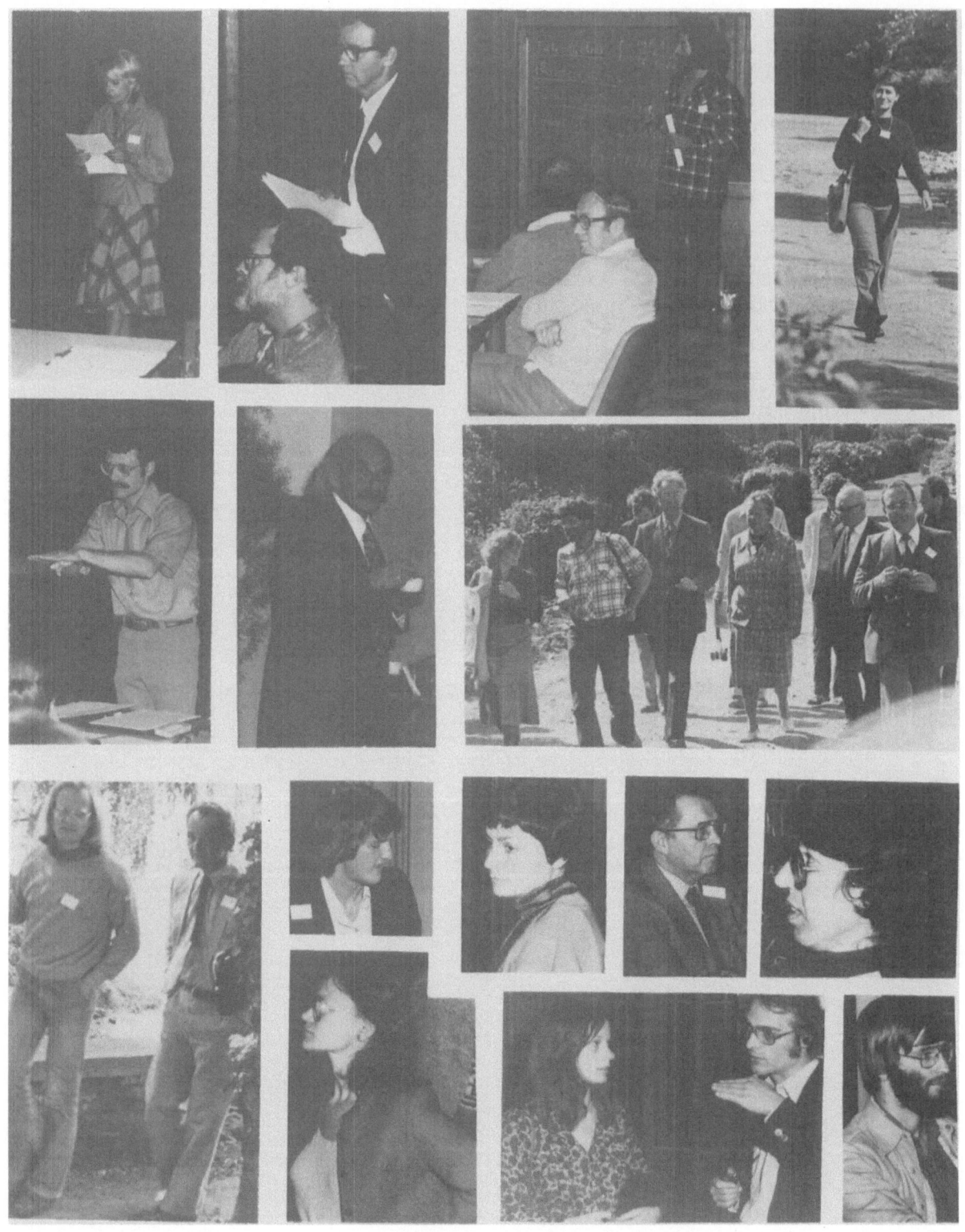

Top row: N. Lair; J. Green and C. E. King; P. Clément and T. Nogra07day: C. Ricci.
Second row: R. L. Wallace; W. Kunicki-Goldfinger; H. Fankhauser, M. Schlüter, W. Koste, M. De Ridder, J. Donner, K. Preissler and R. Gründstrom.
Third row: J. P. Nilssen and A. Herzig; K. Martens; C. Wattiez; P. Leentvaar; E. Lubzens.
Bottom row: E. Wurdak; J. Pensaert and F. Fiers; J. Sinkeldam.

Top row: T. W. Snell, L. May and R. W. Epp; R. Pourriot and A. Ruttner-Kolisko; L. A. Kutikova.
Second row: W. Koste and M. Coussement; general view of dining room; R. Shiel.
Third row: H. Dumont and W. Kunicki-Goldfinger; N. Angeli and M. C. Pizay; A. Cornillac and P. L. Starkweather.
Fourth row: J. C. Rostan, N. Leimeroth, A. Amsellem, M. Schlüter, W. Kunicki-Goldfinger and K. Preissler; N. Walz; P. Sorgeloos; W. Hofmann.
Bottom row: J. J. Gilbert; P. Turner and C. Hussey; B. Pejler.

A PERSPECTIVE ON AGING IN ROTIFERS*

Charles E. KING & M. R. MIRACLE

Department of Zoology, Oregon State University, Corvallis, Oregon 97331 USA
Department de Zoologia, Universidad de Valencia, Spain

* We gratefully acknowledge the aid and advice of William R. Rice in conduct of the regression analysis. This work was supported by grants from the National Science Foundation and National Institutes of Health to CEK.

Keywords: aging, senescence, rotifers, life histories, population dynamics

Abstract

Most research on aging in rotifers has been performed with populations, not with individuals. As a consequence, the dependent variable in these studies is usually either mean lifespan or rate of survivorship. After a brief consideration of the literature published since the last major review (King, 1969), the results of a series of experiments are presented. Males and females of three genetically distinct clones of *Brachionus plicatilis* were used for a factorial life table analysis at three different temperatures. The results of these experiments indicate several potential problems in using populations to study the aging process of individuals. These problems derive from the fact that lifespan is only one component of fitness, and its relative duration may not reflect the evolutionary success of the clone. That is, lifespan is free to vary in response to both stochastic and deterministic events without significantly reducing fitness. Under these conditions, neither mean lifespan nor pattern of survivorship will provide meaningful data on the determinants of individual senescence.

Introduction

Although aging occurs at the level of the individual, most studies of senescence in rotifers have been performed using populations. The rationale is a simple one. Laboratory populations are groups of individuals, each of which is born, matures, senesces, and dies. If the genetic composition of the population, or the physiological characteristics of the environment are changed, rates of maturation and senescence are also expected to change. Mean lifespan can, therefore, be used to characterize the aging process.

Our purpose is to suggest that this rationale may lead to inapplicable experiments and erroneous conclusions when applied to the aging process. The paper will be divided into two sections: a brief consideration of the major literature published since the last review of aging in rotifers (King, 1969), and a presentation of some experiments that bear on the occurrence of aging in rotifers. Throughout this paper the term 'aging' will be used synonomously with 'senescence'.

Part I. Review

Aging studies using rotifers date from the first quarter of the present century. Most of the early research (e.g. Finesinger, 1926) correlated lifespan and gross environmental parameters such as temperature and pH. In general these studies confounded the aging process with metabolic activity; the fact that rotifers tend to live longer at lower temperatures does not provide much insight to the physiological basis of the aging process.

Jennings & Lynch (1928) carried out the first substantial and controlled study of aging in this group. Using a single clone of the rotifer *Proales sordida,* the lifespan of offspring was shown to vary inversely with maternal age. As an individual female ages, an increased frequency of abortive (non-hatching) eggs is produced and the progeny of old females show more variability in both rate of development and fecundity. In addition, as an individual female ages her eggs become increasingly more variable in both size and shape. Jennings and Lynch's explanation for the maternal effect on survivorship and fecundity of offspring was that individuals developing from small

13

Hydrobiologia 73, 13-19 (1980).   *0018-8158/80/0731-0013$01.40.*
© *Dr. W. Junk b.v. Publishers, The Hague.*

eggs were physiologically stressed compared to those developing from large eggs. This effect, since it was observed within a single clone, was attributed to environmental rather than genetic factors. Today it is apparent that these differences might also be explained by age-related variation in gene activity, particularly in regulatory genes. This latter effect is also environmental, but the relevant environmental parameters are correlates of age.

A major contribution to the study of aging in rotifers was made by Lansing (1942a, 1942b, 1947, 1948, 1954) who established isogenic lines ('orthoclones') differing only in maternal age. Using this protocol, maternal age becomes an independent experimental variable. Lansing found that orthoclones established from young females (pediaclones) had expanding patterns of survivorship (mean lifespan increased in each generation), whereas the lifespan of old orthoclones (geriaclones) decreased until extinction occurred. This effect was therefore both transmissible and cumulative from one generation to the next. It was also reversible since individuals obtained from young females in a geriaclone lived longer than their parents. Lansing determined that the division between increasing and decreasing survivorship patterns occurred at the age of growth cessation. After that point he suggested, a cytoplasmic 'aging factor' starts to accumulate. On the basis of earlier microincineration studies, either intracellular or extracellular calcium was proposed as the aging factor.

Lansing's conclusions on maternal effects were confirmed and extended by King (1967) who also found that aging patterns could be modified by both food quantity and quality (different types of food). In addition, King conducted an extensive analysis of the relationship between age-specific fecundity and survivorship patterns. Several other papers relating to this analysis appeared between 1940 and 1968. These papers, and the entire topic of aging in rotifers were critically reviewed by King (1969) and will not be considered herein.

Since 1969, several promising leads have developed but, unfortunately, none appears to have been pursued. Meadow & Barrows (1971a, 1971b) analyzed relationships among different culture media, food types, and age-specific lactic and malic dehydrogenase activities in the bdelloid rotifer *Philodina acuticornis*. Although bdelloids also show the Lansing effect and are suitable for several types of aging studies, the complete lack of sexual reproduction in this group renders them inappropriate for most types of genetic manipulation.

Maternal age has recently been demonstrated by Rou-

gier & Pourriot (1977) to control (or correlate with) the production of sexual (mictic) females by amictic females. These workers established young and old orthoclones of *Brachionus calyciflorus* and assayed progeny type (mictic or amictic) in both daily renewed and unrenewed media. A high proportion (~95%) of the offspring of young females was found to be mictic, whereas most (~91%) offspring derived from old females were amictic. The fact that the same result was found in both renewed and unrenewed media for the young orthoclone suggests that the effect was not due to the accumulation of an inducing agent in the medium. However, when progeny from old orthoclone females were studied in unrenewed media, lifespan, fecundity, and proportion of mictic progeny all decreased. Thus the influence of maternal age on offspring is subject to modification by the external environment.

Both Lansing (1942b) and Sincock (1974, 1975) demonstrated that rotifers start to accumulate calcium when they complete growth. Rotifers reared on low calcium media were found to have longer lifespans than those reared on high calcium media. Uptake of calcium[45] was in direct proportion to the calcium content of the media. Further, mean lifespans were increased by brief treatments with a variety of ion chelating agents that were shown to reduce mean calcium content. Unfortunately, Sincock (1975) did not use orthoclones. Moreover, the absence of variance measures on either lifespan or calcium content make this result difficult to interpret.

Lifespan in rotifers can be altered in predictable directions by (a) change in temperature, (b) change in pH, (c) change in photoperiod, and (d) change in nutrition. All of these factors have marked effects on the physiological state of the individual, thus their influence on lifespan is not surprising. What is intriguing is the finding that a strong relationship exists between calcium metabolism and lifespan. Sincock (1975) demonstrated that treatment of rotifers with ion chelators removed bound calcium and lengthened lifespan. These results led him to conclude that calcium itself is the determinant of lifespan. Alternatively, we suggest that calcium accumulation may be either correlated with an unknown physiological factor and not causally related to the aging phenomenon, or act as a triggering device for initiation of the aging process.

The last investigation to be considered in this brief review is that of Snell & King (1977), who examined relationships between reproduction and lifespan. An analysis of the survivorship and fecundity schedules of 1,774 individuals of the rotifer *Asplanchna brightwelli* revealed that long-lived individuals spent the largest part of their lifespan in the reproductive phase, distributed reproduction evenly over many age classes, and reproduced at relatively low rates. In contrast, short-lived individuals spent the largest part of their lifespan in the prereproductive phase, concentrated reproduction in a few age classes, and reproduced at a high rate producing many offspring late in life. . Results of this study, therefore, are consistent with the view that reproduction is deleterious to future survival.

With the application to rotifers of electrophoretic analysis of isozymes, genetic studies of aging are now possible. These techniques, and the relevant background of rotifer genetics, are considered in King (1977, 1979) and King & Snell (1977). Several types of approaches should now be feasible in this group. For instance, rotifers would seem to be ideal organisms to use in testing the theory of senescence developed by Medawar (1957) and Williams (1957). The basic idea proposed in these papers is that senescence results from deleterious mutations that decrease postreproductive survival. Since selection cannot act directly on such genes, they will accumulate in the gene pool of the population by mutation pressure alone.

Actually, the effects of the deleterious genes need not be confined to the postreproductive phase of the life cycle. Arguments framed in terms of reproductive value in individuals suggest that mortality due to senescence may begin any time after the onset of reproduction.

The interesting idea is that selection of a pleiotropic gene that confers an advantage at one age and a disadvantage at another age depends on both the magnitudes of the effects and time of their action. It is expected, for example, that genes which increase reproduction in later age classes would be favored by natural selection as long as mean fitness of individuals carrying these genes is greater than that of individuals bearing alleles without such effects. Selection experiments using *Brachionus plicatilis* designed to test this theory are currently being conducted in King's laboratory.

Part II. Measures of Aging in Populations

A. Mean Lifespan Analysis

Mean lifespan is measured using a group of individuals, usually members of a cohort. Interpretation of the significance of this measure is dependent upon both the amount of variation in lifespan and the source(s) of mortality. In the experiments to be reported we examined the effects of temperature on lifespan. Our objectives were to determine whether males and females had the same patterns of survivorship, and also to determine whether mortality rates varied between temperatures in a similar way for genetically distinct clones of the same species.

Three clones of *Brachionus plicatilis* were used in the following study: 'LA' from LaJolla, California, USA; 'MC' from McKay Bay, near Tampa, Florida, USA; and 'SP' from Castellon, Spain. As indicated in Table 1, at both 20° and 30° the mean body length of clone SP is about 40% larger than LA and MC.

Experimental individuals were derived from cultures of each clone that had been maintained for several generations at each of the experimental temperatures, 20°, 25°, and 30°C. To further assure physiological adaptation to experimental conditions, for a week prior to initiation of the experiments both density and food level were maintained so that the rotifers were rapidly growing. Parents of the individuals used in the experiments were 3-5 days old. All work was performed with vitamin-supplemented Guillard-Ryther (1962) 'Ala' medium at pH 8.0 and a salinity of 26-28 parts per thousand.

Survivorship curves based on experiments performed at 25°C are presented in Fig. 1. Because male rotifers do not feed, rates of male survivorship were uniformly lower than those of females. In addition, both females and males of clone SP had higher survivorship rates than either clone LA or clone MC.

Table 1. Body length (± SE) of males and females belonging to three different clones of *B. plicatilis* at 20° and at 30°C. Each value is based on 50 individuals.

Clone		Temperature 20°	Temperature 30°
LA	females	194 ± 2.3	168 ± 1.3
	males	116 ± 0.7	105 ± 0.9
MC	females	199 ± 2.4	188 ± 1.3
	males	115 ± 1.0	100 ± 1.0
SP	females	278 ± 3.5	253 ± 2.5
	males	142 ± 1.6	133 ± 2.1

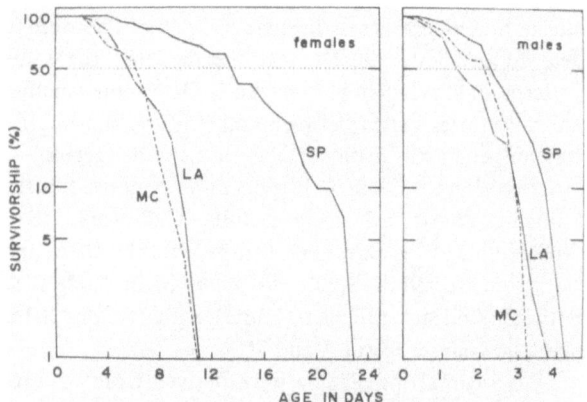

Fig. 1. Female and male survivorship at 25°C of three clones of *Brachionus plicatilis*. Female curves are based on a minimum of 30 individuals, male curves on a minimum of 40 individuals.

Variation among clones in mean lifespan was also observed at both 20° and 30°C, as shown in Fig. 2. Substantial genetic differences exist between all three clones as suggested by the body size data in Table 1, and as demonstrated by electrophoretic analysis of isozymes (to be presented elsewhere). These differences are not surprising when viewed in the context of the geographical variation represented by the three clones.

Of more interest are the patterns of variation in mean lifespan at the three temperatures. Rotifer lifespans are normally expected to be inversely related to environmental temperature (King, 1968). Although clone SP meets this expectation, clones MC and LA do not. In particular it should be noted that there is no significant

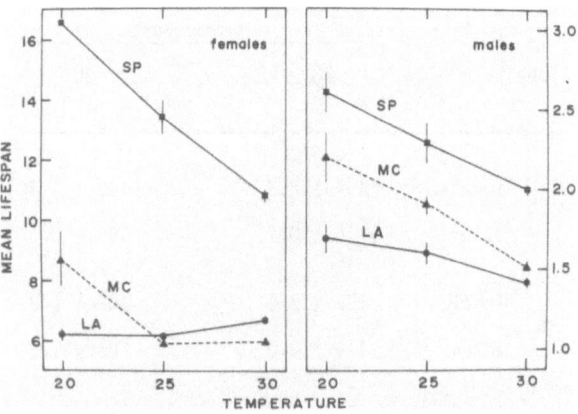

Fig. 2. Mean (± SE) lifespan at each of three temperatures for male and female *B. plicatilis* belonging to 3 different clones (standard errors for those points lacking vertical indicators are smaller than the symbols).

difference between mean female lifespans at 25° and 30°C for clone MC (by analysis of variance). The same conclusion holds at all three temperatures for clone LA.

Survivorship constitutes only one aspect of fitness in rotifers. To obtain a more complete view, fecundity parameters must also be considered. We have therefore used our life table data to calculate instantaneous growth rates (r) for each clone and temperature. These values, presented in Fig. 3, help to explain the different patterns of survivorship. In particular, note that clone SP has both the highest growth rate and longest mean lifespan at 20°. At 30°, clone SP still has the longest mean lifespan, but it has the lowest growth rate. These results clearly suggest that SP is a 'cool-water' clone, whereas LA and MC are 'warm-water' clones. Two general conclusions may be deduced from these results:

I. Lifespan Comparisons Among Different Populations in the Same Environment

Biologically meaningful results can seldom be obtained by directly comparing the quantitative performance of a series of clones (or species) in a single environment when these clones (or species) are adapted to a diversity of different environments.

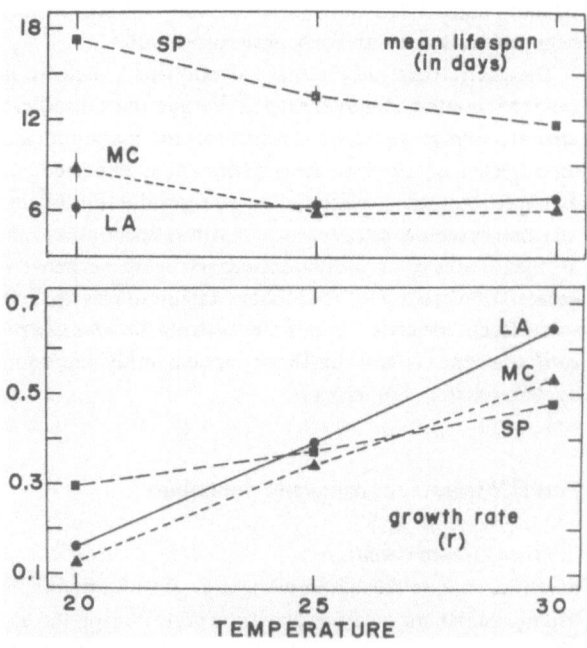

Fig. 3. Comparison between mean lifespans (top panel) and instantaneous growth rates (bottom panel) calculated from the amictic female life tables.

II. Lifespan Comparisons Made with the Same Population in Different Environments

Biologically meaningful results can seldom be obtained by directly comparing the quantitative performance of a single clone (or species) in a series of environments when the environmental spectrum is greater than the adaptive range of the clone (or species).

B. Survivorship Curve Analysis

To study the effects of aging using survivorship analysis, it seems obvious that mortality should be related to senescence. However, since we seldom know the cause of individual deaths when studying populations, a relationship between senescence and mortality is implicitly assumed. In this section we will consider the validity of this assumption.

Two hypothetical survivorship curves are graphed in Fig. 4. Curve A illustrates the 'negative-skew' rectangular pattern in which there is a high rate of survival through the young and middle age classes, and a high mortality rate in the old age classes. This high mortality rate is the result of age-specific factors associated in one way or another with senescence.

In contrast, curve B illustrates a pattern in which there is a constant death rate. If 'B' had been graphed on semilogarithmic axes, it would form a straight line. In curve B therefore, the probability of death is independent of age (i.e., not due to senescence). Because of this, it is not possible to use cohort survivorship curves of this form to study the process of aging.

Survivorship curves, even when taken under laboratory conditions, often tend to be somewhat irregular and intermediate between types 'A' and 'B'. Since 'A' can be used to study aging, but 'B' can not, it would be useful to have an objective measure of the form of survivorship. The simplest way to achieve this measure is to perform a nonlinear regression analysis of the form:

$$l_x = a_0 + a_1 x + a_2 x^2$$

In such an analysis, the sign and magnitude of the quadratic coefficient, a_2, yields the desired information. If a_2 is significant (by analysis of variance) and negative in sign, survivorship is of the general form of curve A in Fig. 4. If a_2 is significant and positive, survivorship is of type B. If a_2 does not significantly differ from 0, an intermediate curve between types A and B (similar to curve C in Fig. 4) is indicated. The characteristics of these patterns of survivorship are further considered in Table 2.

In summary, if a_2 is negative and significant, it is possible to use both mean lifespan and the entire survivorship curve to study aging. If a_2 is positive and significant, the probability of death in a given age class is independent of age and no relevant data can be obtained from the survivorship curve on the process of aging. Interpretation of

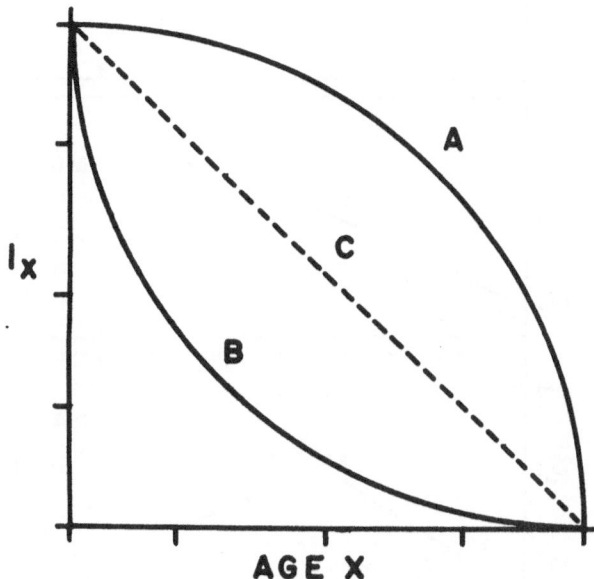

Fig. 4. Hypothetical survivorship curves: A, negative skew in which probability of death increases with age; B, positive skew in which the probability of death is independent of age; C, intermediate pattern in which the number of individuals dying per unit time is uniform. Both axes are on arithmetic scales.

Table 2. Comparison of survivorship in curves A and B of Fig. 4. Coefficient a_2 is from the quadratic term of the regression equation presented in the text.

	Curve	
	A	B
Survivorship Rate		
a. early	high	constant
b. late	low	constant
Lifespan variance	low	high
Sign of a_2	neg	pos
Prob of death	$= f(age)$	$\neq f(age)$

curves that have a quadratic term with a sign of zero must be made with caution.

A regression analysis has been performed on each of the 18 survivorship curves obtained in our study of *Brachionus plicatilis*. Signs of the curves were considered to be '0' unless they were significantly positive or negative by analysis of variance. Results of this analysis are given in Fig. 5. Note that both males and females of clones LA and MC have positive quadratic terms at 20°. Obviously, these conditions should not be used to study aging in the two clones.

Note also in Fig. 5, that LA males at 25° and LA females at 30° have positive values of a_2. For the females at 30°, mortality appears to be related to variation in fecundity patterns of the type noted by Snell & King (1977). The same explanation is obviously inadequate for males. Although these data do not permit an identification of mortality sources, it is important to remember that while all females of a given clone are genetically identical, the males are not identical. Males are haploid and each individual represents an independent sample of the maternal genome. If the parental clone has a substantial level of heterozygosity, variation from one male to another should also be quite high. In addition, the haploid constitution of the male is expected to reduce its level of genetic homeostasis.

Fig. 6 presents the sign of a_2 as a function of average daily reproductive effort (net reproduction/mean lifespan). Clone SP has both reproductive effort (net reproduction/mean lifespan). Clone SP has both negative a_2 and intermediate reproductive efforts in all three temperatures. As noted earlier, MC and LA are 'warm-water' clones. For this reason, physiological effects of the 20° environment produced a high variation in mean lifespan and a low mean reproductive effort. The result of this environmental stress is a positive a_2. At 25°, there is less thermal stress for MC and LA, the variance in lifespan decreases, and the a_2 values of both clones change to 0. Note in Fig. 6 that a_2 for clone MC remains at zero at 30°, whereas a_2 becomes positive for LA. Accompanying this change in sign of a_2 for LA is a 120% increase in R_0 (from 5.0 to 11.0). The same change in temperature produced only a 60% increase in R_0 (from 4.7 to 7.5) for MC, and a_2 remained at 0. We therefore suggest that the positive value of a_2 for LA at 30° reflects physiological stress induced by very high rates of egg production.

These data reveal substantial changes in survivorship

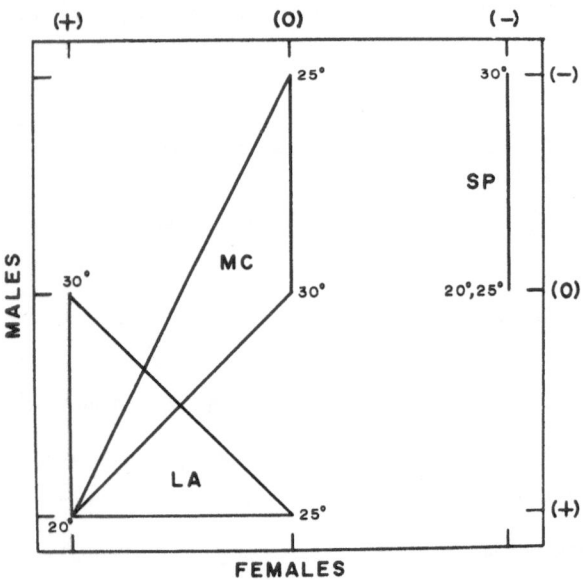

Fig. 5. Sign of the quadratic term, a_2, calculated from the male and female survivorship curves of each clone at the three temperatures.

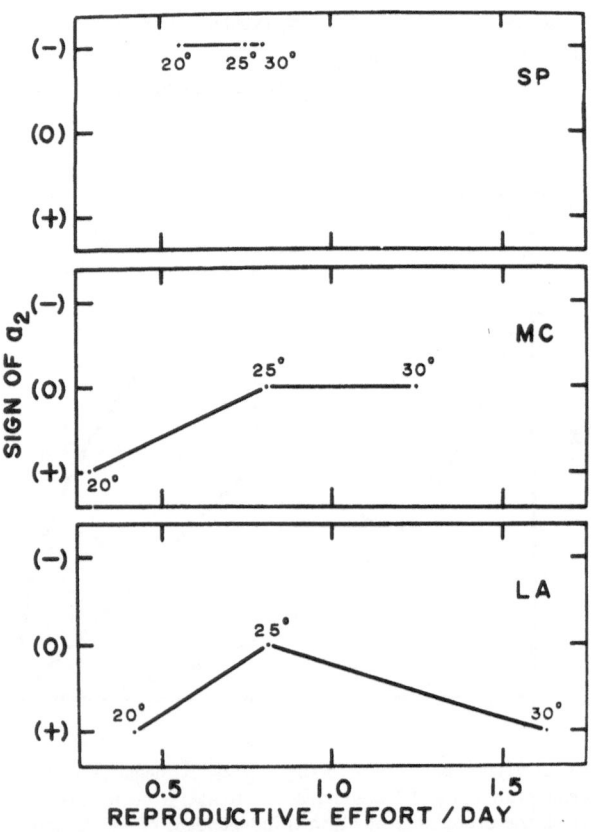

Fig. 6. Sign of a_2 as related to average reproductive effort per day of life (R_0/mean lifespan).

pattern from one clone to another and from one environment to another. In some of these combinations it is clearly inappropriate to use a life table survivorship analysis to study the aging process in rotifers. Clearly, even when one is dealing with a genetically homogeneous population in a presumed homogeneous environment, some factors remain uncontrolled. The most important of these is likely to be maternal environment during egg formation.

Discussion

We have shown that one of the three clones, SP, has substantially higher rates of survivorship and reproduction at 20° than the other two. At 30°, this clone again has the highest rate of survival, but its rate of population growth is lower than the other two clones. Thus although lifespan is one component of fitness, it does not always accurately reflect the composite fitness of the population. For this reason there can be substantial variation in lifespan (from either stochastic or deterministic sources) that does not affect the ability of a clone to exploit its environment. When these circumstances exist it is difficult to assert that mean lifespan or pattern of survivorship is a biologically meaningful dependent variable.

In a similar way, lifespan analyses in inappropriate environments are likely to be meaningless. The net result of such research is the same as that from experiments performed under uncontrolled or poorly defined conditions: lifespan variances increase and survivorship departs from the negative skew rectangular ('physiological') pattern expected in more appropriate environments.

Our purpose in this paper has not been to decry use of population analysis to study phenomena occuring in individuals. Instead we have attempted to identify and illustrate some of the hazzards involved with such studies. The most fundamental of these is that any complex process – including senescence – may have multiple determinants and different individuals in a population may be subjected to different stresses. In this respect it is important to recognize that while death is an event, dying is a process.

References

Finesinger, J. E. 1926. Effect of certain chemical and physical agents on fecundity and length of life and on their inheritance in a rotifer Lecane (Dystyla) inermis (Bryce). J. Expl. Zool. 44: 63-94.

Guillard, R. R. L. & Ryther, J. H. 1962. Studies on marine planktonic diatoms. I. Cyclotella nana Hustedt and Detona confervacae (Cleve) Gran. Can. J. Microbiol. 8: 229-239.

Jennings, H. S. & Lynch, R. S. 1928. Age, mortality, fertility and individual diversities in the rotifer Proales sordida Gosse. I. Effect of the age of the parent on characteristics of the offspring. J. Expl. Zool. 50: 345-407.

King, C. E. 1967. Food, age, and the dynamics of a laboratory population of rotifers. Ecology 48: 111-128.

King, C. E. 1969. Experimental studies of aging in rotifers. Expl. Gerontology 4: 63-79.

King, C. E. 1977. Genetics of reproduction, variation, and adaptation in rotifers. Arch. Hydrobiol. Ergebn. Limnol. 8: 187-201.

King, C. E. 1979. The genetic structure of zooplankton populations. In: Kerfoot, W. C. (ed.), The Structure of Zooplankton Communities. New England University Press. in press.

King, C. E. & Snell, T. W. 1977. Sexual recombination in rotifers. Heredity 39: 357-360.

Lansing, A. I. 1942a. Some effects of hydrogen ion concentration, total salt concentration, calcium and citrate on longevity and fecundity of the rotifer. J. Exp. Zool. 91: 195-211.

Lansing, A. I. 1942b. Increase of cortical calcium with age in the cells of a rotifer, Euchlanis dilatata, a planarian, Phagocata sp. and a toad, Bufo fowleri, as shown by the microincineration technique. Biol. Bull. 82: 392-400.

Lansing, A. I. 1947. A transmissible, cumulative and reversible factor in aging. J. Gerontol. 2: 228-239.

Lansing, A. I. 1948. Evidence for aging as a consequence of growth cessation. Proc. Nat. Acad. Sci. 34: 304-310.

Lansing, A. I. 1954. A nongenic factor in the longevity of rotifers. Ann. New York Acad. Sci. 57: 455-464.

Meadow, N. D. & Barrows, C. H. 1971a. Studies on aging in a bdelloid rotifer: I. The effect of various culture systems on longevity and fecundity. J. Exp. Zool. 176: 303-313.

Meadow, N. D. & Barrows, C. H. 1971b. Studies on aging in a bdelloid rotifer: II. The effect of various environmental conditions and maternal age on longevity and fecundity. J. Gerontology 26: 302-309.

Medawar, P. B. 1957. The uniqueness of the individual. London: Meuthen.

Rougier, C. & Pourriot, R. 1977. Aging and control of the reproduction in Brachionus calyciflorus (Pallas) (Rotatoria). Exp. Gerontology 12: 137-151.

Sincock, A. M. 1974. Calcium and aging in the rotifer Mytilina brevispina var redunca. J. Gerontology 29: 514-517.

Sincock, A. M. 1975. Life extension in the rotifer Mytilina brevispina var redunca by the application of chelating agents. J. Gerontology 31: 2-7.

Snell, T. W. & King, C. E. 1977. Lifespan and fecundity patterns in rotifers: the cost of reproduction. Evolution 31: 882-890.

Williams, G. C. 1957. Pleiotropy, natural selection, and the evolution of senescence. Evolution 11: 398-411.

EXPERIMENTAL OBSERVATIONS ON MATERNAL REPRODUCTIVE RATE AND OFFSPRING CHARACTERISTICS

C. RICCI

Istituto di Zoologia, Università degli Studi, via Celoria 10, Milano, Italy

Keywords: Rotifera, Bdelloidea, life table, fecundity, aging

Abstract

Starting from a clone of *Philodina roseola* (Rotifera, Bdelloidea) six homogeneous animals have been reared. From them three different age cohorts have been studied in order to determine their life tables. Cohorts and parental life schedules are compared and discussed. A sort of correlation between offspring fecundity and life span and maternal rate of egg production could be established.

Introduction

Since Lansing's papers (1947, 1954) numerous studies have been performed to point out maternal age effect on orthoclones of different animals (i.e. Wattiaux, 1968; King, 1969; Beguet, 1972; Rougier & Pourriot, 1977). Furthermore, maternal aging seems to affect both survivorship and fecundity of the first generation. Jennings & Lynch (1928), when working on individuals produced in three different stages of the reproductive period of *Proales sordida* (Rotifera, Monogononta), had noted a decreasing mean life span as well as a decreasing but more variable mean fecundity related to increasing parental age. Lansing (1947), using the same experimental design on *Philodina citrina* (Rotifera, Bdelloidea), pointed out a decreasing life span and fertility related to increasing maternal age, but from his data in series 1°, a small increase of F_1 life span and fertility when the central line (from 11 days old mothers) is considered, can be observed. Lastly Beguet & Brun (1972), considering young and old series from a group of parental hermaphrodite nematodes *(Caenorhabditis elegans)*, had observed a significant reduction in fecundity of old versus young lines.

In the present paper, attention is focused on the effect of parental rate of egg production to offspring characteristics, such as fecundity and longevity.

Material and methods

Philodina roseola Ehr. (Rotifera, Bdelloidea) clone A78 has been reared under laboratory conditions: $24 \pm 0.5°C$, daily renewed culture medium consisting of natural spring water with a low mineral content (Panna water) and a suspension of commercial fish food (details of the culture technique in Ricci, 1976). 15 eggs have been isolated randomly in glass cells and bred according to the life table experimental design, so that clonal characteristics have been tested. From this cohort six eggs laid by a female during its maximum reproductive activity period have been isolated and allowed to develop, and the hatched animals established the parental stock, originating three different series according to the procedure followed by Lansing (1947). Each series was composed of approx. thirty eggs, having been isolated when the parental stock was 3-4, 7 and 12-13 days old; the hatched animals were studied as is usual for life table experiments. The same experimental schedule has been followed using another clone of the same species, *P. roseola* N75. The founder animals of both strains were collected from the same environment, an activated sludge tank, in different years and in different months, April 1978 and November 1975. All tested eggs of both experiments were viable.

Results and discussion

Age specific fecundity values (m_x) fit on a bell-shaped

Hydrobiologia 73, 21-25 (1980).   *0018-8158/80/0731-0021$01.00.*
© *Dr. W. Junk b.v. Publishers, The Hague.*

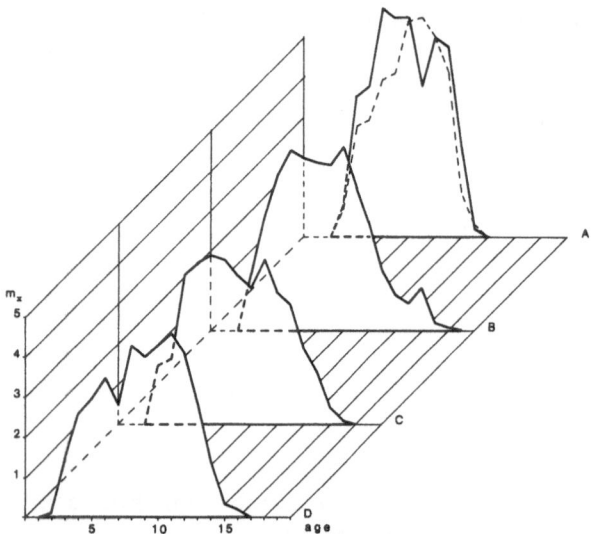

Fig 1. Egg curve of (A) clonal (dotted line) and parental (solid line) stock, of individuals hatching from eggs laid by young (B), mature (C) and old (D) parents. Age in days, $m_x = m_x l_x$ = age specific fecundity rate.

curve for both clonal and parental cohorts of P.r.A78, with no tail for aging (the tail constituted of low m_x values has been observed in similar experiments on the same organism (Ricci, 1978)). This can be ascribed, in part, to the absence of deaths in the reproductive period, so that in this experiment the m_x values are equal to the $m_x l_x$ values. This fact can bias the age effect on the oldest cohort which has been started when the mothers were still relatively prolific ($m_{12} = 3.25$). However (Fig. 1), a light shift of the reproductive activity toward a younger egg production can be observed when considering four to twelve days old cohorts. This shift could be the same as the initial trend of

the young and old orthoclones of Lansing (1954). On the contrary, despite the Author's observations, the oldest cohort doesn't seem to achieve the maximum rate of egg production earlier.

As it may be noted from Table 1, the differences in mean fecundity among the three aged cohorts seem not significant at all. The same seems to hold true for mean life span and for the survivorship curves, characterized by an almost rectangular shape, with no death occurring during reproductive period and never before the age of 18. Therefore no trend similar to Lansing's findings seems to arise from these data: this fact could be ascribed to the uncommon feature of the maternal egg curve; on the other hand, offspring fecundity curves exhibit the same shape.

However, particular attention must be given to the standard deviation values both for fecundity and longevity. Jennings & Lynch (1928) found an increase in variability of offspring fecundity related to parental aging; moreover, it seems that progeny of mature parents exhibits higher fecundity than progeny of juvenile or aging parents. In fact, when the offspring mean fecundity values are superimposed on the maternal egg curve and plotted against the parent age (Fig. 2), a very light trend of the offspring fertility can be foreseen. The animals hatched from eggs laid during the higher fertility period of the parents appear to have higher fertility, while the old cohort shows a wider range of variability.

The same can be foreseen when the offspring mean longevity is considered (Fig. 3): progeny of mature parents displays longer life span, and offspring of aging animals exhibits a light increase in variability. Even if slightly, the progeny characteristics, such as mean longevity and fecundity, seem to follow the trend of parental reproductive activity: the higher the fertility of parents, the better

Table 1. *P.r.*A78 cohorts values (mean ± standard deviation).

	IND. NUMBER	MEAN FECUNDITY	MEAN LONGEVITY	r
CLONE	15	35.75 ±2.83	24.75 ±3.84	0.548
PARENTAL	6	41.75 ±2.06	24.75 ±1.71	0.590
JUVENILE	30	39.74 ±3.72	26.48 ±2.59	0.586
MATURE	30	40.52 ±3.33	26.78 ±2.85	0.558
OLD	31	39.96 ±4.00	26.54 ±3.24	0.573

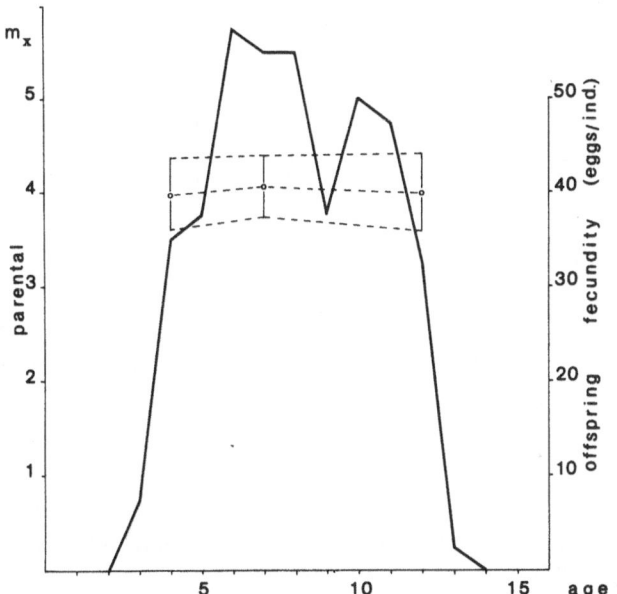

Fig. 2. Different age offspring mean fecundity values superimposed on maternal egg curve and plotted against mothers age. The bars represent standard deviation.

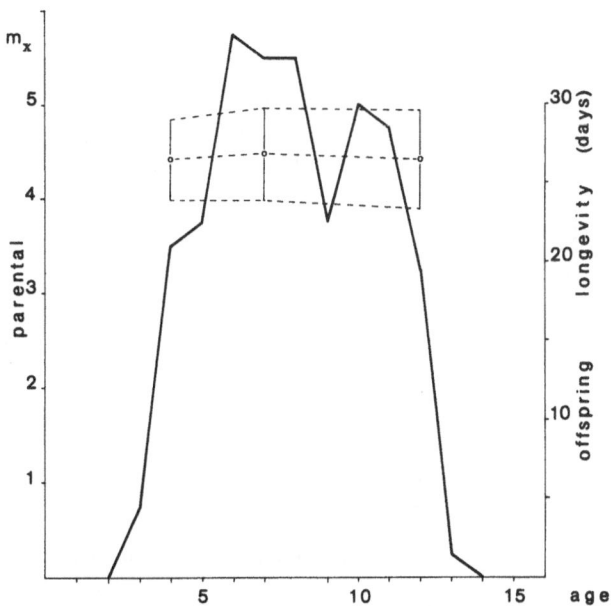

Fig. 3. Different age offspring mean life span values superimposed on maternal egg curve and plotted against mothers age. The bars represent standard deviation.

the ability of the offspring to survive and to reproduce. The term ability and not fitness of offspring has been used because the intrinsic rate of natural increase for the mature cohort (Table 1), calculated through Lotka's formula, is somewhat lower than, but generally very close to, the r values of young and old cohorts. The life histories of the three lines are very similar and such an high r value is more affected by developmental time decrease than by the net fecundity increase (Snell, 1978): small variations in developmental time according to maternal age should be subject of further investigation.

Till now, no difference in egg shape and dimensions related to parental aging has been found by the author,

while Jennings & Lunch (1928) have ascertained a direct relation between variation in egg size and maternal age.

To test the hypothesis of a correlation between offspring characteristics and maternal age, a second experiment, using a different clone of *Philodina roseola* (Pr N75), has been carried out under the same laboratory conditions and by the same experimental procedure. The results are summarized in Table 2. In this case the parental egg curve presents the tail when mothers become older, so that the aging effect is more evident. The tail is mainly due to decreasing m_x values.

The central cohort started from 7 days old mothers, and shows the highest fecundity, while the old cohort, from 13

Table 2. *P.r.*N75 cohorts values (mean ± standard deviation).

	IND. NUMBER	MEAN FECUNDITY	MEAN LONGEVITY	r
PARENTAL	6	43.67 ±5.89	18.83 ±2.04	0.677
JUVENILE	18	36.44 ±4.68	19.33 ±4.38	0.709
MATURE	19	41.84 ±3.98	21.47 ±4.68	0.791
OLD	19	31.25 ±10.84	20.50 ±7.12	0.680

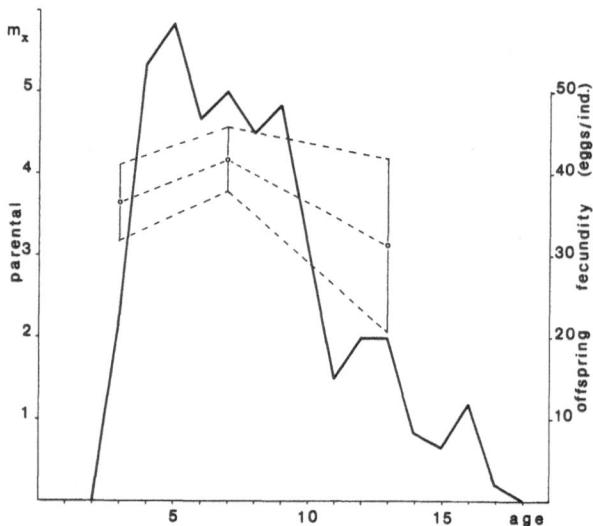

Fig. 4. The same as in Fig. 2 for *Philodina roseola* N75 clone experiment.

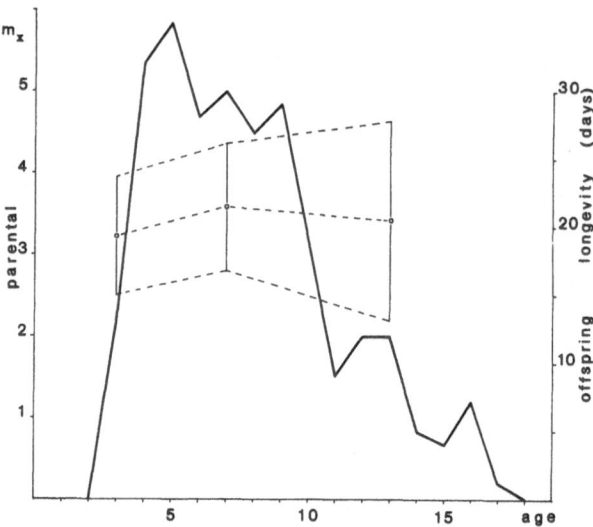

Fig. 5. The same as in Fig. 3 for *Philodina roseola* N75 clone experiment.

days old mothers, displays a wider range of variability for egg production (Fig. 4). Mean life spans exhibit the same pattern (Fig. 5). The longevity and fecundity behaviour for the three cohorts seems to support the hypotesis af a correlation between offspring characteristics and parental egg curve. Yet, offspring fitness, as measured by intrinsic rate of natural increase, seems to follow the maternal rate of egg production: the best fitness is reached by animals hatched from eggs laid by mothers in their most fertile age. This would mean that higher fertility period eggs wil give birth to fitter animals.

On the other hand, old age reproduction seems to increase the variability of offspring life span and net reproduction. Surely, the cause of this fact is a nongenic factor, as already pointed out by several papers dealing with maternal age effect. Old age reproduction provides, however, a variable progeny, which again will affect its own offspring as the present data suggest.

Such a variability could be very important to an animal that is characterized by a parthenogenetic and ameiotic reproduction. It is a possible answer to the question posed by Laughlin (1965) about the possible advantages of older age reproduction.

In conclusion, while younger age reproduction is the most effective way to future growth of a population, as stated by most Authors, the whole of the recorded data seems to suggest that middle age reproduction maximizes offspring fitness and that senile reproduction could be a source of variability for subsequent generations.

Summary

The aim of the present paper is to point out the effect of parental rate of egg production on offspring fecundity and life span.

Following the life table experimental design three different lines from young, mature and aging individuals have been tested in order to determine their mean life spans and fecundities. Comparing the offspring cohorts to the parental rate of egg production, a relationship can be ascertained. In particular, higher fecundity and longevity occur when the cohort has hatched from eggs laid by mature, actively reproductive parents, while a doubtfully lower, but certainly more variable fecundity and longevity are typical of cohorts originated from aging parents.

References

Beguet, B. 1972. The persistence of processes regulating the level of reproduction in the hermaphrodite nematode Caenorhabditis elegans, despite the influence of parental ageing, over several consecutive generations. Expl. Geront. 7: 207-218.

Beguet, B. & Brun, J. L. 1972. Influence of parental aging on the reproduction of the F_1 generation in a hermaphrodite nematode Caenorhabditis elegans. Expl. Geront. 7: 195-206.

Jennings, H. C. & Lynch, R. S. 1928. Age, mortality, fertility, and individual diversities on the Rotifer Proales sordida Gosse. I. Effect of age of the parent on characteristics of the offspring. J. Expl. Zool. 50: 345-407.

King, C. E. 1969. Review article. Experimental studies of ageing in Rotifers. Expl. Geront. 4: 63-79.

Lansing, A. I. 1947. A trasmissible, cumulative, and reversible factor in ageing. J. Geront, 2: 228-239.

Lansing, A. I. 1954. A nongenic factor in the longevity of Rotifers. Ann. N. Y. Acad. Sci. 57: 455-464.

Laughlin, R. 1965. Capacity for increase: a useful population statistic. J. anim. Ecol. 34: 77-91.

Ricci, C. 1976. Nota preliminare sull'allevamento di un Rotifero Bdelloideo. Atti Soc. ital. Sci. nat. Museo Civ. Stor. nat. Milano. 117: 144-148.

Ricci, C. 1978. Some aspects of the biology of Philodina roseola (Rotifera). Mem. Ist. Ital. Idrobiol., 36: 109-116.

Rougier, Cl. & Pourriot, R. 1977. Aging and control of the reproduction in Brachionus calyciflorus (Pallas) (Rotatoria). Expl. Geront. 12: 137-151.

Snell, T. W. 1978. Fecundity, developmental time, and population growth rate. Oecologia, 32: 119-125.

Wattiaux, J. M. 1968. Cumulative parental age effects in Drosophila subobscura. Evolution, 22: 406-421.

ABOUT A TRANSMISSIBLE INFLUENCE THROUGH SEVERAL GENERATIONS IN A CLONE OF THE ROTIFER NOTOMMATA COPEUS EHR.

P. CLÉMENT & R. POURRIOT

University Lyon I, Laboratoire d'Histologie, 69622 Villeurbanne, France
and E.N.S. 46 rue d'Ulm, Laboratoire de Zoologie, 75230 Paris, France

Keywords: cytoplasmic inheritance, reproduction cycle, crowding, rotifer, *Notommata copeus*

Abstract

Over ten years, the individuals of an orthoclone of *Notommata copeus* gradually lost their sensitivity to photoperiod. This loss is transmissible but reversible in three generations. It is endogenous but not chromosomal. The hypothesis is forwarded that it is induced by external factors, and quite possibly crowding.

Introduction

In *Notommata copeus,* mixis is induced by long photophases (Pourriot, 1963). Nevertheless, in inducing light-conditions, other factors influence the rate of mixis, in particular a high population density (n/v) causes an unfavourable mass-effect (Clément & Pourriot, 1975). In total darkness (LD = 0 : 24) or in short photophases (L ≤ 12h), the females reproduce parthenogenetically. Two clones are maintained in such conditions in our laboratory: the S69 since 1969 and the S74 since 1974.

Recently, an influence of the age of the grandmothers on the mixis rate in their F_2 has been showed in the S74 clone (Clément & Pourriot, 1979).

Similar experiments has been realized on the S69 clone (experimental schedule in Fig. 1; constant experimental conditions: temperature = 23°C, LD = 18 : 6; *Spirogyra* 918 as food, population density = 1 ♀ / 10 ml). A similar rhythmic influence has been observed in the F_2 progeny of three grandmothers, while the F_2 progeny of the two other grandmothers is almost entirely amictic (Fig. 2). Most of the parental females born of these two grandmothers do not react more to the photoperiod.

These results raise two questions: 1- is the almost total disappearance of the photoperiod sensitivity irreversible? 2- what is the origin of its disappearance? The results obtained on a clone born of one of these grandmothers, D, are presented and the possible mechanisms involved will be discussed.

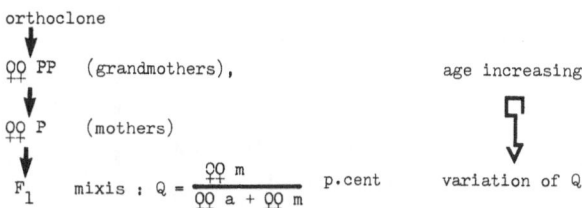

Fig. 1. Experimental schedule for the study of the influence of age in grand-mothers.

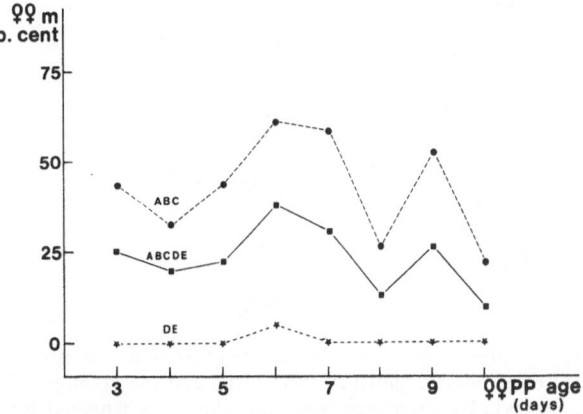

Fig. 2. Variations of the mictic rate in relation to the age of five grand-mothers (A to E).

Hydrobiologia 73, 27-31 (1980).   0018-8158/80/0731-0027$01.00.
© Dr. W. Junk b.v. Publishers, The Hague.

Material and methods

A. Mass-culture conditions

The females of *N. copeus* are a subclone 'De' of the S69 clone. They are born of descendants of the female D (Fig. 2). This sub-clone is cultivated in darkness or in short photophases (LD = 8 : 16), at 18-20°C. Regularly, about 15 young females, corresponding to about the 10th offspring of each parental female, are isolated in a fresh medium to ensure the maintainance of an orthoclone. The population density of the culture varies, thus, from about 15 to about 150 females in 20 ml of medium (spring water from Volvic).

The rotifers are fed with two strains of *Spirogyra* : *Sp. varians* from the culture collection of Cambridge (G.B.) and *Sp.* 918 from the culture collection of Indiana (U.S.A.).

One or two generations before the beginning of the experiments, the orthoclone is placed at 23°C.

B. Experimental conditions (Figure 3)

The experiments are performed on some selected offspring in three successive generations.

1° First generation. The young parental females (9 ♀♀P a to i, Tabl. 1) are isolated in 10 ml of fresh medium, unrenewed during the experiment (about 12 days). They are placed in constant and precise experimental conditions: t = 23°C, LD = 16 : 8 (fluorescent light, 'day-light' tube, E = 200 μw cm^{-2}), *Spirogyra* 918 as food.

These females become mature in two days and then begin to lay eggs. The eggs hatch two days later: the young females F_1 are isolated until their maturity in order to know their type, mictic or amictic, according to the egg size. The percentage of mictic females in the F_1 is thus established every day; and then, at the end of the experiment, the total percentage of mictic F_1, Q, for each female is known. This last percentage, Q, will be the only to be considered in the present work.

2° Second generation. In the F_1 generation of the previous experiment, the amictic females born on the third day of egg laying of the mothers (about the age of the orthoclone) are isolated in 10 ml of fresh medium and placed in the same experimental conditions.

Their offspring are collected every day in order to know the mixis rate as before.

3° Third generation. The same experimental procedure is followed as in the second generation.

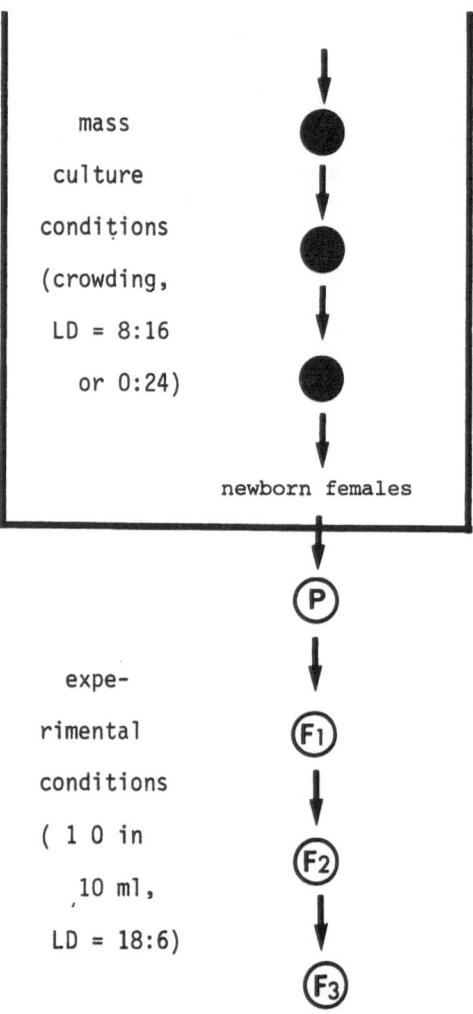

mass
culture
conditions
(crowding,
LD = 8:16
or 0:24)

newborn females

(P)

experimental
conditions
(1 0 in
10 ml,
LD = 18:6)

(F₁)

(F₂)

(F₃)

Fig. 3. Succession of the experimented generations: the mothers (P) of F1 are the grand-mothers of F2 and the grand-grand-mothers of F3. The nearest ascendants which were in mass culture conditions are the grand-mothers for F1, the grand-grand-mothers for F2, ...

Results

The analysis of the mean percentages of mictic females produced by the parental females in each of the three generations (Table 1), shows that in entirely identical experimental conditions, the photoperiod reaction of the parental females varies: the mictic rate increases regularly from the first (3.8%) to the second (28.5%) and to the third generation (42.3%).

Fig. 4 shows the dispersion of the obtained Q percentages and their evolution during the three generations.

Table 1. Percentages of mictic females produced by each mother isolated in 10 ml, along three generations.

Below each column of percentages are the ratios $\dfrac{♀♀\,m}{♀♀\,m + ♀♀\,a}$ obtained per generation in the progeny of one initial female.

♀P	F_1	F_2	F_3
a	19.2	33.3	
		67.9	
	5/26	26/49	
b	8.0	56.3	60.7
		67.7	59.1
		45.5	38.1
		44.0	72.7
			50.0
	2/25	55/99	56/102
c	6.7	11.1	
		18.8	
		16.7	47.6
		45.0	39.1
		33.3	
	2/30	30/115	19/46
d	0		
	0/32		
e	0	45.2	
		42.9	
		30.4	
		21.7	
		36.4	
	0/27	50/138	
f	0	0	17.4
		8.3	71.4
		0	64.3
		0	68.0
	0/26	3/120	39/77
g	0		
	0/28		
h	0	0	15.4
			32.1
			26.1
			7.7
		0	42.3
			30.4
			45.5
	0/24	0/54	64/196
i	0		
	0/26		
TOTAL	9/240 = 3.8%	164/575 = 28.5%	178/421 = 42.3%

29

Thus, a completely amictic response (Q = 0; Fig. 4, column 1) prevails in the first generation, but is not observed in the third. In contrast, the medium or high mictic percentages (Q ⩾ 40%) do not appear in the first generation but prevailed in the third.

Three different cases can be observed in the fluctuations of the mictic reactions in the F_1, F_2, F_3 progeny of each parental females (a, b, c, e, f, h): 1- a similar evolution as in the total results, i.e. Q increases regularly in each generation (female c); 2- a fast increase from the first to the second generation and then a stability of the responses from the second to the third (female b); and 3- a steadiness of the mictic reactions from the first to the second generation as well as an increase in the third (female h).

Discussion and conclusion

The loss of sensitivity to photoperiod in *Notommata copeus* appears to be reversible and non-chromosomal. The sensitivity to photoperiod reappeared in the second or the third generation when the females were isolated in 10 ml of medium. Therefore, the disappearance of this sensitivity cannot have a genetic origin. This could be expected from the fact that in parthenogenetical reproduction, all the females of the same clone are genetically identical.

The phenomenon of loss of sensitivity to the period is endogenous since females tested in the three generations were placed in experimental conditions as uniform as possible where the exogenous factors which influence the mixis rate in *N. copeus* (see review in Clement *et al.*, 1976) were constant. The observed variations must, thus, have an endogenous origin.

The disappearance of the mictic reaction is very likely induced by exogenous factors in the ancestry and the appearance of females temporarily less sensitive or unsensitive to the photoperiod is determined by the culture conditions of the clone preceding the experiment (Table 1).

Without excluding a possible effect of the light-conditions, we think that the crowding is the most important factor. A mass effect on the mixis rate related to high population densities, has been previously shown in *N. copeus* (Clement & Pourriot, 1973; 1975).

This crowding effect is probably due to the production of substance(s) ('DFm') released by the females in the medium. In crowding conditions, the DFm substance(s) would cause the synthesis of an unknown substance ('X') by the vitellarium which would be transfered into the

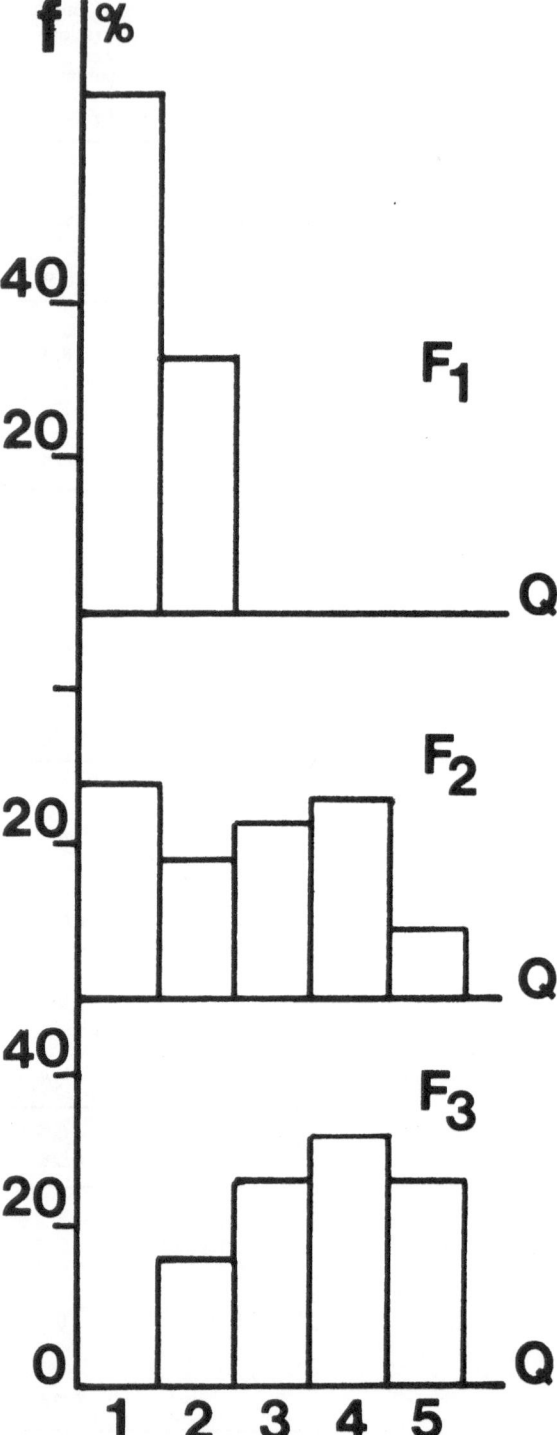

Fig. 4. Frequencies (f) of the mictic responses (Q, cf Table 1) per generation.
Column 1: Q = 0%
Column 2: Q < 20%
Column 3: Q = 20 to 40%
Column 4: Q = 40 to 60%
Column 5: Q > 60%.

cytoplasm of the egg. During embryogenesis, this substance will be distributed in different cells of the rotifer, in particular in its ovocytes.

When the crowding conditions are maintained, each ovocyte would receive more X-substance: its influence would increase.

If the rotifers are placed in low population density conditions, the ovocytes would only contain the part of the X-substance inherited from their mother. During embryogenesis, this substance would be then distributed among all the ovocytes of the future female and a such dilution effect would explain the gradual disappearance in a few generations of the influence of the X-substance.

In conclusion, this work gives a new illustration of a phenomenon of non chromosomal heredity involving a transmission along several generations of probably cytoplasmic substances and showing a cumulative and reversible effect.

The phenomenon described here is close to the Lansing effect (Lansing, 1947; King, 1968); but whereas the latter is a consequence of endogenous physiological modifications, the former is initially determined by exogenous factors. A similar exogenous origin has been assumed by Rougier & Pourriot (1977) for the influence of the age of the grandmothers on both the net reproduction rate and the mixis rate in *Brachionus calyciflorus*.

Besides these influences related to aging, cytoplasmic transmission has been observed by Birky (1964) for the pigments of the cerebral eye cupule of *Asplanchna* and by Birky & Gilbert (1972) for the transmission of labelled α-tocopherol in the same rotifer. In other content, Ruttner-Kolisko (1969) has also suggested the preponderance of maternal effects in the crossing over of two *Brachionus* species.

Rotifers, which by their parthenogenetic reproduction can form clones, are an appropriate material to understand the importance of hereditary non chromosomal phenomena which possibly exist in other animals but are difficult to detect due to the extreme variability of the genome.

Summary

Experiments were made on an orthoclone of the rotifer *Notommata copeus* by placing the experimental females in rigourously identical experimental conditions during three successive generations. The observed variations show the gradual disappearance of influences which affected the generations previous to the experiment (loss of sensitivity to photoperiod). It is postulated that a such transmissible and reversible influence on several generations, endogenous but not chromosomal, is initially induced by external factors, likely crowding.

Acknowledgements

We wish to thank Dr. Gerschenfeld for his help to translate the manuscript in english.

References

Birky, C. W. 1964. Studies on the physiology and genetics of the rotifer Asplanchna. I. Methods and physiology. J. expl. Zool. 155: 273-288.

Birky, C. W. & Gilbert, J. J. 1972. Vitamin E as an extrinsic and intrinsic signal controlling development in the rotifer Asplanchna: uptake, transmission and localization of [^{3}H]α-tocopherol. J. Embryol. expl. Morphol. 27: 103-120.

Clément, P. & Pourriot, R. 1973. Mise en évidence d'un effet de masse et d'un effet de groupe dans l'apparition de phases de reproduction sexuée chez Notommata copeus (Rotifère). C.r. Acad. Sci. sér. D, 277, 22: 2533-2536.

Clément, P. & Pourriot, R. 1975. Influence du groupement et de la densité de population sur le cycle de reproduction de Notommata copeus Ehr. (Rotifère). I Mise en évidence et essai d'interprétation. Arch. Zool. expl. gén. 116: 375-421.

Clément, P. & Pourriot, R. 1979. Influence de l'âge des grandparents sur l'apparition des mâles chez le Rotifère Notommata copeus Ehr. Int. J. Invert. Reprod. 1: 89-98.

Clément, P., Rougier, C. & Pourriot, R. 1976. Les facteurs exogènes et endogènes qui contrôlent l'apparition des mâles chez les Rotifères. Bull. Soc. Zool. Fr. 101, Suppl. 4: 86-95.

King, C. E. 1969. Experimental studies of ageing in Rotifers. Expl. Geront. 4: 63-79.

Lansing, A. I. 1947. A transmissible, cumulative and reversible factor in aging. J. Geront. 2: 228-239.

Pourriot, R. 1963. Influence du rythme nycthéméral sur le cycle sexuel de quelques Rotifères. C.r. Acad. Sci. Paris, sér. D, 256: 5216-5219.

Rougier, C. & Pourriot, R. 1977. Aging and control of reproduction in Brachionus calyciflorus (Pallas) (Rotifère). Expl. Geront. 12: 137-151.

Ruttner-Kolisko, A. 1969. Kreuzungsexperimente zwischen Brachionus urceolaris und B. quadridentatus, ein Beitrag zur Fortpflanzungsbiologie der heterogenen Rotatoria. Arch. Hydrobiol. 65: 397-412.

WORKSHOP ON CULTURE TECHNIQUES OF ROTIFERS

R. POURRIOT

Ecole Normale Supérieure, 75230 Paris Cedex 05, France

Keywords: Rotifers culture, clone

Culture techniques of Rotifers can be considered of two types according to whether the number of associated species is known or not (Dougherty *et al.,* 1960). The cases of monoxenic (one associated species), xenic (unknown number of associated species) and clonal cultures will be successively examined in the following.

Monoxenic cultures

A review of corresponding techniques will be found in Gilbert (1970) and details of procedures in the latter as well in Aloia & Moretti (1973). To carry out monoxenic cultures, the following requirements are necessary:

1. An axenic food-source, either algal or bacterial.

Among the most common nutritive species, our choise will be restricted to species that can be cultivated and obtained in axenic cultures.

2. Axenic Rotifer embryos to initiate the cultures.

They can easily be obtained by washing and setting to hatch in a sterile medium either old embryos from amictic females (Gilbert, 1970) or resting eggs treated in a 0.1 or 1‰ solution of hypochlorite (Nathan & Laderman, 1959; Pourriot, 1965).

Handling and culture maintenance techniques are similar to those usually used in microbiology such as autoclaved media, sterilized tools (micropipettes), manipulation under a sterile hood equipped with glove-boxes and germicidal lamps (U.V.). These constraints on monoxenic culture techniques will of course restrict the possibilities of experimentation.

Initiation and maintenance of monoxenic cultures of phytophagous or bacteriophagous Rotifer species does not present difficulties when an axenic food source is available.

Permanent monoxenic cultures in non-defined or organic medium have been reported by Bazire (1953), rearing *Epiphanes senta* on *Euglena gracilis* or *Polytoma caeca,* and by Dougherty *et al.* (1960) rearing Bdelloids. *Lecane inermis* has also been obtained in axenic cultures with dead material as food (Dougherty, 1963).

In some cases, the monoxenic culture can even be maintained on a defined inorganic medium[1]. We did so over the year for two species *Brachionus urceolaris* feeding on *Chlorella pyrenoïdosa* (Pourriot, 1965) and *Epiphanes brachionus* feeding on *Haematococus pluvialis* (Pourriot, unpublished). The algae were cultivated in a synthetic inorganic medium (KNO$_3$: 100 mg; Ca(NO$_3$)$_2$, 4H$_2$O: 100 mg; K$_2$HPO$_4$: 40 mg; MgSO$_4$.7H$_2$O: 30 mg; oligoelements: 1 ml; bidistilled water: 1000 ml) and the Rotifers in a dilution of the same.

This shows that the nutritive requirements of the two species are entirely satisfied without addition of any other dissolved substance.

A strain of *Brachionus calyciflorus* has been maintened with *Euglena gracilis* as food by Gilbert (1970) using an inorganic medium with a citrate chelating agent and vitamins B$_1$ and B$_{12}$ added.

Maintenance of predatory species, i.e. of a 3 levels foodchain (as bacteria-*Paramecium-Asplanchna* or alga-*Brachionus-Asplanchna*) can be achieved as well, following the same aseptic procedures. But as the number of manipulations increases, risks of contamination are higher (Aloia & Moretti, 1973).

Xenic cultures

Initiation and maintenance of xenic Rotifer cultures are simpler. However, microbiological techniques are still to be used, if possible, to restrict contamination of algal or Rotifer cultures. For instance, contamination by fungi or Protozoa may kill the culture or prevent its development.

The equipement needed for such cultures is easily available: Pasteur pipettes for handling the algae, microdroppers for handling the Rotifers under a microscope; cotton-plugged test tubes or petri dishes for Rotifer cul-

---

[1] For a review of culture media for Rotifers, see King & Snell, 1978.

Hydrobiologia 73, 33-35 (1980). 0018-8158/80/0731-0033$00.60.
© *Dr. W. Junk b.v. Publishers, The Hague.*

tures. Depression slides (used in tissue-cultivation with 3 depressions (1 cm in diameter by 0.5 in depth) proved to be very useful to follow isolated females.

The most important requirement to succeed in Rotifer cultures is to provide the animals with a suitable food source. Most failures results from

– problems in the food-source maintenance (as Chrysophyceans or Dinoflagellates cultures)

– technical difficulties in the realization of a proper detritus flow of controlled origin

– our ignorance of their exact diet.

Difficulties to maintain easily permanent cultures of species as common as *Asplanchna priodonta* or *Keratella cochlearis* is due to a misappreciation of their food requirements since some limited success (Pourriot, 1965; Pourriot & Hillbricht-Ilkowska, 1969; Pejler, 1977) showed no methodological obstacle.

Sometimes, the food sources used in a Rotifer culture appear to be surprising, such as a strain of *Ephiphanes senta* fed on *Paramecium* (Kreiskott, 1958), while they had been cultured for a long time on various flagellates which they obviously prefer (De Beauchamp, 1928; Pourriot, 1965, p. 125, 171), or *Dicranophorus forcipatus* known as a carnivore preying on other Rotifers (Pourriot, 1965), which has been fed on bacteria (Erben, 1978)!

Clonal culture

In an experimental framework, it is of course advisable to maintain clonal cultures, where the parthenogenetic offspring keep their parents characters. The clone is initiated by an amictic female, isolated from a natural population or by a resting egg. Maintenance of parthenogenetic offspring only, might cause problems when sexual and parthenogenetic reproduction are simultaneous (such as in *Brachionus*). When resting eggs do not spontaneously hatch under culture conditions, clone isolation is safe. On the contrary, one will have to separate as early as they appear, amictic females from resting eggs in the cultures. This will require regular and sometimes frequent time-consuming observations.

When there is no sexual reproduction (gamogenesis) or when it appears only under well-controlled conditions (as photoperiod in *Notommata copeus*) there is no problem.

Discussion

Participants: P. Clément, H. J. Dumont, J. J. Gilbert, C. E. King, B. Pejler, A. Ruttner-Kolisko, T. W. Snell, P. Sorgeloos, P. L. Starkweather.

In the discussion following Dr. Pourriot's introduction the following points were made:

1. A clone is characterized by parthenogenetic reproduction, in contrast to an inbred line which involves bisexual reproduction. Apart from somatic mutations all individuals of a clone should be genetically similar.

2. Some species of rotifers appear to be highly selective in their feeding habits. The size of the food particle is important, but some rotifers are selective within the suitable size range and may prefer a particular microorganism even when others of a similar size are present. This selectivity varies greatly between rotifer species.

3. The possibility of rotifers having the capacity to ingest and metabolize dissolved substances needs investigation.

4. In chemostat cultures there may be considerable fluctuations in the rotifer population, and the final death of the culture is sometimes caused by very high algal concentrations.

5. Photoperiod is most important in regulating the reproductive behaviour of rotifers in culture. By appropriate changes in the photoperiod it is possible in some species to regulate the occurrence of mixis and to control the hatching of resting eggs (this theme is developed elsewhere in the symposium, pp. 51-54).

6. It may be possible to preserve clones by means of dried or frozen resting eggs, or in the case of some bdelloids by freezing the adults.

7. There were no reports of previously uncultured species being brought into culture. Some of the commonest species such as *Keratella cochlearis* are difficult to culture and require further research to bring them into continuous culture.

References

Aloia, R. C. & Moretti, R. L. 1973. Sterile culture techniques for species of the Rotifer Asplanchna. Trans. am. microsc. Soc., 92: 364-371.

Bazire, M. 1953. Cultures pures d'Hydatina senta. Premiers résultats. C. r. Acad. Sci. Paris, 236: 855-857.

Beauchamp, P. M. de, 1928. Coup d'oeil sur les recherches récentes relatives aux Rotifères. Bull. biol. Fr. Belg., 62: 52-125.

Dougherty, E. C. 1963. Cultivation and nutrition of Micrometazoa. III. The minute Rotifer Lecane inermis. J. expl. Zool., 153: 183-186.

Dougherty, E. C., Solberg, B. & Harris, L. G. 1960. Synxenic and attempted axenic cultivation of Rotifers. Science, 132: 416.

Erben, R. 1978. Effects of some petrochemical products on the survival of Dicranophorus forcipatus O. F. Müller (Rotatoria) under laboratory conditions. Verh. int. Ver. Limnol., 20: 1988-1991.

Gilbert, J. 1970. Monoxenic cultivation of the Rotifer Brachionus calyciflorus in a defined medium. Oecologia, 4: 89-101.

King, C. E. & Snell, T W. 1978. Culture media (natural and synthetic): Rotifera. In Diets, Culture media, Food supplements. Publ. chem. Rubber Comp.; 71-75.

Nathan, H. A. & Laderman, A. D. 1959. Rotifers as biological tools. Annls. N.Y. Acad. Sci., 77: 96-101.

Pejler, B. 1977. Experience with Rotifer cultures based on Rhodomonas. Arch. Hydrobiol. Beih., 8: 264-266.

Pourriot, R. 1965. Recherches sur les Rotifères. Vie Milieu, Suppl., 21: 224 p.

Pourriot, R. & Hillbricht-Ilkowska A. 1969. Recherches sur la Biologie de quelques Rotiferes planctoniques. Bull. Soc. zool. Fr., 94: 111-118.

METHODS FOR OBTAINING AN AXENIC CULTURE OF HABROTROCHA ROSA DONNER 1949

K. PLASOTA, M. PLASOTA & W. J. H. KUNICKI-GOLDFINGER

Institute of Microbiology, University, of Warsaw, Nowy Swiat 67, 00-046 Warsaw, Poland

Keywords: axenisation, rotifers

Abstract

A culture of the bacterivorous rotifer *H. rosa* was separated from other Eucaryota in activated sludge and then purified by passages from most representatives of the bacterial microflora. A fully axenic culture was finally obtained by use of a lysing buffer and certain antibiotics.

Introduction

H. rosa Donner, 1949 (family Habrotrochidae) feeds mainly on bacteria. Its food is transformed by the mastax into spherical 'pellets' which are digested in a syncytial stomach. In this connection the choice of food for cultivation purposes and axenisation procedure should be based on microbiological techniques, treating the bacterial microflora as an equiponderant component of the system rotifer-rotifer food.

Material and methods

H. rosa was isolated from activated sludge. A culture of the rotifer free from other Eucaryota was obtained by transferring individual specimens with sterile preparation needles to a synthetic medium without any carbon source (to inhibit bacterial growth) composed of: NH_4Cl – 60 mg, $MgSO_4.7H_2O$ – 2 mg, K_2HPO_4 anh. – 50 mg, NaCl – 1000 mg, H_2O dist. – 1000 ml, pH 7.0, supplemented with crystaline penicillin (1000 u/ml) and *Micrococcus* sp. Every three weeks the culture was renewed by introducing 30 ml rotifer inoculum into 300 ml fresh medium. Cultures were maintained in 500 ml flasks, without aeration. The medium for passages was prepared similarly in 20 ml volumes in 150 ml flasks. Sterility was ensured by irradiation of the working chamber with UV light. Passages were made by transferring the rotifers with a sterile needle. The rotifers were fed bacteria (*Micrococcus* sp.) grown in nutrient broth and centrifuged three times with washes of the rotifer medium to remove all traces of the nutrient broth before being introduced into the rotifer culture. This procedure was repeated every 14 days. Axenisation of the rotifer culture by combined action of several antibiotics was carried out in a sterile chamber. A dense culture of rotifers from a passage was poured into a Petri plate and washed with sterile medium with or without antibiotics. The fluid was poured out of the plate leaving the rotifers fixed to the bottom. Bacteria of the genus *Pseudomonas* were removed with the use of cell wall lysing buffer (Gilleland, Stinnett & Eagon, 1974) composed of: K_4HPO_4 – 0.04 M, KH_2PO_4 – 0.022 M, $(NH_4)_2SO_4$ – 0.007 M, $MgSO_4.7H_2O$ – 0.5 mM, glucose – 0.03 M, pH 8.6 and Tris + EDTA buffer.

Scheme of washes and axenisation:

1. two washes with pure medium
2. EDTA + Tris
3. Pseudomonas lysing buffer — 20 min.
4. EDTA + Tris + 2000 units penicillin/ml
5. pure medium — 15 min.
6. 40 μg nebcin/ml — 60 min.
7. pure medium — 15 min.
8. 100 μg chlorotetracycline/ml — 90 min.
9. pure medium — 5 min.
10. 5000 μg carbenicillin and 40 μg/ml chlorotetracycline — 20 h
11. pure medium — 30 min.
12. passage with needle to rotifer medium containing 1000 units penicillin/ml; feeding with *Micrococcus* sp. was carried out after 24 hours.

Purity controls of the axenic culture were made very three days by spreading fluid from the culture for bacterial growth into plates with nutrient agar, nutrient agar with penicillin (1000 units/ml), blood agar and malt agar. The concentration of the antibiotics and time of exposure were determined in preliminary experiments on the sensitivity of the rotifers and the bactericidal activity of the preparation employed.

Hydrobiologia 73, 37-38 (1980). 0018-8158/80/0731-0037$00.40.
© Dr. W. Junk b.v. Publishers, The Hague.

Results and discussion

Activated sludge as the environment of *H. rosa* contains, besides Rotatoria, a large number of eucaryotic organisms including different groups of protozoans and nematodes. These organisms were successfully removed by use of the synthetic medium described under 'Methods' with addition of a bacterial culture isolated from the primary culture as a food source. Single rotifer specimens were transferred to this medium with a sterile needle resulting first in microcultures and then in larger cultures in Erlenmeyer flasks. Most of the species of the bacterial microflora were removed in the next stage which involved cultivation in serial passages. Single specimens were transferred with a sterile needle to a synthetic medium without carbon source. *Micrococcus* sp. was introduced as a source of food and the culture was stabilized by the addition of 1000 units penicillin/ml, which favourably affected the development of the culture. This procedure was repeated every two weeks (time needed for propagation of the rotifers) by transferring single specimens to fresh medium.

The use of conditions preventing infection with microorganisms and stringent control of the culture with respect to bacteriological purity resulted in the 'loss' of most of the bacterial species accompanying the rotifers. After twelve passages the rotifer culture contained only two species from the initial activated sludge besides the bacteria introduced into the medium. After preliminary tests with over 20 antibiotics we chose nebcin, chlorotetracycline and carbenicillin for further experiments. The antibiotics were potent bactericidal agents but were also toxic to the rotifers. None of the antibiotics was 100% bactericidal and always forms from the *Pseudomonas* group survived. This led us to employ initial lysis of the bacterial cells with EDTA and Tris and a cell wall lysing buffer for *Pseudomonas* (Gilleland, Stinnett & Eagon, 1974). This procedure was followed by the use of antibiotics. The simultaneous use of several antibiotics and short exposure time of the rotifers to their action resulted in full axenisation of the culture. A sufficient number of rotifers survived to allow its renewal.

At this stage it is possible to introduce food sources into the culture (selected strains of bacteria or certain species of yeast with small cells) and to conduct studies which require the removal of uncontrolled and variable factors linked with the unknown microflora.

Summary

A method involving the use of *Pseudomonas* cell wall lysing buffer and antibiotics (to most of which *Pseudomonas* is resistant) was elaborated for axenisation of rotifer cultures derived from activated sludge. It should be stressed that this method, particularly with respect to the types of antibiotics used, cannot be regarded as universal but has been adapted for a particular set of microorganisms. Its adaptation to other systems requires testing of the sensitivity of the bacterial microflora.

References

Gilleland, H. E. Jr, Stinnett, J. D., Eagon, R. 1974. Ultrastructural and chemical alteration of the cell envelope of Pseudomonas aeruginosa, associated with resistance to ethylenediamine-tetraacetate resulting from growth in Mg^{2+} deficient medium. J. Bact. 117: 302.

SOME PROBLEMS IN THE EMBRYOGENESIS OF HABROTROCHA ROSA DONNER 1949

K. PLASOTA & M. PLASOTA

Institute of Microbiology, University of Warsaw, Nowy Swiat 67, 00-046 Warsaw, Poland

Keywords: life span, number of eggs laid, spiral cleavage, epibolic gastrulation, organogenesis

Abstract

Several parameters connected with the biology of *H. rosa* were investigated under laboratory conditions: average life span (20 days) divided into three characteristic stages, mean number of eggs laid (30 eggs) and average time of egg development (31.5 hours). Ontogenesis was studied (until the stage of early organogenesis) and a spiral type of cleavage and epibolic gastrulation were observed. The paper also presents data on the origin of the digestive system and sex cells.

Introduction

Problems connected with early ontogenesis – which is the subject of part of this study – have been studied in only few rotifer species mostly of the Monogonont group (Tannreuther, 1920; Beauchamp, 1965). In these forms a spiral type of cleavage, and gastrulation of typically epibolic type (*Ploesoma* sp.) or resembling this type (*Asplanchna* sp.) with some reservations (Beauchamp, 1965), were observed. However, data on the development of the Bdelloidea are lacking.

Material and methods

A culture of *H. rosa* was maintained in laboratory conditions in 500 ml flasks in a synthetic medium composed of: NH_4Cl -60 mg, $MgSO_4.7H_2O$ -2 mg, KH_2PO_4 anh. -20 mg, K_2HPO_4 anh. -50 mg, NaCl -1000 mg, H_2O dist. -1000 ml, pH 7.0, supplemented with crystaline penicillin (1000 u/ml). The rotifers were fed with bacteria (*Micrococcus* sp.). Every three weeks the culture was renewed by introducing 30 ml rotifer inoculum into 300 ml fresh medium. Studies on single specimens were conducted at

room temperature in July in small watch-glasses into which single eggs were transferred. Observations on single rotifers were made continuously and eggs laid were removed with a micropipette. To avoid aging of the medium, part of the fluid was removed daily and replaced with a suspension of bacteria in fresh medium. In studies on time of egg development and the early stages of cleavage the specimens were maintained in a hanging drop. The development stages were examined with the use of the following techniques: 1. direct observations of unstained material (sometimes in phase contrast) or similar observations of thermally fixed (+ 60° C) material; 2. observations on material fixed with alcohol and made translucent with glycerol; 3. preparations (lightly squashed) fixed in Bouin's fluid and stained with iron-hematoxylin or stained without fixing by the RVD method (R. V. Dippel, 1955; Plasota & Plasota, this volume).

Results and discussion

A. Observations on the biology of reproduction of *H. rosa*. By examining several specimens in microcultures (Table 1) we found that the average life span under laboratory conditions was approximately 20 days. However, some of the rotifers died earlier, probably due to disorders in the egg laying process since the eggs were not laid one after the other but accumulated in the body of the specimen (2-3 eggs). Such specimens are rather frequent in cultures and the development of the eggs within the dead body is quite normal.

The life of an individual consists of three stages (Table 1): a period before maturation (2-3 days), egg laying period (6-9 days) and a period of full biological activity unaccompanied by egg laying (10-14 days). The number of

Hydrobiologia 73, 39-41 (1980).   0018-8158/80/0731-0039$00.60.
© Dr. W. Junk b.v. Publishers, The Hague.

Table 1. Data on life span and reproduction of *H. rosa*.

| specimen | number of days of life | | | life span | number of eggs laid |
	before egg laying stage	in egg laying stage	after egg laying stage		
1	3	6	10	19	24
2	2	6	12	20	22
3	3	9	10	22	35
4	2	5	–	7°	20
5	2	9	14	25	33
6	2	8	2?	12?	34

° Dead rotifer, filled with three unlaid eggs.

eggs laid during the life of a single specimen averages 30 and may reach 6 eggs per day. This observation is compatible with cytological observations which have shown that the highest number of oocytes in the ovaries of rotifers entering the stage of sexual activity is also about 30. Eggs are very large compared to the dimensions of a rotifer – approximately ¼ of its body length. At the moment of laying, their shell is still soft. When the egg passes through the cloaca it becomes constricted (Fig. 1). After several minutes it hardens and acquires its characteristic surface structure.

Under laboratory conditions the average time of egg development from laying till hatching is 31.5 hours and is fairly constant (± 1 h). The process of hatching takes only 2-2.5 minutes. We also found that some of the eggs – particularly those laid by a rotifer shortly after its maturation – may become arrested at an early stage of development and die. Moreover, completely formed embryos may also be unable to hatch if they overstay (by several hours) in the egg shell.

B. Selected problems in the ontogenesis of *H. rosa*. Our observations extended until the stage of early organogenesis. The course of early ontogenesis in the examined representative of the Digonont – that is the first two divisions resulting in the four blastomere stage, indicates the occurrence of a cleavage pattern of the spiral type (Fig. 1).

Two further divisions lead after about two hours to the 16 blastomere stage in which after the separation of the micromeres four macromeres can be distinguished. In the

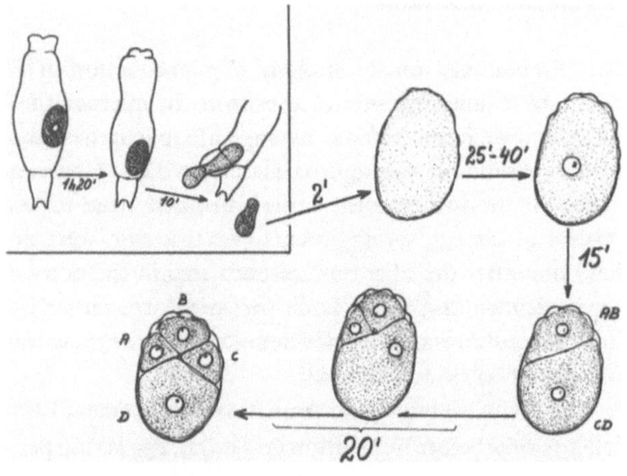

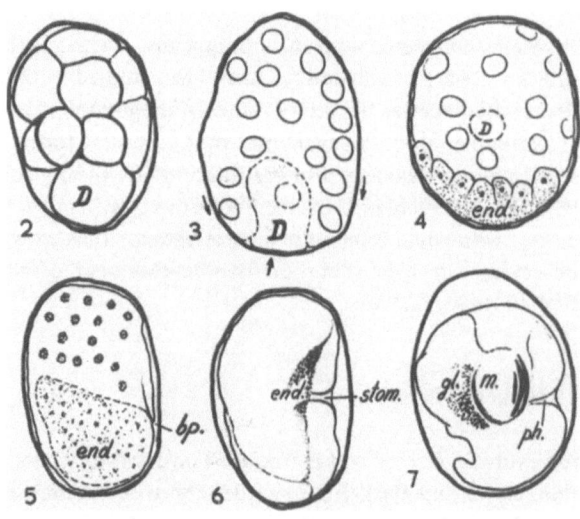

Fig. 2. Further two divisions leading to 16 blastomere stage; visible macromere D.

Fig. 3. Beginning of genital line; visible macromere D.

Fig. 4. Differentiation of endoderm (end.).

Fig. 5. Gastrulation (bp. – blastopore).

Fig. 6. Cavitation of stomodeum (stom.).

Fig. 7. Development of digestive system (ph. – pharynx, gl. – stomach glands, m. – mastax).

Fig. 1. Early ontogenesis in *H. rosa* from egg-laying stage to 4 blastomere stage. The times of the individual stages are given.

course of these divisions the largest macromere D becomes
evident (Fig. 2). It moves inwards (Fig. 3) and gives rise to
the genital line (vitellocytes and oocytes). The cells formed
from divisions, of the other macromeres (Fig. 4) – which
seem to be the presumptive endoderm proper – are subject
to epiboly and move under and are covered by the micromeres; a remnant of this process is a residual blastopore
(Fig. 5). Soon afterwards, at the cost of the ectodermal
micromere material, a cavity of the stomodeum is formed
(Fig. 6) from which the pharynx and mastax originate.
The endodermal cells form the inner gut and its glands
(Fig. 7).

Summary

The average life span of *H. rosa* under laboratory conditions is approximately 20 days and divides into three
stages. In the period of full biological activity (6-9 days)
the specimen lays an average of 30 eggs. The development
of the egg from the moment of laying until hatching of a
young rotifer lasts 31.5 ± 1 hour. The cleavage is of the
spiral type. The macromeres give rise to the sex cells and
endoderm. Gastrulation is of the epibolic type. The anterior part of the alimentary tract (pharynx and mastax)
forms from the ectodermal micromere material and the
inner gut and the stomach glands from the endoderm.

References

Beauchamp, P. de. 1965. Rotifers. In: Traité de Zoologie (ed.
    Grassé, P.), Masson et Cie, Paris, 4/3: 1225-1379.
Dippel, R. V. 1955. A temporary stain for Paramecium and other
    ciliate protozoa. Stain Technol., 30: 69-75.
Plasota, K. & Plasota, M. 1980. The determination of the chro-
    mosome number of Habrotrocha rosa Donner, 1949. Hydro-
    biologia.
Tannreuther, G. W. 1920. The development of Asplanchna
    ebbesbornii. J. Morphol., 33: 389-437.

THE DETERMINATION OF THE CHROMOSOME NUMBER OF HABROTROCHA ROSA DONNER, 1949

K. PLASOTA & M. PLASOTA

Institute of Microbiology, Univ. Warsaw, Nowy Swiat 67, 00-046 Warsaw, Poland

Keywords: Digononts, Bdelloids, chromosome number

Abstract

The number of chromosomes in *Habrotrocha rosa* Donner 1949 has been estimated at 2n(?) = 14. Chromosomes were stained and photographed by the RVD method (acetocarmine + fast green). Observations were made on oocytes and cleavage eggs.

Introduction

In undertaking caryological studies on the Rotifera Bdelloidea, it is necessary to keep in mind a certain peculiarity. This group, which embraces the species studied here can, according to some authors (Hsu, 1956a, b; Birky & Gilbert, 1971) be treated as forms which, in the course of evolution (conversion to exclusive parthenogenetic mode of reproduction) forfeited chromosome homology. They have, however, retained a 'diploid' number of chromosomes, further designated 2n(?), which does not necessarily has to be even. This state naturally excludes the occurrence of normal meiosis due to the absence of synapsis between homologous chromosomes. But Bdelloidea have retained some rudiments of this process, i.e. two maturation divisions during oogenesis (Hsu, 1956a, b). The latter may indicate that the Bdelloidea do not derive from the Monogononta through simple loss of the ability to produce mictic females in their life cycle. Information on the chromosome number in Rotifera is not only incomplete but also often contradictory. In the Monogononta, Tauson (1924, 1927) gives n = 12 for *Asplanchna intermedia,* but Whitney (1929) gives n = 13 for *A. sieboldi,* and Storch (1924) and Robotti (1975) give n = 8 or n = 7, respectively for *A. priodonta.* In Digononta Hsu (1956a, b) gives 2n(?) = 13 for the two species *Habrotrocha tridens* and *Philodina roseola.*

Material and methods

A culture of rotifers was concentrated by centrifugation and the pellet was transferred to a protein-coated glass slide. The rotifers were then squashed under a cover-glass which was removed by freezing the preparation at -20° C. The preparation was then dried and either fixed or immediately stained.

Chromosomes were visualized by one of the following staining techniques: 1. fixing with Bouin's fixative, followed by iron-hematoxylin 1.0%; 2. orcein according to Keyl & Keyl, 1959; 3. stain RVD (R. V. Dippel, 1955) consisting of: 10.5 ml acetocarmine and 4.5 ml of 45% acetic acid, 1.0 ml 1N HCl and 1.0 ml fast green; 30 minute staining not preceded by fixing. The nuclear material is stained red and the cytoplasm green.

Results and discussion

Chromosomes were visualized in the nuclei of maturating sex cells and in cleavage eggs. The result obtained for *Habrotrocha rosa* by all methods was 2n(?) = 14.

The best results were obtained with RVD staining (Fig. 1) which yielded a two-tone image in the preparation and swelling of the chromosome matrix. The latter is of particular importance since the chromosomes of the Bdelloids are very small, with dimensions at the limit of the resolving power of the light microscope.

Summary

The species *Habrotrocha rosa* is characterized by a chromosome number 2n(?) = 14 as shown in both maturating sex cells and blastomer nuclei in the course of cleavage.

Hydrobiologia 73, 43-44 (1980).   0018-8158/80/0731-0043$00.40.
© Dr. W. Junk b.v. Publishers, The Hague.

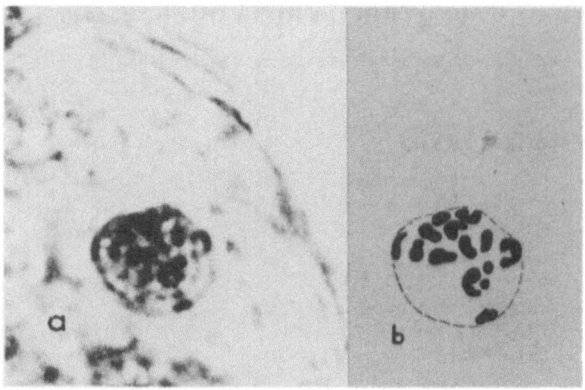

Fig. 1. Chromosomes in the nucleus (prophase) in early cleavage cells of *H. rosa*.
1a. RVD 100x immersion
1b. optical observations, correcting micrograph.

References

Birky, C. W. & Gilbert, J. J. 1971. Parthenogenesis in Rotifers: The control of sexual and asexual reproductions. Amer. Zool., 11: 245-266.

Dippel, R. V. 1955. A temporary stain for Paramecium and other ciliate protozoa. Stain Technol., 30: 69-75.

Hsu, W. S. 1956a. Oogenesis in the Bdelloidea rotifer Philodina roseola Ehrenberg. Cellule 57: 283-296.

Hsu, W. S. 1956b. Oogenesis in Habrotrocha tridens (Milne). Biol. Bull., 111: 364-374.

Keyl, H. G. & Keyl, J. 1959. Die cytologische Diagnostik der Chironomiden. Arch. Hydrobiol., 56: 43-57.

Robotti, C. 1975. Chromosome complement and male haploidy of Asplanchna priodonta Gosse 1850 (Rotatoria). Experientia, 37: 1270-1271.

Storch, O. 1924. Die Eizellen der heterogonen Rädertiere. Zool. Jb., 2, 45: 309-404.

Tauson, A. O. 1924. Die Reifungsprozesse der parthenogenetischen Eier von Asplanchna intermedia Huds. Z. Zellforsch. Gewebelehre 1: 57-84.

Tauson, A. O. 1927. Die Spermatogenese bei Asplanchna intermedia Huds. Z. Zellforsch. Mikrosk. Anat., 4: 652-681.

Whitney, D. D. 1929. The chromosome cycle in the rotifer Asplanchna amphora. J. Morphol. Physiol., 47: 415-433.

MASS CULTURE EXPERIMENTS WITH BRACHIONUS RUBENS

M. SCHLÜTER

Kernforschungsanlage Jülich GmbH, Abt. für Algenforschung und Algentechnologie, D-4600 Dortmund 1 Germany

Keywords: rotifers, Brachionus rubens

Abstract

In order to develop the optimum conditions for a mass culture of *Brachionus rubens*, eight strains of phytoplankton were tested as food for the rotifers. The optimum food concentration as well as the concentration of algal medium tolerated by *R. rubens,* and the influence of nitrite, sodium chloride, extreme pH-values and low oxygen concentrations on the reproduction of *B. rubens* were determined.

Introduction

Rotifers of the genus *Brachionus* are a suitable food for the fry of fish and other aquatic animals (Howell, 1974; Theilacker & Mc Master, 1971). The technique of rearing rotifers was investigated in laboratory cultures to work out the basis for cultivating rotifers on a technical scale.

Seven series of experiments were performed in which eight strains of green algae from the collection of the 'KFA Institut für Algenforschung' in Dortmund, Germany (Fig. 1) were tested as food for *Brachionus rubens* and the optimum food concentration as well as the concentration of algal medium, tolerated by *Brachionus rubens* were determined. Moreover the influence of nitrite, NaCl, extreme pH-values and low oxygen concentrations on the reproduction of *B. rubens* were determined.

Materials and methods

The rotifers were fed with algae grown chemostatically in glasstubes with 300 ml culture volume. The algae were incubated in a light thermostat at 30°C under constant illumination with 40 W coolwhite neon tubes and aerated with a 1% CO_2 in air mixture. The optical density of the algal-culture ranged from 0.3-0.4, measured at 560 nm in a Medico-photometer. The inorganic nitrate medium N8 (Soeder, Schulze & Thiele, 1967) was used as culture medium. The algal cultures remained free from contamination from other algae, but were not axenic.

The rotifer culture medium (inorganic Medium after Halbach & Halbach-Keup, 1974) was inoculated with 5 animals/ml in 200 ml glass tubes and incubated for six days at 20°C without illumination. Except for the algal strain experiment they were fed with *Scenedesmus costato-granulatus* from the chemostat. The density of the rotifers was measured by counting a definite culture volume (1.0 or 0.2 ml) under 10-fold magnification.

Results

Algal strain experiment
The algae were grown in pure cultures and removed from the culture fluid by centrifugation before being fed to the rotifers. The concentration of algae was adjusted to give an optical density of 0.08 in the rotifer cultures. *Scenedesmus costato-granulatus, Kirchneriella contorta* and *Chlorella fusca* gave the best results, *Chlamydomonas* sp., *Chlorella vulgaris* and *Scenedesmus acuminatus* were unsuitable (Fig. 1).

Food concentration
The rotifer cultures were supplemented with algal concentrations ranging from 0-100 mg algal dry weight/l. The highest rotifer density after six days of incubation was obtained with 70 mg algae/l (Fig. 2). At higher concentrations the algae cause blockage of the filter system of the rotifers.

Hydrobiologia 73, 45-50 (1980).   0018-8158/80/0731-0045$01.20.
© Dr. W. Junk b.v. Publishers, The Hague.

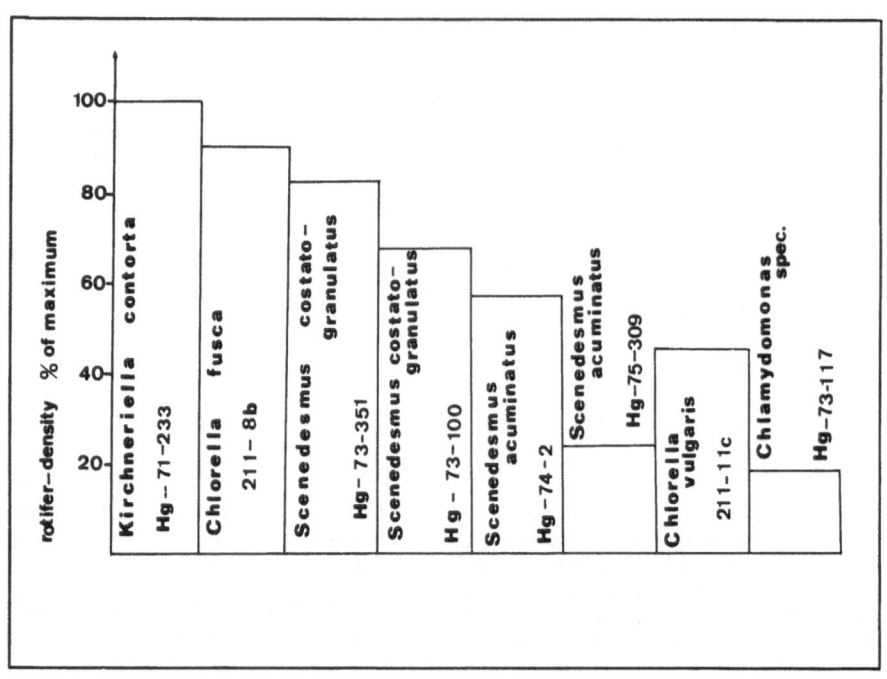

Fig. 1. Block diagram, showing the relative suitability of eight algal
strains as food for Brachionus rubens.

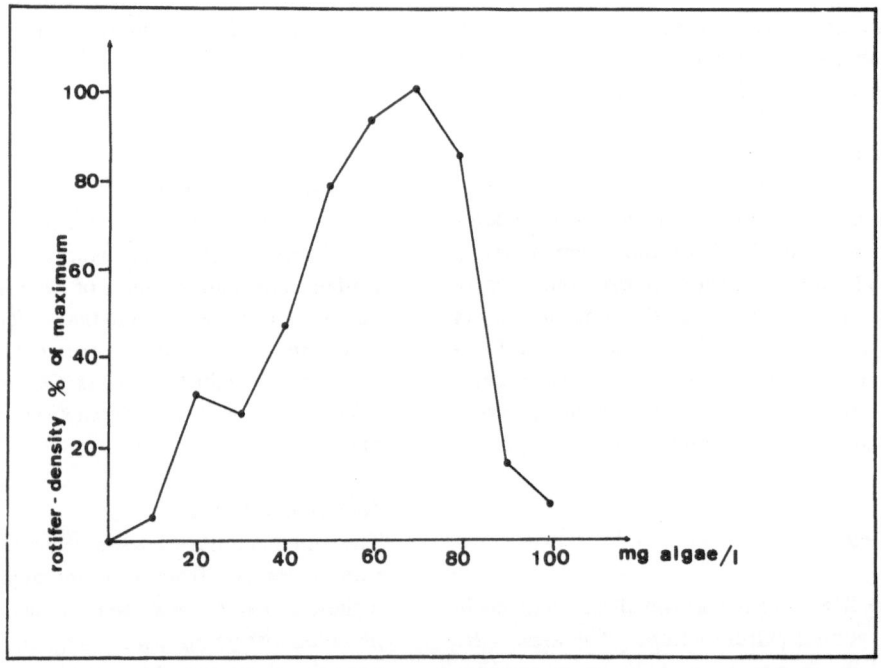

Fig. 2. Influence of the concentration of the food organism (Scenedesmus costato-granulatus) on
the reproduction of Brachionus rubens.

Tolerance of the rotifers to algal medium

According to Halbach & Halbach-Keup (1974, algal media are toxic for rotifers. For mass cultures, there is a great advantage in being able to feed the algae directly with their culture medium to the rotifers. In order to test the toxicity of the algal medium, a series of mixtures of algal and rotifer-media were prepared (0-100% algal-medium), and the density of the rotifer population after six days of incubation with algal supplement was determined. A population density of 50% of the control was obtained in a medium containing 80% algal medium. *B. rubens* did not survive in 100% algal medium (Fig. 3).

Tolerance to nitrite

In rotifer mass cultures, nitrite may accumulate in the medium as a toxic metabolite. In order to test the toxicity of nitrite ranging concentrations of the $NaNO_2$ (0-50 ppm NO_2-N) were prepared in rotifer medium. The series was inoculated with rotifers from NO_2-free-medium as well as with rotifers adapted to nitrite from a medium containing 12 ppm NO_2-N. In the first series a 90% inhibition of the reproduction rate of *B. rubens* was caused by concentrations of 10 ppm NO_2-N. A concentration of 20 ppm NO_2-N, only caused a 50% inhibition with the NO_2-adapted rotifers (Fig. 4).

Tolerance to sodium chloride

Rotifer medium was enriched with NaCl in a range from 0-8 g/l to determine the highest salt concentration in which the reproduction of *B. rubens* is not inhibited. Inhibition of the reproduction-rate was found at concentrations over 2 g NaCl/l, the highest salt concentration in which animals survived for more than 12 days was 4 g NaCl/l (Fig. 5).

Influence of pH

The pH of the rotifer cultures was adjusted to values between pH 3 and pH 11. After six days of incubation the highest rotifer densities were found at pH 6-8. No rotifers survived at values above pH 9.5 and below pH 4.5 (Fig. 6).

Influence of the oxygen concentration

The rotifer cultures were aerated with a mixture of air and N_2 in constant proportions in order to obtain a constant O_2-concentration in the culture medium. No inhibition of the reproduction rate of *B. rubens* was observed above a concentration of 1.15 mg O_2/l. At 0.72 mg O_2/l the rotifers ceased to reproduce and the cultures died off within five days (Fig. 7).

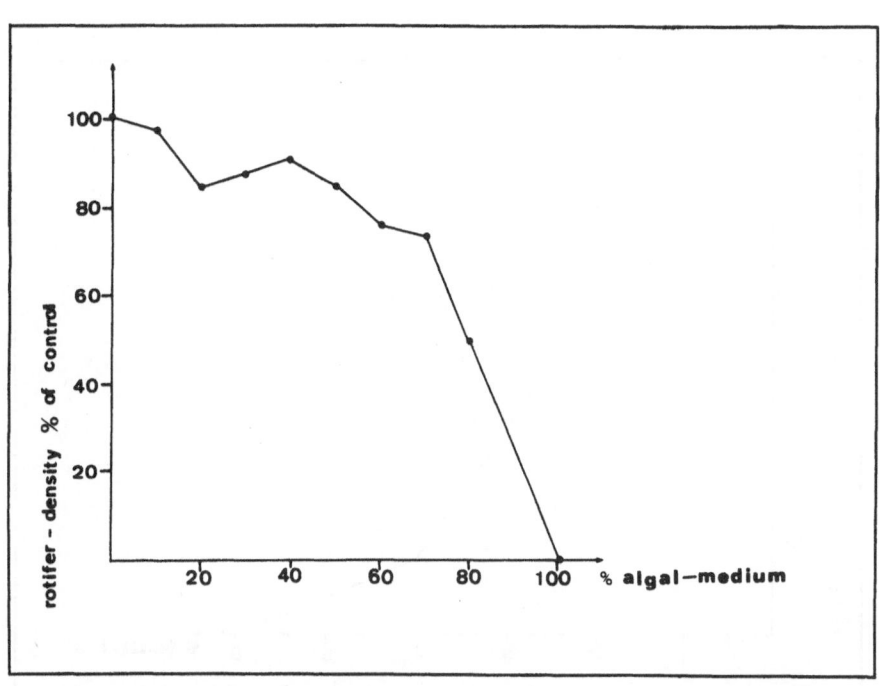

Fig. 3. Population growth of Brachionus rubens cultures in relation to the concentration of the algal medium N8 (Soeder).

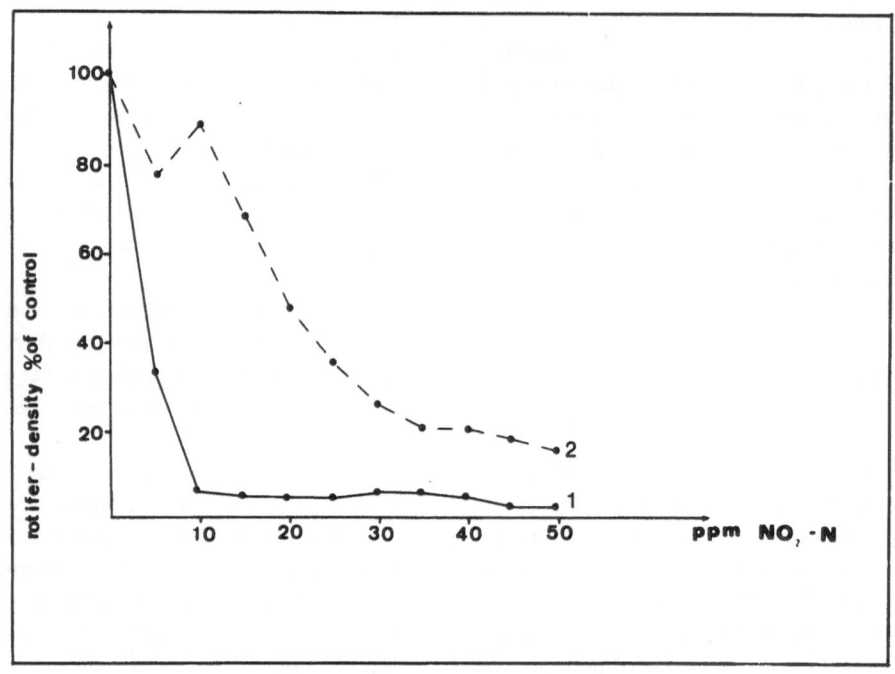

Fig. 4. Influence of NaNO₂ on the reproduction of Brachionus rubens.
unadapted rotifers: 1
NO₂⁻-adapted-rotifers: 2

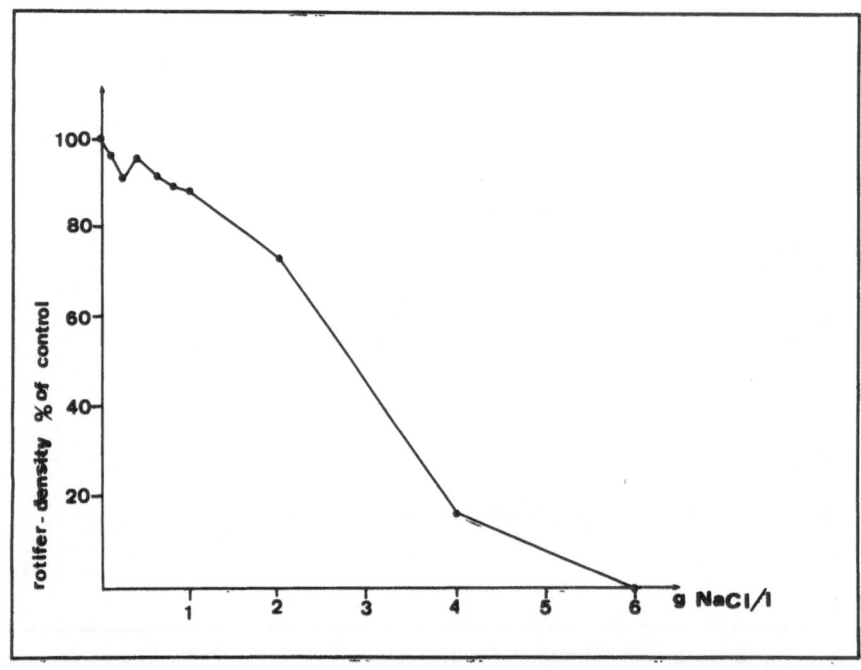

Fig. 5. Influence of the NaCl-concentration on the reproduction of Brachionus rubens.

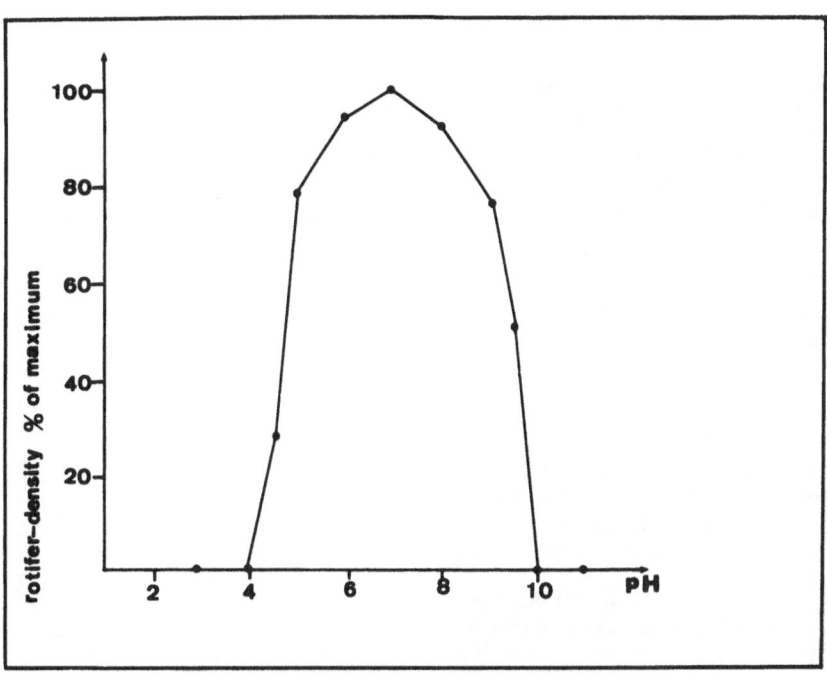

Fig. 6. Population growth of Brachionus rubens cultures in relation to the pH-values of the culture medium.

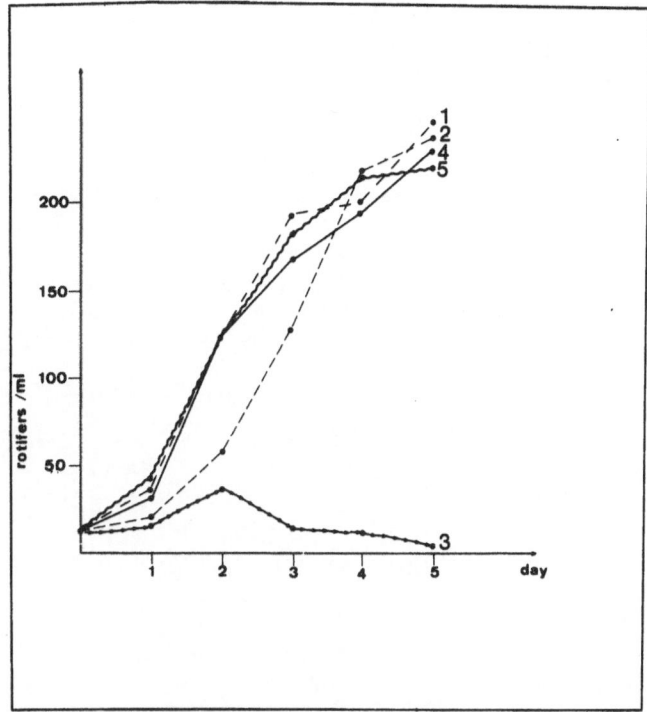

Fig. 7. Influence of the O_2-concentration in the culture medium on the reproduction of Brachionus rubens.

8.1 $\pm$ 0.41 ng O_2/l: 1
2.3 $\pm$ 0.19 mg O_2/l: 2
0.72 $\pm$ 0.03 mg O_2/l: 3
4.1 $\pm$ 0.11 mg O_2/l: 4
1.15 $\pm$ 0.18 mg O_2/l: 5

Discussion

The results show that *B. rubens* is a suitable organism for a mass culture. Several phytoplankton species can be used by the rotifers as a source of nutrition in a medium consisting of 80% algal medium and 20% rotifer medium This concentration supports sufficient algal growth. Salt, nitrite and pH-values are tolerated in a wide range of concentrations and oxygen concentrations as low as 1 or 2 mg O_2/l do not effect the reproduction of *B. rubens*.

References

Halbach, U. & Halbach-Keup, G. 1974. Quantitative Beziehungen zwischen Phytoplankton und der Populationsdynamik des Rotators Brachionus calyciflorus Pallas. Befunde aus Laboratoriumsexperimenten und Freilanduntersuchungen. Arch. Hydrobiol. 73: 273-309.

Howell, B. R. 1973. Marine fish culture in Britain VII. A marine rotifer, Brachionus plicalilis Müller, and the larvae of the mussel Mytilus edulis, L. as food for larval flatfish. J. Cons. int. Explor. Mer. 35: 1-6.

Soeder, C. J., Schulze, G. & Thiele, D. 1967. Einfluß verschiedener Kulturbedingungen auf das Wachstum in Synchronkulturen von Chlorella fusca Sh. et Kr. Arch. Hydrobiol. Suppl. 127-171.

Theilacker, G. H. & Mc Master, 1971. Mass culture of the rotifer Brachionus plicatilis and its evaluation as a food for larval anchovies. Marine Biology 10: 183-188.

HATCHING OF BRACHIONUS RUBENS O. F. MULLER RESTING EGGS (ROTIFERS)

R. POURRIOT, C. ROUGIER & D. BENEST

Ecole Normale Superieure, 75230 Paris Cedex 05 France

Keywords: Dormancy, hatchability of resting eggs, rotifers, *Brachionus rubens*

Abstract

After a dormant period at low temperature (5°C) and darkness, hatching of *Brachionus rubens* resting eggs is induced by an increase of temperature (10-22°C) in presence of light.

Introduction

Little is known about the stimulus which initiates development of rotifer resting eggs. Besides scattered observations, the only works are those of Nipkow (1958, 1961) and Bogoslovsky (1963, 1967, 1969) on resting eggs from plankton samples. The former showed the necessity of a long period of dormancy before hatching, i.e. 6 months for *Brachionus calyciflorus*. The latter observed variations of hatching ability, partly related to morphological differences and determined by the physiological status of the mictic females (Bogoslovsky, 1963). For a review of dormancy characteristics, see Gilbert (1974).

No information being available concerning clones under experimental controlled conditions, we put to a test resting eggs from our laboratory clones. This paper will present our first results concerning *B. rubens*.

Material and methods

Culture conditions

All the resting eggs are isolated from a clone of *B. rubens;* the initiating female was obtained from a farm pool in Chartainvilliers (Eure et Loire, France) in 1977 (clone referred to: Ch 77).

The mass-culture is maintained in darkness at 15°C provided with euglenoid *Phacus pyrum* or chlorococcale *Chlorella pyrenoïdosa* as a food supply.

Since resting eggs do not spontaneously hatch under culture conditions in the first weeks after their laying, the clone continuance is ensured by placing, at short enough time intervals (i.e. less than a month), some amictic females in a fresh medium. Commercial spring water (Volvic) was used for both culture and experimental mediums.

Conditions prevailing during resting eggs formation

As both exogenous and endogenous factors induce dormant egg production (review in Clément *et al.,* 1977 and Pourriot & Rougier, 1976 for *B. rubens*), in *Brachionus,* control of mixis in mass-culture is not simple. Resting eggs for our experiments were collected when a lot of mictic females were observed; conditions of resting eggs formation are indicated in Table 1.

A few resting eggs shown some abnormal features (in shape, in size or in absence of the inner cavity) and were eliminated.

Experimental conditions

For experiments a, b, c, and e, each lot of dormant eggs has successively been submitted to different temperature and light conditions (see Fig. 1 and 2).

In experiment d, the initial lot of 180 resting eggs was divided in two uneven lots d_1 (30 r.e.) and d_2 (150 r.e.), and d_1 was then spread into 5 equal parts (d_{11} to d_{15}). Conditions for each lot are indicated in Fig. 2.

A upper case 'R' by an arrow indicates a transfer of the resting eggs to a fresh medium, eggs are kept in the medium where they have been laid otherwise.

Temperature and light conditions for experiment e are LD = 16:8 and 23°C respectively, All experiments were performed in controlled environmental chambers except for the last stages of b and c were respectively:

b, 21 ± 2°C, LD = 14.30 : 9.30 (04.28.'78)

c, 21 ± 2°C, LD = 16 : 8 (07.03.'78).

All controlled photoperiods in environmental chambers were LD = 16 : 8.

Hydrobiologia 73, 51-54 (1980).   0018-8158/80/0731-0051$00.80.
© Dr. W. Junk b.v. Publishers, The Hague.

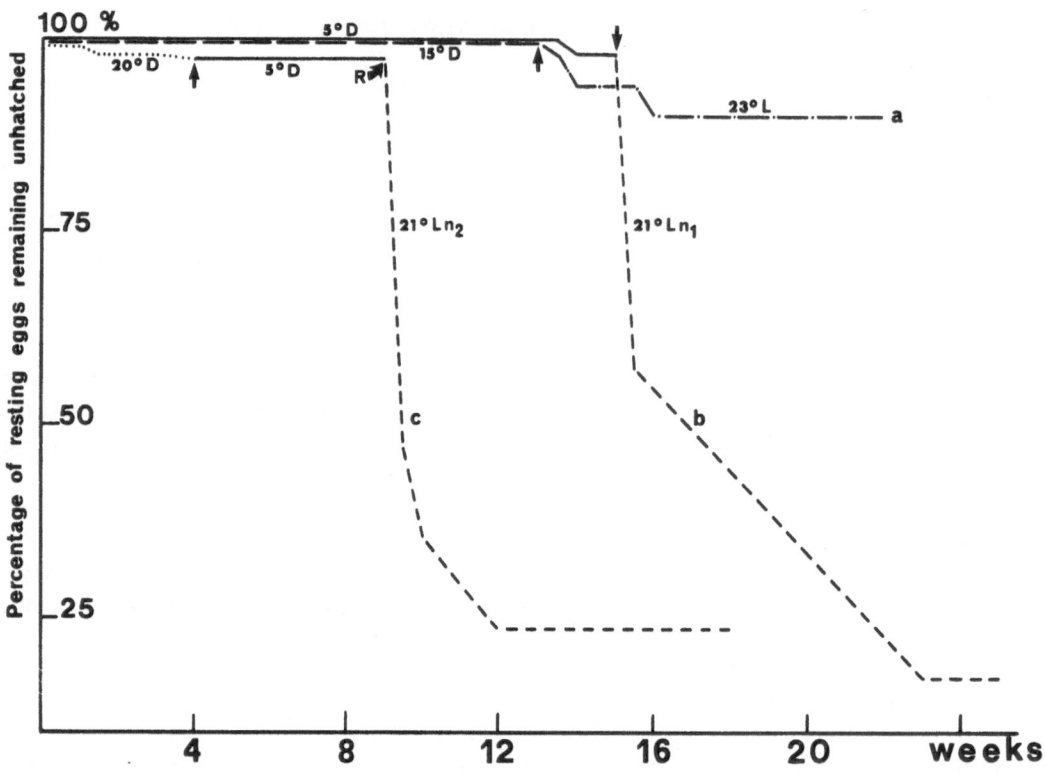

Fig, 1. Decrease of three lots of resting eggs by hatching (in percent). in different light and temperature conditions.

Results and discussion

Results of experiments a, b, c and d are reported in Figures 1 and 2. Experiment e does not differ greatly from the first part of experiment c: hatching starts two weeks after isolation of resting eggs; occasional hatchings occur until the 5th week, where hatching rate reaches 7% and remains constant for 4 more weeks.

1. Most resting eggs must undergo an obligatory dormancy: whatever light and temperature conditions could be, there is no or very little hatching the first month following laying.

2. No hatching or nearly, occurs when resting eggs are kept (three months or more) in darkness at temperature below 15°C.

3. An increase in temperature immediately after egg formation, whatever light conditions may be, does not promote hatching (cf exp. e and first part of c).

4. An increase in temperature with light, following a period at intermediate temperatures (10 or 15°C) in darkness, induces but a low percentage (< 15%) of hatching, even when resting eggs are placed in a fresh medium (exp. a and d_2).

5. A high proportion of dormant egg hatching is observed

Table 1. Conditions of resting eggs formation (f.m.f.: fertilized mictic females).

exp.	N r.e,	dates of formation	temp.	LD	origin	food supply
a	48	02.20-24.1978	15°	0:24	40 f.m.f.	Chlorella pyrenoidosa
b	100	12.22.'77-01.10.'78	15°	0:24	mass-cult.	Phacus pyrum
c	96	04.14-28.1978	20°	0:24	id.	id.
d	180	04.12-19.1978	10°	0:24	id.	Chlorella pyrenoidosa
e	110	03.10-23.1978	15°	0:24	id.	id.

when they are maintained in darkness at low temperature (5°C) during 1 to 3 months and then placed at higher temperatures (10, 14, 18, 22°) with light (exp. d₁).

6. Light is a key factor to obtain a high rate of hatching (compare exp. d_{11}, d_{12} with d_{13} to d_{15}); light intesity and photoperiod have little or no importance (cf very different light conditions in b and c), when higher than minimum values which have to be determined.

7. Bogoslovsky (1963) suggested that the female physiology, determined by food quality and quantity, influences the duration of dormancy. Nevertheless, in experiments b, c and d, resting eggs are produced at different temperatures and feeding conditions, but the hatching process is about the same.

Low to moderate hatching rates (< 35%) obtained by Bogoslovsky (1969) over several months for *B. rubens* resting eggs maintained in ambient conditions (somewhat imprecise) can be explained by the lack of stimulus promoting fast hatching.

Sporadic hatchings, spread over time, were observed by us as well. These sporadic hatchings, out of the major hatching process, could be due to the coincidence of the short periods of illumination necessary for observations with a phase of sensitivity or to some endogenous factors (genotypic) independant from morphological characteristics.

A possible explanation for the observations of Bogoslovsky and ours is the following: hatching, at least for a

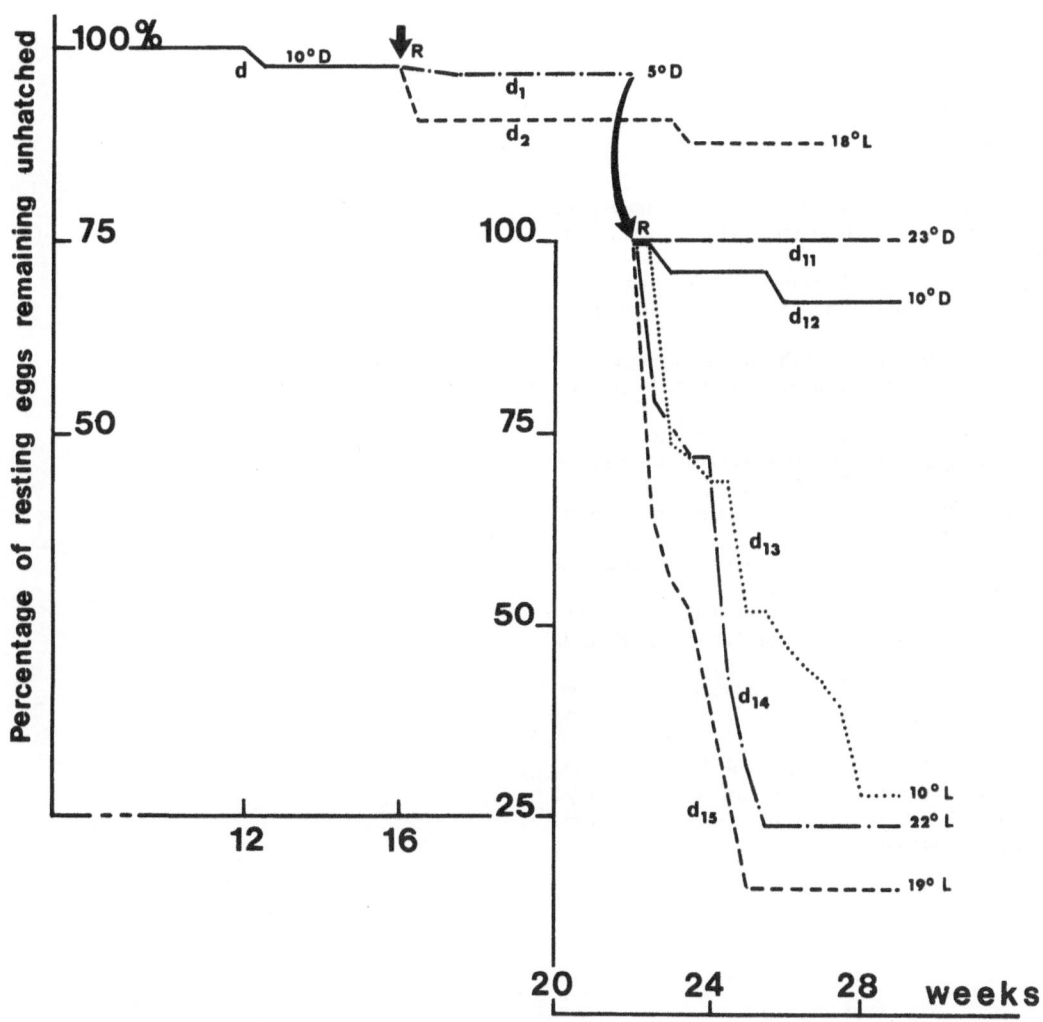

Fig. 2. Decrease of a stock of resting eggs by hatching (in percent), in various conditions (changes are indicated by arrows).

part of the resting eggs, will be due to two different processes: a fast one if the dormant eggs undergo suitable stimulus and a slower one, if they don't.

From the experiments described herein, we conclude that, for *B. rubens,* resting eggs hatchability is increased by a dormant period in darkness at low temperature (conditions occuring when they are in the sediments in winter) followed by illumination and an increase in temperature (spring and summer conditions). In a biotope, this process promotes a fast colonization when external conditions are suitable.

Acknowledgements

We wish to thank Miss S. Des Clers for her help to translate the manuscript in English.

References

Bogoslovsky, A. G. 1963. Material to the study of resting eggs of Rotifers. Communication 1. Byul. Mosk. Obshch. Ispat. Prir. 68: 50-67.

Bogoslovsky, A. G. 1967. Material to the study of resting eggs of Rotifers. Communication 2. Byul. Mosk. Obshch. Ispat. Prir. 72: 46-68.

Bogoslovsky, A. G. 1969. Material to the study of resting eggs of Rotifers. Communication 3. Byul. Mosk. Obshch. Ispat. Prir. 74: 60-79.

Clement, P., Rougier, C. & Pourriot, R. 1977. Les facteurs exogènes et endogènes qui contrôlent l'apparition des mâles chez les Rotifères. Bull. Soc. Zool. Fr. 101, supp.[4]: 86-95.

Gilbert, J. J. 1974. Dormancy in Rotifers. Trans. Am. microsc. Soc. 93: 490-513.

Gilbert, J. J. 1977. Mictic female production in Monogonontes Rotifers. Arch. Hydrobiol. Beih. 8: 142-155.

Nipkow, F. 1958. Beobachtung bei der Entwicklung des Dauereies von Brachionus calyciflorus Pallas. Schweiz. Z. Hydrol. 20: 186-194.

Nipkow, F. 1961. Die Rädertiere im Plankton des Zürich Sees und ihre Entwicklungsphasen. Schweiz. Z. Hydrol. 23: 398-461.

Pourriot, R. & Rougier, C. 1976. Influence de l'âge des parents sur la production de femelles mictiques chez Brachionus calyflorus (Pallas) et B. rubens Ehr. (Rotifères). C. r. Acad. Sci. Paris. sér. D, 283: 1497-1500.

INDUCTION OF SEXUAL REPRODUCTION AND RESTING EGG PRODUCTION IN BRACHIONUS PLICATILIS REARED IN SEA WATER

Esther (wajc) LUBZENS, Rachel FISHLER & Viviane BERDUGO-WHITE

Israel Oceanographic & Limnological Research, Tel-Shikmona, P.O.B. 8030 Haifa, Israel

Keywords: Rotifer resting eggs, application to mariculture

Abstract

Brachionus plicatilis raised in our laboratory in sea water reproduces asexually even under high crowding conditions (at least 40 individuals per ml). Amictic females were induced to produce mictic females, males and resting eggs by reducing the concentration of the sea water culture medium. Mictic females and males appeared predominantly among the progeny produced by the amictic females during 4 days following their transfer into 25% sea water. Resting eggs appeared first 5-12 days after the onset of the experiment. Following the disappearance of males, the culture consisted of amictic females.

Resting eggs produced by the method described above may be preserved for at least three months at -14°C or by desiccation at room temperature. Under the appropriate experimental conditions, resting eggs hatch into amictic females. Since *B. plicatilis* is one of the most commonly used food sources of fish larvae in aquaculture, the methods reported here may offer an easy and versatile way of preserving rotifer culture stock to be used on demand.

Introduction

One of the main aims in growing zooplankton for fish hatcheries is to become independent of the immediate success or failure of the zooplankton culture.

The rotifer *Brachionus plicatilis* is one of the most commonly used organisms in aquaculture, serving as an important food source for fish larvae. An adequate supply of food for fish larvae depends, therefore, on a successful culture of rotifers. One of the possible ways of ensuring a good supply of rotifers would be the building up of reserve stocks and their use when the demand rises. This goal can possible be achieved by: (a) finding methods for inducing rotifers to produce resting eggs in large quantities; (b) finding ways for preserving and hatching the resting eggs.

It has been previously reported that changes in environmental conditions, such as an increase in crowding, cold-shocks, decrease in food quantity and changes in photoperiod (see review by Gilbert, 1974) may induce production of males. Ito (1960) reported that *B. plicatilis* collected from eel ponds (Cl 2-3‰) was induced to produce large numbers of resting eggs when transferred from high chlorinity (18‰) culture media into lower chlorinity media.

Brachionus plicatilis was raised in our laboratory for the past seven years in sea water (38‰) and fed on a variety of marine species of algae. Under these conditions, rotifers reproduced asexually even under high crowding conditions (40 or more individuals per ml). Induction of sexual reproduction was achieved by reducing the salinity of the sea water culture medium. Large numbers of resting eggs could be thus produced and various methods for their preservation and hatching were tested under laboratory conditions.

Material and methods

Brachionus plicatilis was collected in the summer of 1972 from sea water fish ponds near Dor, Israel. The rotifers were since then bred in filtered (0.45 μm) sea water and fed on various species of marine algae. Rotifers used for the experiments reported here were fed on *Chlorella stigmatophora*. Sea water dilutions were carried out with double distilled water.

The appearance of females carrying three or more eggs, mictic females, males and resting eggs was followed in three experimental set-ups:

55

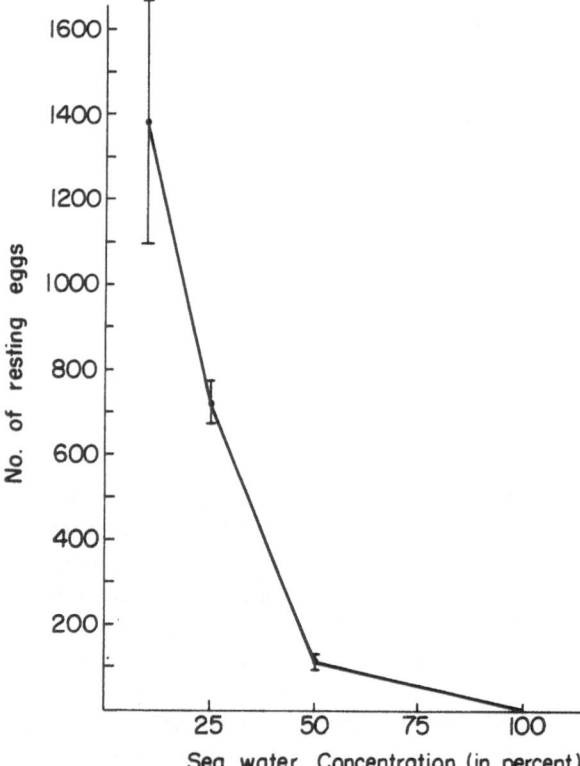

Fig. 1. The number of resting eggs (Mean ± S.E.M.) produced by 50 amictic *B. plicatilis* females, three weeks after their transfer into various sea water concentrations.

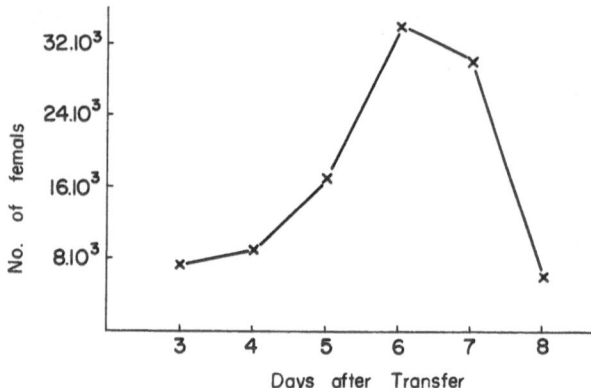

Fig. 2. The average number of females carrying three or more eggs, produced during 8 days by 2000 amictic females transferred into 25% sea water.

1. Fifty or 2000 rotifers were transferred from 100% sea water (SW) into beakers containing 100 ml and 2 litres of diluted sea water, respectively. The number of females carrying three or more eggs and the number of resting eggs were counted at intervals for up to three weeks from the onset of the experiment (see Figs. 1, 2 and 4).

2. Ten amictic females were transferred from 100% SW culture into small petri dishes containing 2 ml of 25% SW culture medium. These adult amictic females were transferred daily into new petri dishes, leaving their progeny behind in the old culture dish. The progeny produced during the first day following the transfer was termed 1st day progeny, the progeny produced during the second day after the transfer was termed 2nd day progeny, etc. This procedure was replicated 8 or 10 times in Experiments 1 and 2, respectively, as reported in Table 1. The appearance of mictic females, males and resting eggs was recorded at daily intervals.

3. Same as in (2) above, except that single amictic females were transferred. The appearance of mictic females, males and resting eggs was followed in females that carried no eggs or one egg at the time of transfer (Experiments 3 and 4, respectively, in Table 1) or on females that carried two or three eggs 24 hours after the transfer (Experiments 5 and 6, respectively, in Table 1). Observations on the appearance of males (Fig. 3) were carried out by recording the presence of males in the petri dishes containing the original transferred amictic females and those of their progeny from 1-8 days after the transfer (1st to 8th day progeny).

Counted batches of resting eggs were frozen in distilled water (-14°C) or left to dry at room temperature and kept in a desiccator. Resting eggs were hatched at room temperature in 100% SW containing *C. stigmatophora* and were followed at intervals for three weeks.

Results and discussion

Production of resting eggs

B. plicatilis raised in our laboratory for several years in sea water (38‰) was induced to produce resting eggs by reducing the concentration of sea water in the culture medium. The number of resting eggs produced increased with the decrease in concentration (Fig. 1). Attempts to culture these rotifers in media containing less than 10% SW were unsuccessful. Ito (1960) obtained maximal numbers of resting eggs in cultures of 7.3‰ chlorinity. The induction of resting egg production, in our laboratory, was independent of the density of the transferred rotifers (Table 1). In addition, feeding rotifers reared in 100% SW with *C. stigmatophora* previously transferred into 25%

Table 1. The time of appearance of the first mictic females, males and resting eggs, after transferring amictic females from 100% sea water into 2 ml of 25% sea water culture medium

	Experiment no.	No. of replicates	Mictic females Time (days)	Mictic females No. of observations	Males Time (days)	Males No. of observations	Resting eggs Time (days)	Resting eggs No. of observations
Females transferred in groups of 10 individuals into petri dishes (30°C)	1	8	2	2	2	2	5	3
	2	10	2	5	3	5	5	3
Females transferred singly into petri dishes* (24-26°C)	3	8	4	2	4	2	11	1
	4	10	7	3	7	3	12	2
	5	10	5	2	5	2	-	-
	6	10	4	1	5	1	-	-

*In Experiment 3 and Experiment 4, females carried no eggs or one egg, respectively, at the time of transfer.
In Experiment 5 and Experiment 6, females carried two or three eggs, respectively, 24 hours after their transfer into 25% sea water.

SW for extended periods of time did not induce sexual reproduction (unpublished results). Rotifers also produced resting eggs, although in smaller numbers, if fed on *Platymonas* and *Phaeodactylum* in diluted sea water media. Most probably, *Chlorella* tolerated lower salinities better than the other two algal species.

The events leading to the production of resting eggs involved the appearance of large numbers of females carrying three or more eggs shortly after the amictic females were transferred from 100% SW into 25% SW (Fig. 2). These females were seldom observed in 100% SW cultures. Females carrying three or more large eggs ($130\ \mu \times 85\ \mu$ in diameter) gave rise to amictic or mictic females, while those carrying several small eggs ($90\ \mu \times 75\ \mu$ in

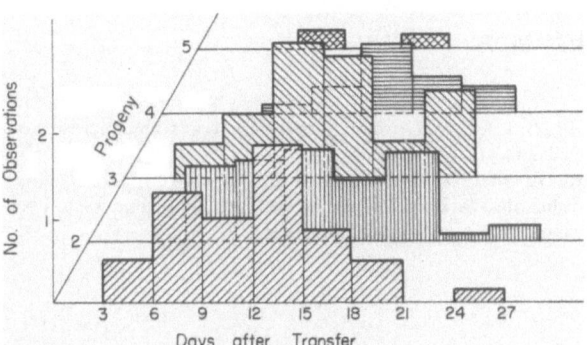

Fig. 3. A histogram showing the average number of petri dishes in which males were observed in the progeny produced by individual amictic females during the first 5 days following their transfer from 100% to 25% sea water. Results were averaged for four different experiments and normalized for the number of replicates and number of observation periods made in each experiment.

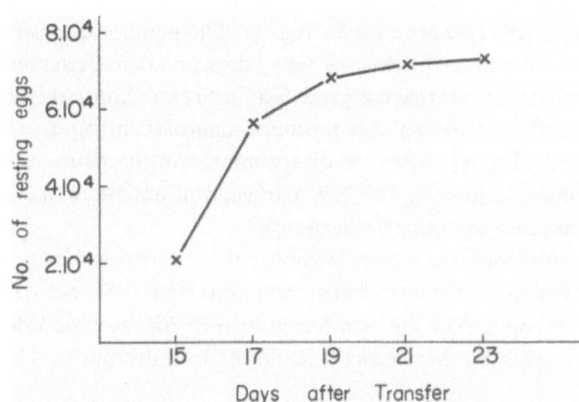

Fig. 4. The average number of resting eggs produced during 23 days by 2000 amictic females transferred into 25% sea water.

Table 2. The effect of the preservation method on hatching of resting eggs.

Method	Duration of preservation (weeks)	No. of eggs preserved	Hatching (%)	
Freezing -14°C	1	100	81±6	(3)
	2	100	70	(2)
	3	100	81	(2)
	4	166	78±13	(6)
	6	100	75	(2)
	8	200	67	(3)
	12	200	85	(1)
	16	200	50	(2)
Desiccation	2	200	100	(2)
	3	200	80	(1)
Desiccation and sonication	25	200	100	(1)

diameter) produced only male offspring. Mictic females were also produced by amictic females carrying one or two large eggs (Table 1). The first mictic females carrying small eggs appeared 2-7 days after the transfer into 25% SW was made (Table 1). This was closely followed by the appearance of males and later by that of resting eggs.

Males appeared among the progeny of between 90-100% of the females transferred into 25% SW. This was revealed by transferring single females into petri dishes containing 25% SW (see Material and methods). Most of the males appeared in the progeny of eggs that were formed by the amictic females during 1-4 days after their transfer into 25% SW (progeny 1-4 in Fig. 3). The number of newly produced males increased for 15 days and then decreased with their eventual complete disappearance. Similarly the number of resting eggs produced increased for up to 19 days (Fig. 4). After the disappearance of the males, the rotifer culture in 25% SW consisted of amictic females that carried one or two large eggs.

Although the factors leading to the production of males by amictic females transferred into 25% SW are yet unknown, they are transferred to the progeny produced closely after the transfer is made and later disappear.

Hatching of resting eggs
In contrast to the report of Ito (1960), resting eggs pro-

duced in 25% SW did not hatch readily in media containing sea water of reduced concentrations. This permitted the harvesting of large numbers of resting eggs from the bottom of the breeding vessel.

Resting eggs may be preserved for at least 12 weeks by freezing at -10°C without signifcant loss of viability (Table 2). Eggs dried and kept desiccated at room temperature retained their viability for up to 3 weeks. Hatching of desiccated eggs preserved for 25 weeks was facilitated by sonification at low energies.

Summary

Large numbers of resting eggs of the rotifer *Brachionus plicatilis* were obtained by transferring amictic females from 100% sea water into 25% sea water culture medium. This resulted from the appearance of males in the progeny produced, by almost all the transferred amictic females, during the first four days following the transfer. The induction of sexual reproduction was mainy due to the change in the salinity of the culture medium. Increasing or decreasing the density of rotifers reared in various sea water concentrations or changing the algal species fed to the rotifers had no effect on this phenomenon.

The resting eggs produced were preserved for 12 weeks at -14°C or for up to 3 weeks in desiccated form at room temperature, without loss of viability.

The methods reported here may offer an easy and versatile way of preserving rotifer culture stocks to be used on demand in fish hatcheries.

References

Gilbert, J. J. 1974. Dormancy in rotifers. Trans. am. microsc. Soc. 98: 490-513.
Ito, T. 1960. On the culture of mixohaline rotifer Brachionus plicatilis, O. F. Müller in the sea water. Rep. Fac. Fish. Pref. Univ. Mie 3: 708-740 (in Japanese).

THE EFFECT OF SOME BIOTIC AND ABIOTIC FACTORS ON THE FERTILITY OF PLANKTONIC ROTIFER SPECIES

S. RADWAN

Academy of Agriculture, Department of Zoology and Hydrobiology, Akademicka 13, 20-934 Lublin, Poland

Keywords: Rotifers, Poland, environmental factors, fertility

Abstract

The influence of some biotic and abiotic factors on the fertility of some planktonic rotifers was studied. Trophic factors, such as community structure and the amount of nannoplankton food available was of considerable influence to rotifer fertility, especially among sedimentators, while abiotic factors had a much smaller influence.

Introduction

Experiments were carried out on planktonic rotifers that carry their eggs and occur in abundance at least part of the year. They were: *Asplanchna priodonta, Brachionus angularis, Filinia maior, Keratella cochlearis, K. cochlearis tecta* and *K. hiemalis, Pompholyx sulcata.*

Fertility was calculated as a fertility coefficient E, according to the formula

$$E = \frac{N_e}{N} \cdot l^{-1}$$

which expresses the ratio of females with eggs (N_e) to all the females of that species (N) per unit volume (1).

Morphometric-ecological characteristics of study sites

The experiments were carried out during 1971-1972 in three small lakes of Eastern Poland, differing by depth, area, and trophic level. Eutrophic, polymictic Lake Bikcze (85 ha) has a maximum depth of 3.5 m; dystrophic, polymictic Lake Brzeziczno (7.5 ha) has a maximum depth of 2.5 m, while a-mesotrophic, dimictic Lake Piaseczno (85 ha) has a depth of 39 m. In the two shallow lakes, thermal and oxygen conditions showed small differences between the surface and bottom layers. In the deeper lake thermal and oxygen stratification was formed during both summer and winter; the epilimnion reached down 6 m, the metalimnion extended between 6-12 m, and the hypolimnion between 12-39 m.

Each lake had a distinctive nannoplankton. In dystrophic Lake Brzeziczno different algal groups succeeded each other in the nannoplankton. Blue-green algae predominated in early Winter and Cryptomonads in late Winter. Chlorococcales were perennial (*Chlorella* sp., *Chlorococcus* sp.) but their greatest biomass was reached in Summer and Autumn. In this lake, the nannoplankton formed a very important fraction of the phytoplankton (Brzek *et al.,* 1975).

In the eutrophic Lake Bikcze the main year-round nannoplankters pertained to the genus *Cryptomonas; Cryptomonas pusilla* was the dominant species. In summer *Koliella longiseta* and *Elacatothrix lacustris* also occurred. The maximum of the nannoplankton in the total phytoplankton biomass was from Winter till early Summer, and the minimum was reached in late Summer.

In mesotrophic Lake Piaseczno the year-round, dominant nannoplankters were *Cryptomonas pusilla, Monoraphidium minutum, Chlorella vulgaris, Mallomonas caudata.* Species from the genera *Ankistrodesmus* and *Oocystis* were also observed. In the epilimnion of this lake the maximum fraction of nannoplankton in the total phytoplankton biomass was in Winter-Autumn and late Summer (Wojciechowska, 1976).

Hydrobiologia 73, 59-62 (1980).   0018-8158/80/0731-0059$00.80.
© Dr. W. Junk b.v. Publishers, The Hague.

The influence of throphic factors on the fertility of rotifers

My results suggest that a considerable influence on the reproduction of Rotifers, especially sedimentators, is exerted by the structure and the amount of nannoplankton biomass in each lake.

The ferility coefficient of the eurytopic *Keratella cochlearis* in dystrophic Lake Brzeziczno and in the epilimnion of the mesotrophic Lake Piaseczno showed a neat negative correlation with the nannoplankton biomass in spring and autumn and a weak one in Summer. Here, the highest values of the fertility coefficient were found in Spring and Autumn and the lowest ones in Summer. In the deeper layers of Lake Piaseczno there was generally a positive correlation between fertility and nannoplankton biomass. In eutrophic Lake Bikcze a positive correlation was found

in Winter and Spring, and a negative one in Summer and Autumn (Fig. 1, 2).

A very high reproduction was found in *Brachionus angularis*, occurring abundantly in Lake Brzeziczno. The maximum reproduction corresponded with the highest nannoplankton biomass which, in this lake, occurred in Summer (Fig. 1). It coincided with a shift in both the phytoplankton composition and in the relation of nannoplankton biomass to the microplankton biomass. In Winter and Spring nannoplankton biomass to the microplankton biomass. In Winter and Spring nannoplankton had been dominating the phytoplankton biomass, and even more clearly so in eutrophic Lake Bikcze; *Cryptomonas, Trachelomonas, Chlorella* and *Scenedesmus* were the commonest genera. But in Summer and early Autumn the nannoplankton biomass decreased as microplankton,

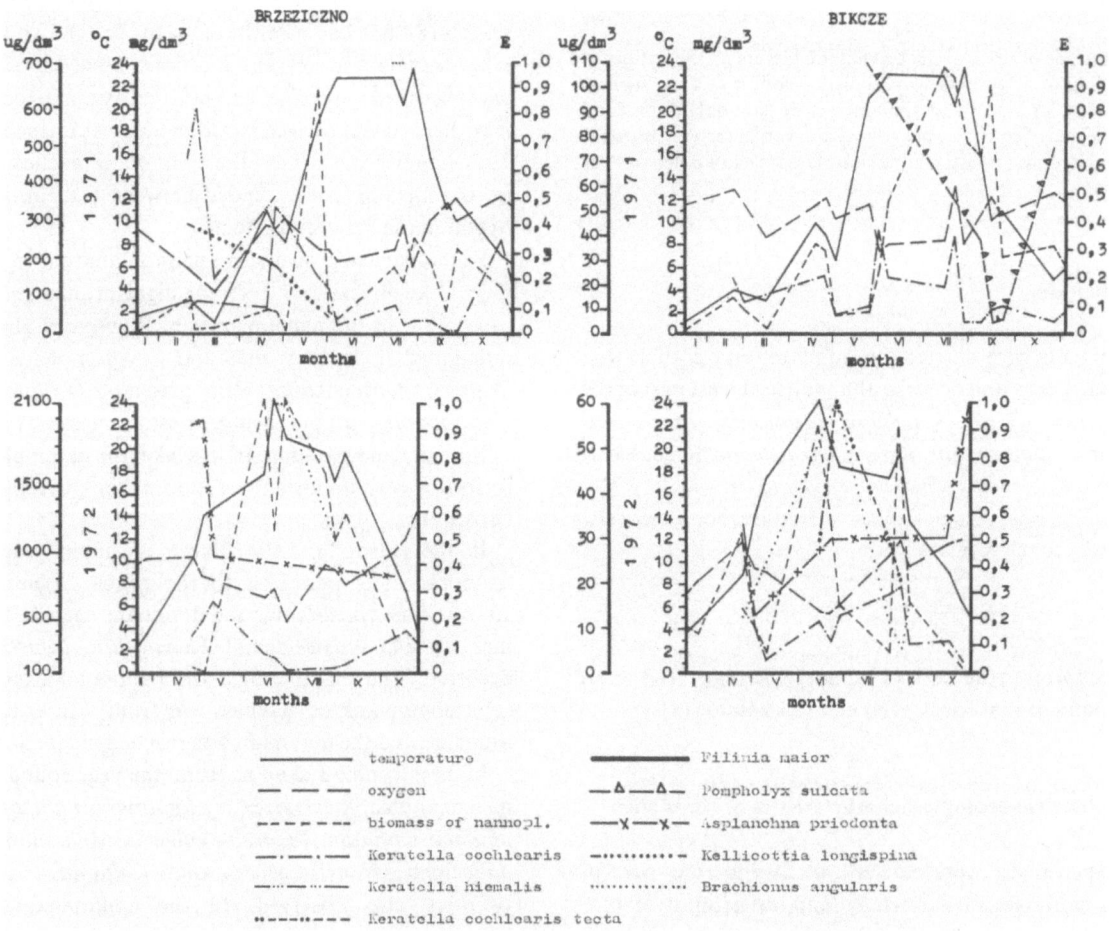

Fig. 1. Correlation between fertility coefficient E of some rotifers species and environmental factors in lakes Bikcze and Brzeziczno.

in which blue-green algae predominated increased: *Apha-nothece clathrata, Microcystis aeruginosa* and *M. incerta* appeared in great abundance. Excessive growth of blue-green algae inhibits the development of most sedimen-tators by eliminating most of the small algae which are the basic food of the phytophagous sedimentators (Edmond-son, 1965; Pourriot, 1965; Dumont, 1977).

In addition, it has been suggested that toxic substances released by blue-green algae can directly limit the fertility of planktonic rotifers (Dumont, 1977).

Cold stenothermal *Kellicottia longispina* occurred in numbers in Lake Piaseczno only. Its fertility coefficient was lowest in Winter and early Spring. A clear positive correlation with the nannoplankton fraction ($< $ 10 μm) was found in the hypolimnion only, while in the epi-and metalimnion correlations were variable. Their fluctua-tions were, however, closely connected with changes in nannoplankton biomass and community structure. Dur-ing peak reproduction of *K. longispina, Cryptomonas pusilla,* and the slightly smaller *Monoraphidium minu-tum, Chlorella vulgaris* and *Mallomonas caudata* made up the main nannoplankton biomass. These small algae were thus consumed in great numbers by *Kellicottia*

longispina. Conversely, as soon as the microplankton biomass predominated over the nannoplankton biomass, the fertility of this species sharply sank. Dumont (1977), too, attracted attention to the high correlation between the occurence of *Kellicottia* and the abundance of the genus *Cryptomonas.*

In Lakes Bikcze and Brzeziczno, for unknown reasons, *Kellicottia* occurred only sporadically and it was not possible to relate its presence to that of phytoplankters.

Another cold stenothermal species, *Filinia maior,* re-sponded similarly to changes in nannoplankton biomass.

The value of the fertility coefficient in *Asplanchna pri-odonta* is interesting because it is negatively correlated to fertility in *Keratella cochlearis* (Fig. 1, 2), suggesting that *Asplanchna priodonta* is feeding on *Keratella* (Radwan & Popiolek, 1976). *Asplanchna priodonta,* above a certain abundance treshold, thus appears to regulate the population density of some sedimentators.

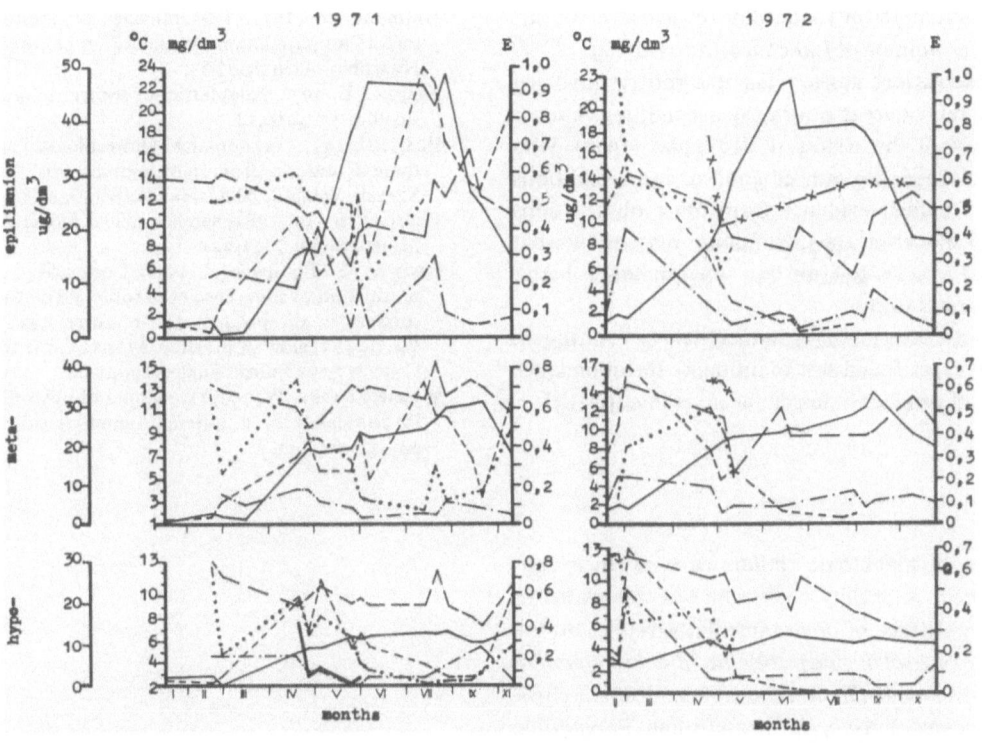

Fig. 2. Correlation between fertility coefficient E of some rotifers species and environmental factors in lake Piaseczno.

The effect of abiotic factors on the fertility of rotifers

Abiotic factors had only a weak influence on the fertility of the rotifers studied here. Still, a relation can exist between the reproduction of some species, especially cold stenothermal ones, and temperature and oxygen levels in Winter (King, 1972; Hofmann, 1977).

My results illustrate this point. Such species as *Asplanchna priodanta* and *Pompholyx sulcata* reached highest fertility when the temperature was about 10°C (Spring and Autumn), whereas good growth in *Keratella cochlearis* took place over a considerable range of water temperatures. It reached high fertility both below 10°C and above 20°C, but the highest values coincided with high temperatures (Fig. 1, 2).

Summer species such as *Keratella cochlearis tecta* and *Brachionus angularis* reached maximum fertility at a water temperature af about 18°C.

Cold stenothermal species *(Kellicottia longispina* and *filinia maior)* should show positive correlations between Fertility and low water temperature, as is seen in the meta- and hypolimnion of deep Piaseczno Lake (Fig. 2).

Some relation was found between fertility in the cold stenothermal rotifers *Kellicottia longispina* and *Pompholyx sulcata,* and the oxygen content of the water. They reached maximum fertility while low oxygen levels occurred in the hypolimnion of Lake Piaseczno (Fig. 2).

These observations suggest that the rotifers involved can live and reproduce in poor oxygen conditions but are at variance with the results of Hofmann (1977), who believes that these species need good oxygen conditions for growth. In fact, evidence from other observations (Pejler, 1956; Pourriot, 1965; Hofmann, 1977) shows that stenothermal species tolerate low oxygen levels better than eurythermal forms.

Other abiotic factors (pH, conductivity, Ca^{++}, nitrogen, phosphorus) were found not to influence the abundance and fertility of pelagic rotifers in the lakes investigated.

Results show that structure and amount of nannoplankton biomass influence rotifer reproduction considerably, especially among sedimentators. This relation is perennial in mesotrophic and dystrophic lakes, but limited to Winter and Spring in a eutrophic lake.

The population abundance of small *Keratella* was found to determine the fertility of its predator, *Asplanchna priodonta.*

Abiotic factors showed a weaker influence on the fertility of rotifers. Only temperature and oxygen content influenced the fertility of some species, especially cold stenothermal ones.

References

Brzek, G., Kowalczyk, C., Lecewicz, W., Radwan, S., Wojciechowska, W. & Wojciechowski, I. 1975. Influence of abiotic environmental factors on plankton in lakes of different trophy. Pol. Arch. Hydrobiol. 22: 123-139.

Dumont, H. J. 1977. Biotic factors in the population dynamics of rotifers. Arch. Hydrobiol. Beih. 8: 98-122.

Edmondson, W. T. 1965. Reproductive rate of planktonic rotifers as related to food and temperature in nature. Ecol. Monogr. 35: 61-111.

Hofmann, W. 1977. The influence of abiotic environmental factors on population dynamics in planktonic rotifers. Arch. Hydrobiol. Beih. 8: 77-83.

King, C. E. 1972. Adaptation of rotifers to seasonal variation. Ecology 53: 408-418.

Pejler, B. 1957. Taxonomical and ecological studies on planktonic Rotatoria from Northern Swedish Lapland. Kungl. Svensk. Vetensk. Akad. Handl., Ser. 6: 1-68.

Pourriot, R. 1965. Recherches sur l'écologie des rotifères. Vie Milieu, Suppl. 21: 1-224.

Radwan, S. Popiołek, B. 1977. Comparison of biomass and production of some species of rotifers Rotatoria in spring and summer in the pelagic zone of three lakes in the Leczna-Włodawa region. Ann. Univ. Mariae Curie Skłodowska, Sec. C 32: 277-291 Polish, English summary.

Wojciechowska, W. 1976. Dynamics of phytoplankton biomass in two lakes of a different limnological character. Ekol. pol. 24: 237-252.

Summary

Only species of planktonic rotifers carrying their eggs were examined. A fertility coefficient was calculated to express the influence of environmental variables on the reproduction of *Asplanchna priodonta, Brachionus angularis, Filinia maior, Kellicottia longispina, Keratella cochlearis, K. cochlearis tecta, K. hiemalis* and *Pompholyx sulcata.*

ASPECTS OF THE FEEDING BEHAVIOR AND TROPHIC ECOLOGY OF SUSPENSION-FEEDING ROTIFERS

Peter L. STARKWEATHER

Department of Biological Sciences, University of Nevada, Las Vegas, Las Vegas, Nevada 89154 U.S.A.

Keywords: rotifers, feeding, diet, food selectivity, field, regulation, chemostat

Introduction

While the evidence is not yet adequate to provide a general assessment of rotifer feeding in the dynamics of planktonic communities, there can be no question but that the feeding biology of this group is a rich and exciting field for freshwater ecologists and natural historians. Two recent reviews, those of Dumont (1977) and Pourriot (1977), provide ample evidence for this assertion, together summarizing some 70 years of research on rotifer feeding behavior. This paper is a further discussion of certain aspects of the feeding behavior and feeding ecology of free-swimming, suspension-feeding rotifers. I will deliberately omit consideration of sessile or predacious genera since the feeding biology of these forms has been reviewed very recently; sessile rotifers are examined by Wallace (1980) and the predacious genus *Asplanchna* by Gilbert (1979).

The comments below relate to four areas of rotifer trophic ecology in which significant contributions have been recently made. The mechanisms by which members of the phylum regulate the quantity and quality of their diets are not fully understood. Accordingly, I review below the substantial literature describing the patterns of rotifer feeding rates relative to changes in food cell density. In addition, there is new work quantitatively examining the degree of food selectivity demonstrated by brachionids. The second section summarizes this work. As a third category, I have reviewed the literature on *in situ* determinations of rotifer feeding rates. This material is largely the result of experimentation between 1978 and the present. The final section of the current review considers the integration of chemostats with rotifer feeding biology. Experimental study with this technique appears likely to foster rapid progress in rotifer feeding, energetics and population dynamics.

Quantitative regulation of feeding rates

Perhaps the most obvious aspect of dietary regulation in rotifers is the modification of feeding rates with respect to food cell density. Since food consumption is the initial step in grazer productivity, the patterns of change of clearance and ingestion rates are of central importance in the trophic and population ecology of these animals. In addition, quantifying the potential grazing of a rotifer species or population permits evaluation of the impact that the rotifers may have on the food assemblage, both as a discrete taxonomic group and as a component of the larger suspension-feeding community.

Two terms that are conventionally used to describe the quantitative feeding activities of rotifers, and other grazing zooplankton, are clearance (or filtering) rate and ingestion rate. Ingestion rate is a straightforward unit, referring to the biomass or number of cells consumed by an animal (or per unit weight of an animal) in an interval of time. Clearance rate is ideally an estimate of the volume of water processed by an animal (or unit weight, as above) in a time interval during which it is engaged in feeding. In practice, the measured clearance rate conforms more exactly to Erman's (1956) usage, i.e. the volume of water which contains the biomass or number of cells actually ingested by the animal per unit time. This departure from the ideal is due to the fact that suspension-feeders are seldom 100% efficient in the removal of particles from water which they have processed, and is especially striking

63

Hydrobiologia 73, 63-72 (1980).   *0018-8158/80/0731-0063$02.00.*
© *Dr. W. Junk b.v. Publishers, The Hague.*

in the case of rotifers which readily reject particles already collected or behaviorally exclude particles before processing by the principal feeding structures.

Erman (1956) determined the clearance and ingestion rates of *Philodina roseola* and *Brachionus urceolaris* var. *rubens* (*B. rubens*) fed either *Scenedesmus acuminatus* or *Chlorella vulgaris*. For both foods, *P. roseola* clearance rates were constant (abour 1 μl animal^{-1}h^{-1}) over a cell density range between 10^3 and $> 10^7$ cells ml^{-1}. For the brachionid, on the other hand, clearance rates varied from less than 1 μl animal^{-1}h^{-1} to over 7 μl animal^{-1}h^{-1}, with the lowest rates obtained at the two extremes of food density used and the maximal rate at about 10^4 cells ml^{-1}. Pilarska (1977) found a similar pattern for *B. rubens* feeding on *C. vulgaris,* with a maximal clearance rate of 11.3 μl animal^{-1}h^{-1} near 10^4 cells ml^{-1}. The rotifers in Pilarska's study achieved ingestion rates which approximated 700 cells animal^{-1}h^{-1} (somewhat lower than those in Erman's work) and which became density-independent at greater than 3×10^6 cells ml^{-1}.

Comparable clearance and ingestion rates have been reported for *Brachionus plicatilis*. Ito (1955) reported a clearance rate of slightly greater than 3 μl animal^{-1}h^{-1} (from Doohan, 1973) based on *Synechococcus* consumption; this rate is within the range of other estimates but is certainly quite high considering the elevated cell density (ca. 8×10^6 cells ml^{-1}) under which it was obtained. Hirayama and Ogawa (1972) found *B. plicatilis* clearance rates between 0.18 and ca. 6.0 μl animal^{-1}h^{-1} inversely proportional to cell density *(Chlorella)*. In later work with the same rotifer species but with two different foods *(Chlamydomonas* and *Olisthodiscus),* Chotiyaputta & Hirayama (1978) found clearance rate values higher and lower respectively than those obtained with *Chlorella.* Interestingly, the critical concentrations (critical concentration is defined as the maximal food density at which ingestion rate is directly proportional to food density) differed for the three food types, ranging between ca. 10^5 cells ml^{-1} for *Olisthodiscus* to 2×10^6 cells ml^{-1} for *Chlorella.* Doohan's (1973) clearance rates for *B. plicatilis* varied unsystematically over the tested range of *Dunaliella* densities, averaging close to 1.0 μl animal^{-1}h^{-1}. This fairly low rate might be due to high food density (near 10^6 cells ml^{-1}), inappropriateness of the food alga or a methodological error in the rate determinations (see Starkweather & Gilbert, 1977a).Like Hirayama and his colleagues, Dewey (1976) made feeding rate estimates for *B. plicatilis* feeding on three separate food types. Unlike the earlier work, the different cell suspensions (two species of *Dunaliella* plus

Isochrysis galbana) produced only slight variation in clearance and ingestion rates. Dewey's (1976) clearance rate estimates fell between 1.0 and 9.0 μl animal^{-1}h^{-1} (20°C); the accompanying ingestion rate curves showed critical concentrations near 2×10^5 cells ml^{-1} for the *Dunaliella* spp. but no region of density-independence for consumption of the smaller *I. galbana*.

Feeding rates for *Brachionus calyciflorus* appear to fall in the same general range as those reported for the other brachionid species. Pennington (1941) found clearance rates approximately 0.3-0.5 μl animal^{-1}h^{-1} at food densities (of *Diogenes rotunda*) near 5×10^6 cells ml^{-1}. Erman (1962b) fed *B. caliciflorus* suspensions of *Lagerheimia ciliata* at densities between 2 and 140 μg dry weight ml^{-1}. With densities lower than 60 μg ml^{-1}, clearance rate was constant (2-5.6 μl animal^{-1}h^{-1} for different experiments) resulting in ingestion rate changes in direct proportion to food density. At a higher concentration (140 μg ml^{-1}), clearance rate dropped and ingestion rate became density-independent. Galkovskaja (1953) found *B. calyciflorus* clearance rates which were constant (ca. 1.0 μl animal^{-1}h^{-1}) at cell densities below 5×10^5 cells ml^{-1}, but which declined, as in Erdman's study, at higher food concentrations. Ingestion of the *Scenedesmus* cells increased with algal density until $0.5 - 1.0 \times 10^6$ cells ml^{-1}, above which ingestion rate was constant. Using *Chlorella pyrenoidosa* as the food cell, Halbach and Halbach-Keup (1974) tested *B. calyciflorus* feeding at densities between 0.05 and 5.0×10^6 cells ml^{-1}. Their data show a peak in clearance rate at 5×10^5 cells ml^{-1} (rate: 3.4 ± 0.12 μl animal^{-1}h^{-1}) and a minimal value (rate: 0.59 ± 0.01 μl animal^{-1}h^{-1}) at the highest tested cell concentration. In contrast to other results for brachionids, Halbach & Halbach-Keup (1974) report a decline, rather than a plateau, in ingestion rates found in the most dense cell suspensions.

More recent results for *B. calyciflorus* feeding regulation have been obtained for the rotifers exposed to microfungi and bacteria as well as algae. Starkweather & Gilbert (1976b) compared clearance and ingestion rates based on experiments run with *Euglena gracilis* and *Rhodotorula glutinis* at densities between 0.01 and 1000 μg dry wt ml^{-1}. For both foods, clearance rates varied between 45-50 μl animal^{-1}h^{-1} and < 0.5 μl animal^{-1}h^{-1} at low and high cell concentrations, respectively. Ingestion rates, on the other hand, differed substantially comparing the two foods, with yeast consumption far exceeding that of the alga except at the lowest food densities. The bacterium *Aerobacter (Enterobacter) aerogenese* elicited very low clearance rates (mean values always < 1.0 μl animal^{-1}h^{-1})

which did not vary systematically with cell density (Starkweather *et al.*, 1979). Accordingly, ingestion rates for *B. calyciflorus* feeding on *A. (E.) aerogenese* are distinctly density-dependent, with no obvious plateau in rate within the tested range. The differences among the feeding vs. food density relationships for the three foods equate with similar results of Hirayama & Ogawa (1972) and Chotiyaputta & Hirayama (1978) using *B. plicatilis*. For both species, different feeding rates and different critical concentrations were found using a variety of food cell types. A summary of the effects of food type on critical concentration for 3 species of *Brachionus* is provided in Table 1.

King (1967) measured the feeding rates of the sublittoral brachionid species *Euchlanis dilitata* based on three algal species, *Euglena gracilis*, *E. geniculata* and *Chlamydomonas reinhardti*. In each case, clearance rates declined with increased food density from ca. 10-16 μl animal^{-1}h^{-1} at 1.6 μg dry wt. ml^{-1} to 3-4 μl animal^{-1}h^{-1} at 49 μg ml^{-1}. The two *Euglena* species elicited identical feeding rates while both clearance and ingestion rates differed somewhat from those obtained with *C. reinhardti*.

A number of patterns arise from the results described above. It should be kept in mind, of course, that these patterns may be appropriate only for brachionid rotifers under laboratory conditions. A greater degree of generality must await study of additional families of rotifers.

1. Clearance rates may vary between less than 0.1 μl animal^{-1}h^{-1} to at least 50 μl anumal^{-1}h^{-1} depending upon food type and food density. By far the greatest number of published clearance rates fall between 1 and 10 μl animal^{-1}h^{-1} in the 20-25°C temperature range.

2. Clearance rates generally decrease with increasing food density, either in a progressive and continuous manner or abruptly at a high cell concentration. Some authors report a unimodal pattern of clearance rate vs. cell density with slight reduction in feeding activity at very low food levels. These declines are probably the result of malnutrition of animals maintained on a low food regime for long periods, and are not indicative of an active behavioral response to food shortage (see Starkweather & Gilbert, 1977b).

3. Ingestion rates are extremely variable, with published values from tens to many thousands of cells consumed animal^{-1}h^{-1}. The lowest ingestion rates (on a biomass basis) are found with the largest and smallest tested food types; high food consumption is associated with intermediate cell sizes (*n.b.*: this generalization may fail in particular instances involving selective behavior – see appropriate statements, below).

4. Ingestion rate increases with increased food cell density at low to moderate suspension concentrations. In dense food preparations, rotifer ingestion rates reach a maximal value and remain constant (the 'critical concentration', see Table 1). In some experiments, an ingestion rate plateau was not reached at normal cell densities, reflecting the continuous acceptance of food in proportion to its availability, even if such superfluous consumption is deleterious.

A variety of mechanisms have been proposed to explain the progressive (or occasionally abrupt) reduction in clearance rates with increased food density. The immediate effectors of such changes may be particulate interference with the feeding apparatus (Halbach & Halbach-Keup, 1974), inhibition of ciliary activity as reflected in reduced swimming speeds (Erman, 1956; J. V. Jackson, pers. comm.), increased food rejection frequency or intermittency of feeding effort (Starkweather & Gilbert, 1977b). In addition, Gilbert & Starkweather (1977, 1978) have proposed that the formation of 'screens' by the pseudotrochal cirri of *Brachionus* may dramatically reduce clearance rates in that genus, especially in dense suspensions of large food types.

The above regulatory mechanisms may be ultimately induced by nutritional causes such as reduced assimilation efficiency due to rapid gut passage of food (hence malnutrition) (Halbach & Halbach-Keup, 1974), or direct inhibition by algal, fungal or bacterial toxins or metabolites (Halbach & Halbach-Keup, 1974; Pilarska, 1977). It is also conceivable that rotifers may directly sense cell density, through contact frequency, for instance, and modulate feeding in response to such stimuli. Work on *B. calyciflorus*, examining pseudotrochal screening behavior resulting from starvation or rapid transfers between food types, supports at least a partial role of nutrition in determining feeding activities (Gilbert & Starkweather, 1978).

Selective feeding

As discussed in another section, rotifers exposed to natural assemblages of particulate foods consume a diverse collection of phytoplankton, detritus and bacteria. It is also true, however, that while diets may vary among species, or within a species found in waters of differing character (or at different seasons), many rotifers collect

Table 1. Summary of the influence of food type on the feeding dynamics of 3 species of Brachionus. Critical concentration is the maximal food density at which ingestion rate is directly proportional to cell concentration.

Species	Food type	Critical concentration		Temperature (°C)	Author
		dry weight (μg ml^{-1})	cell density (cells ml^{-1})		
Brachionus calyciflorus	Lagerheimia	100		18.5 - 20.0	Erman (1962b)
	Scenedesmus		1.5×10^6	19.0 - 20.0	Galkovskaja (1963)
	Chlorella		5×10^5	19.0 - 21.0	Halbach and Halbach-Keup (1974)
	Euglena	5-10	$1\text{-}2 \times 10^4$	22.0 - 28.0	Starkweather and Gilbert (1977b)
	Rhodotorula	$\sim$100	$\sim 5 \times 10^6$	"	"
	Aerobacter	>100	$> 10^{12}$	20.0 - 24.0	Starkweather et al. (1979)
B. plicatilis	Chlorella		2.1×10^6	24.0 - 26.0	Hirayama and Ogawa (1972)
	Dunaliella		2×10^5	20.0	Dewey (1976)
	Isochrysis		$>1.2 \times 10^6$	"	"
	Chlamydomonas		1.5×10^5	23.0	Chotiyaputta and Hirayama (1977)
	Olisthodiscus		$\sim 10^5$	23.0	"
B. rubens	Scenedesmus		$> 10^5$		Erman (1956)
	Chlorella		3×10^6	19.6 - 20.4	Pilarska (1977)

and ingest food items in proportions which may substantially depart from their relative availabilities in the surrounding water. Such dietary specialization has made it possible to classify several species groups based on their apparent feeding preferences (Erman, 1962a; Pourriot, 1965).

There are relatively few laboratory studies available which have directly investigated the selective feeding capabilities of suspension feeding rotifers, and those which have are all based on the genus Brachionus. Erman (1962b) demonstrated that Brachionus calyciflorus feeding rates determined with 20 cell types showed great variability (up to 10-fold differences) even though many of the food cells were of similar size. Other workers have found comparable degrees of difference among Brachio-

nus feeding rates based on different foods, but cell sizes or cell shapes were always quite different in those cases (Dewey, 1976; Pilarska, 1977; Chotiyaputta & Hirayama, 1978; Starkweather & Gilbert, 1977b, 1978; Starkweather, 1979). Erman (1962b) was the first to report experiments specifically designed to quantitatively determine the degree of food selection which may occur in simple mixtures of cell types. Each of his mixtures was composed of 100 μg ml^{-1} of Lagerhemia ciliata combined with 100 μg ml^{-1} of one of 13 other algal varieties. Relative to L. ciliata, 2 algae were selectively removed, 4 algae were differentially avoided, and the remaining 7 were ingested at the same rate. The electivity indices for the 13 algae corresponded roughly to the rates at which each cell type was consumed compared to L. ciliata, when each was available in pure suspension. In both sets of comparisons, no clear relationship was found between electivity and cell size or electivity and cell shape.

Erman (1962b) made the additional exciting observation that *B. calyciflorus* could distinguish (as demonstrated by different electivities relative to *L. ciliata*) between conspecific strains of food algae; the results show 4-fold differences between the electivities of paired strains of both *Chlorella vulgaris* and *C. terricola*.

Chotiyaputta & Hirayama (1978) measured the feeding rates of *Brachionus plicatilis* feeding on mixtures of *Chlamydomonas* and *Olisthodiscus*. They found differences in apparent filtering rates comparing the two cell types, with *Chlamydomonas* the preferred food in all cases. In general, the degree of 'selection' of *Chlamydomonas* was inversely proportional to *Chlamydomonas* density but was directly proportional to *Olisthodiscus'* relative concentration. *Olisthodiscus* ingestion was largely unaffected by the presence of the alternative cell type in the feeding mixtures, except with regard to differences in absolute feeding rate compared to values obtained with pure suspensions.

In addition to showing that *B. plicatilis* are able to distinguish between food cells of different species, Chotiyaputta & Hirayama (1978) found that the rotifers had feeding rates on *Chlamydomonas* which differed depending upon the age of the algal culture. Rates obtained with algae growing in exponential culture phase were at least twice those determined with senescent food. Unfortunately, the authors did not test exponential and senescent phase cells in a mixture to determine if the *Brachionus* could selectively consume one type of cell when both were available simultaneously.

Additional direct tests of food selection have been performed using *Brachionus calyciflorus* (Starkweather & Gilbert, 1978, in prep; Starkweather, 1979; Starkweather *et al.,* 1979). In these studies, the rotifers were presented with mixtures of *Euglena gracilis* and *Rhototorula glutinis* or of *R. glutinis* and *Aerobacter (Enterobacter) aerogenese.* In one set of experiments, *E. gracilis* and *R. glutinis* were mixed in equal quantities (by dry weight) but with total biomass varying from 0.1 to 200 μg ml^{-1}. In these and other experiments using mixed food suspensions, the constituent cell types were labeled with different radio-isotopes, permitting the measurement of feeding rates of the two foods simultaneously. The relationships between ingestion rate and food density differed for the yeast and the alga; the former food was always ingested in proportion to its absolute availability while *Euglena* ingestion

varied in a complex pattern relative to total cell concentration (Starkweather, 1979; Starkweather & Gilbert, in prep.). As was true in Chotiyaputta & Hirayama's (1978) study using *B. plicatilis,* electivity values obtained from mixture experiments with *B. calyciflorus* reflected the relative feeding rates obtained for the different foods in pure suspensions.

In other experiments (Starkweather, 1979; Starkweather & Gilbert, in prep.), Starkweather and colleagues mixed two cell types in differing proportions; either keeping the combined density constant, or shifting total density by adding variable quantities of one cell type to fixed amounts of another. Three classes of results were found, also similar to those of Chotiyaputta & Hirayama (1978): 1) selection of a particular food type is usually inversely proportional to the relative abundance of that food in a two-species mixture, 2) total food ingestion may change depending upon the relative abundance of foods in simple mixtures even if the overall availability of food remains the same, and 3) the ingestion of some foods is more affected by the presence of alternative cell types than is the ingestion of other foods. By way of example for this last result, *E. gracilis* consumption by *B. calyciflorus* is dramatically affected by the presence of even small quantities of *R. glutinis* in a mixture, but *R. glutinis* ingestion rates are independent of the quantity of the alga present (Starkweather, 1979).

Field determinations of rotifer feeding

Ecologists concerned with trophic relations of rotifers, while often relying on laboratory observations for fundamental data on feeding behavior, are justifiably uncomfortable when called upon to generalize these results in terms of the activities of natural populations. To reduce this discomfort, and to provide a link between laboratory studies and animal feeding activities *in situ,* three approaches have been taken: gut content analysis, correlation of rotifer and phytoplankton (as well as detritus and bacteria) dynamics, and direct measurement of feeding rates using radioisotope techniques.

Examinations of field-collected rotifers, with special attention to gut contents, have been accumulating for many years. Naumann's (1923) work characterized the diets of seven genera (eight species) including consideration of both the size and qualitative nature of food consumed. Erman (1962a) studied several additional species, stressing the high degree of feeding specialization apparent in many of the forms. Similar studies are available by

Table 2. Field determinations of clearance rates for various suspension-feeding rotifers.

Species	Method	Clearance rates (μl animal^{-1}h^{-1})	Author(s)
Mixed species*	^{14}C-labeled natural seston	0.13 - 1.18	Nauwerck (1959)
Polyarthra dolichoptera	^{32}P(or ^{33}P)-labeled tracer cells	0.02 - 1.69	Bogden, et al. (1980; in prep.)
Polyarthra euryptera	"	0 - 0.36	"
Keratella cochlearis cochlearis (small morph)	"	0.45 - 2.53	"
K. cochlearis cochlearis (large morph)	"	0.29 - 8.12	"
Kellicottia bostoniensis	"	0.97 - 1.31	"
Conochilus dossuarius	"	3.01 - 4.73	Starkweather and Bogden (1980)
Brachionus calyciflorus	^{35}S-labeled tracer cells	1.45	T.M. Frost (pers. commun.)
B. havanaënsis	"	1.15 - 1.56	"
Keratella americana	"	0.50 - 2.09	"
Polyarthra sp.	Labeled tracer cells	0.5 - 1.0	J.F. Haney et al. (pers. commun.)

*species combinations varied during experiment but included: Polyarthra dolichoptera, Synchaeta tremula, S. oblonga, Keratella hiemalis, K. cochlearis, Filinia terminalis and Kellicottia longispina.

Nauwerck (1963), Pejler (1957), Hillbricht-Ilkowska (1972), Infante (1978), and especially by Pourriot (1965), who confirmed the nutritional value of many foods found in guts by performing subsequent laboratory culture experiments. These studies, and many others (see reviews of Pourriot, 1977 and Dumont, 1977) provide a clear indication that rotifers as a group may consume a wide range of natural sestonic foods, but that individual species may show extreme forms of diet specialization even in the presence of heterogeneous particle assemblages.

A variation of gut content analysis has been used by Gliwicz (1969) to determine the food size specificity of

Keratella and *Conochilus*. Gliwicz added finely ground sand to natural seston and examined the animals' guts after a 3 h. feeding interval. The results indicate that *Keratella cochlearis*, *K. quadrata* and *Conochilus uniornis* all have a degree of differential feeding based primarily on food size, but it is not clear how the ingestion of these inert tracers relates to consumption of natural organic particles.

Despite the value of gut content analysis in describing the qualitative aspects of rotifer diet, the technique suffers from its failure to provide information on the rate of food collection or ingestion, and from its inappropriateness for detection of soft-bodied or otherwise easily disrupted food cell types. Such foods are common in rotifer diets (Erman 1962a) and, as pointed out by Pourriot (1977), they are easily broken by the mastax or in the stomach. As a result, many important foods may be missed by such analysis unless identifiable fragments or distinctive cell pigments persist in the digestive tract. This disadvantage is amplified since the assimilation of a food type may be related to the degree to which it is physically disrupted during feeding (Infante, 1973). Thus, it is possible that in some cases the most nutritionally important foods are the least likely to be reliably observed or identified.

The second methodology used to assess the feeding activities of natural rotifer populations has been the statistical correlation of rotifer population parameters with the abundance of particular phytoplankton species on species groups (Nauwerck, 1963; Edmondson, 1965; Zimmerman, 1974; Halbach & Halbach-Keup, 1974; review of Dumont, 1977; reanalysis of Nauwerck's data by Cushing, 1976). This approach rests on the assumption that high birthrates and population sizes of rotifers occur during periods of peak biomass of readily consumed and utilized food types. A high correlation coefficient would logically imply, therefore, that the rotifers eat the particular food or foods in question.

This technique may be misleading in at least two respects. First, algal biomass is a static measure and may not necessarily reflect the actual rate of supply of various food types to the rotifers. The importance of less abundant but more quickly reproducing phytoplankton will be underestimated or missed altogether, especially if the phytoplankton are grazed rapidly (see critique of Lewis, 1977). Secondly, rotifers may not necessarily select the most abundant potential foods; many species of rotifers, as noted in the gut content studies, may preferentially consume rare particles or at least consume foods in proportions not reflecting their relative abundance in the water column.

Correlation techniques, like gut content studies, provide only qualitative indications of what natural rotifer populations are feeding on. To determine the rates at which the animals are removing particles from the surrounding medium requires a more direct quantitative analysis.

Until very recently, the only available *in situ* estimations of feeding rates for suspension-feeding rotifers were those of Nauwerck (1959). He determined clearance rates of mixed species populations feeding on ^{14}C-labeled natural foods from Lake Erken. Values ranged from 0.13 μl animal^{-1}h^{-1} for *Polyarthra dolichoptera* and *Synchaeta tremula* to 1.8 μl animal^{-1}h^{-1} for two *Synchaeta* species plus *Keratella hiemalis*. These rates are of the same order as those found in laboratory studies (Pourriot, 1977 and see Quantitative Regulation, above), but direct comparisons are difficult because of methodological and species differences.

Nauwerck's (1959, 1963) technique has the advantage of quantifying feeding rates on natural particles. However, the procedure estimated feeding activity only on photosynthetically-active plankton and ignores the contribution of detrital particles and heterotrophic cells, which are known to be important dietary constituents for many rotifers (Edmondson, 1957; Erman, 1962a; Pourriot, 1965). Furthermore, the photosynthetic assemblage is unlikely to be labeled uniformly (Rigler, 1971), allowing additional uncertainty about which autotrophic particles constitute the main fractions of the animals' diets. Thus, unless some particle types can be differentially labeled in multiple treatments (as in Saunder's (1963) use of ^{14}C-starch as a heterotroph tracer), or unless a particle fractionation is done before labeling (Bogdan & McNaught, 1975), Nauwerck's technique is of limited value in determining either overall feeding rates or the degree of differential feeding on various food types in the natural seston assemblage.

More recently, Bogdan, Gilbert & Starkweater (1980, in prep.) and Starkweather & Bogdan (1980) have used a modification of Haney's (1971, 1973) method to determine clearance rates of field populations of *Keratella cochlearis* (2 morphs), *Polyarthra dolichoptera*, *P. euryptera*, *Conochilus dossuarius* and *Kellicottia bostoniensis*. This technique requires the introduction of radioactively-labeled tracer particles into an enclosed sample of the rotifer community, a step accomplished with the use of a combination sampling/incubation chamber (see Haney, 1971). Using either one or two tracer particle types for each feeding estimate, Bogdan *et al.* (in prep.) have examined seasonal, diel and vertical variation in the feeding activity of the dominant rotifers in their system. In addition, they have quantitatively documented differential feeding on bacteria, algae, yeast and detritus tracer particles by sympatric rotifer species, suggesting a food-based niche separation operating within the rotifer community (Bogdan *et al.,* 1980; Starkweather & Bogdan, 1980). J. F. Haney *et al.* (pers. comm.) have also used this method for determining clearance rates of *Polyarthra* sp. feeding on *Stichococcus* tracer cells. Their results, while preliminary, are consistant with those of Bogdan *et al.* in terms of the absolute values of the feeding parameters measured. These results, together with those of Nauwerck (1959), Bogdan *et al.* (1980, in prep.), Starkweather & Bogdan (1980) and T. M. Frost (pers. comm.) are summarized in Table 2.

Working in tropical Lake Valencia (Venezuela), Frost (pers. comm.) has measured *in situ* community cleartance rates using a sampling/incubation chamber identical to that used by Bogdan *et al.* (1980). Frost measured the grazing activity of both rotifers (principally *Brachionus* spp.) and microcrustaceans at 6 depths, comparing the calculated clearance rates based on 2 tracers, the bacterium *Aerobacter (Enterobacter) aerogenes* and the yeast *Rhodotorula glutinis.* Regression analysis of clearance rates versus animal population sizes indicated that rotifer density was a better predictor of bacterial consumption than was microcrustacean density, while the reverse was true for grazing rates based on yeast tracers. This suggests that bacteria may be relatively more important food sources for rotifers than for the co-occurring cladocerans and copepods, possibly reflecting a feeding niche separation between the major taxonomic groupings within the Lake Valencia zooplankton.

The use of laboratory-grown tracer cells to estimate *in situ* grazing activities suffers from at least three limitations. First, the technique measures differential clearance rate on only a limited number of sizes or types of radioactive particles, and does not estimate the ingestion of other materials represented in the pool of available foods. Secondly, as noted previously (Starkweather & Gilbert, 1978), the rotifers under study may selectively consume or reject the tracer cells relative to natural seston. Lastly, the addition of substantial quantities of tracers to the enclosed plankton community shifts both the size-frequency distribution and the absolute amount of food available to the rotifers. These changes are very likely to influence the feeding behavior of the suspension-feeders, producing results which may differ from those which occur with unmodified food spectra (see Starkweather, 1979; Starkweather & Gilbert, 1977b).

70

Chemostats and rotifer feeding

All the laboratory experiments described above, except those of Erman (1962a), were performed under static 'batch' conditions; that is, the rotifers and their foods were combined in systems where neither food cells nor cell nutrients were replenished during the course of the experiment. This may allow at least three important changes to occur in experimental vessels during a feeding determination: a) food cell condition (physiological state, growth rate, average cell size) may change as nutrients are depleted and cells are cropped, b) food cell numbers may change, through ingestion losses, settling or cell division, and c) rotifer condition may change via growth, reproduction, mortality or accumulation of metabolities. All of these factors are capable of inducing modifications in rotifer feeding behavior and performance, and all will be accentuated if the feeding determination requires a long period of time.

Chemostats are continuous-flow systems in which flow rates are held constant and biological populations are allowed to achieve steady-state levels controlled by the rate of supply of limiting nutrients. Chemostat techniques permit the culturing of foods appropriate for rotifers and provide supplies of cells of uniform size, shape, physiological state and suspension density. Thus, two of the three difficulties listed above are eliminated in experimental feeding systems which incorporate chemostat food culture and supply. In addition, the direct coupling of rotifer culture vessels to chemostats can produce conditions, after equilibration, which maintain animal populations with essentially constant age structure, physiological state and, presumably, feeding activity.

The use of chemostats in rotifer culture was first described in preliminary terms by M. R. Droop (in Conover, 1970). The system used by Droop combined a chemostat of *Brachiomonas submarina* which was coupled, through a metering dose switch, to a separate vessel containing *Brachionus plicatilis.* The design and construction of the equipment is detailed in a later paper (Droop, 1975).

Several variables relative to rotifer feeding, population dynamics and energetics are affected by the rate at which foods cells are supplied to *B. plicatilis.* Droop (1975) reports substantial differences in absolute food consumption, rate of food consumption and rotifer production with 2-or 3-fold differences in culture dilution rate. Lesser differences in steady-state population size and ecological growth efficiency were detected in the same experimental comparisons. More recently, Droop & Scott (1978) have reported an extensive set of measurements confirming the effects of rate of food supply on *B. plicatilis*. Drawing their results from some 30 steady-state cultures and including systematic measurement of carbon, NO_3-nitrogen and vitamin B_{12} levels from all components of the culture system, the authors were able to describe the population dynamics, energetics and nutrient relation of the algae-rotifer system. Among the principal findings were that culture dilution rate (equivalent to rotifer specific growth rate in this system) has striking impact on rotifer respiration, excretion, and ecological growth efficiency, but has only minor effects on ingestion and assimilation efficiencies.

Relative to the population and reproductive structure of *B. plicatilis* populations in the above chemostat system, Scott (1977) has reported dilution rate influences on the mean numbers of amictic and egg-bearing amictic females per ml. It should be noted, however, that this result was somewhat inconsistent, and that the ratio of amictic to egg-bearing amictic females ml^{-1} changed only very slightly despite increased growth rate and decrease in mean age accompanying higher dilution values.

The only other report of chemostat use in providing food for rotifers is that of Boraas (1979). Boraas fed *Chlorella pyrenoidosa* to *B. calyciflorus* with chemostats of two designs, one similar to Droop's (Droop, 1975; Droop & Scott, 1978) with a two-stage operation segregating algal and rotifer growth vessels, and one where algae and rotifers were mixed in a single vessel. In two stage systems, both rotifers and algae entered steady-state relatively quickly with any population periodicity resembling a critically-damped oscillation. The maximal *B. calyciflorus* population doubling time Boraas measured with this technique was 12.5 h, indicative of the favorable nutritional conditions produced by the chemostat.

With mixed algae-rotifer chemostats, populations were either critically damped or showed 5-6 cycles of population oscillation before steady-state. The initial conditions of the innocula appeared to effect those differences, with low *C. pyrenoidosa* and high *B. calyciflorus* introductions producing rapid, critically damped dynamics and the reverse condition leading to delayed steady-states.

An additional aspect of this work is the incorporation of competing rotifers into one- or two-stage chemostat systems. Boraas (pers. comm.) is currently examining the affects of dilution rate on the outcome of food-based competition between congeneric brachionids. This approach, coupled with indications of size-selective feeding by chemostat-fed rotifers (Boraas, 1979), may lead to useful observations concerning the influence of rate of supply and cell quality in determining rotifer population interactions.

Acknowledgments

This manuscript was prepared while the author was Visiting Assistant Professor at Dartmouth College, Hanover, N. H., supported by N.S.F. research grant DEB 78-02882 (J. J. Gilbert and P. L. Starkweather). Special thanks are due J. V. Jackson who improved portions of the report and Ms. Dorothy N. Edelman who quickly and carefully assembled the manuscript for publication.

References

Bogdan, K. G. & McNaught, D. C. 1975. Selective feeding by Diaptomus and Daphnia. Verh. int. Ver. Limnol. 19: 2935-2942.

Bogdan, K. G., Gilbert, J. J. & Starkweather, P. L. 1980. In situ clearance rates of planktonic rotifers. In this volume, pp. 73-77.

Boraas, M. E. 1979. A chemostat system for the study of rotifer-algal-nitrate interactions. Special Symp. III. Am. Soc. Limnol. Oceanogr. The evolution and ecology of zooplankton communities. W. C. Kerfoot (ed.) in press.

Chotiyaputta, C. & Hirayama, K. 1978. Food selectivity of the rotifer Brachionus plicatilis feeding on phytoplankton. Mar. Biol. 45: 105-111.

Conover, R. J. 1970. Cultivation of plankton populations. Helgoländer wiss. Meeresunters. 21: 401-444.

Cushing, D. H. 1976. Grazing in Lake Erken. Limnol. Oceanogr. 21: 349-356.

Dewey, J. M. 1976. Rates of feeding, respiration, and growth of the rotifer Brachionus plicatilis and the dinoflagellate Noctiluca miliaris in the laboratory. Doctoral Dissertation, Univ. of Washington (U.S.A.), 117 pp.

Doohan, M. 1973. An energy budget for adult Brachionus plicatilis Muller (Rotatoria). Oecologia 13: 351-362.

Droop, M. R. 1975. The chemostat in mariculture. Tenth European Symp. Mar. Biol. (Ostend) 1: 71-93.

Droop, M. R. & Scott, J. M. 1978. Steady-state energetics of a planktonic herbivore. J. mar. biol. Ass. U.K. 58: 749-772.

Dumont, H. J. 1977. Biotic factors in the population dynamics of rotifers. Arch. Hidrobiol. Beih. 8: 98-122.

Edmondson, W. T. 1957. Trophic relations of the zooplankton. Trans. Am. microsc. Soc. 76: 225-245.

Edmondson, W. T. 1965. Reproductive rate of planktonic rotifers as related to food and temperature in nature. Ecol. Monogr. 35: 61-111.

Erman, L. A. 1956. On the quantitative aspects of the feeding of rotifers (Rotifera Phylum) [in Russian]. Zool. Zh. 35: 965-971.

Erman, L. A. 1962a. On the utilization of the reservoirs trophic resources by plankton rotifers. [in Russian]. Byull. Mosk. Obshch. Ispyt. Prir. 67: 32-47.

Erman, L. A. 1962b. The quantitative aspect of nutrition and food selectivity in the plankton rotifer Brachionus calyciflorus Pall. [in Russian, transl. JPRS: 19, 894]. Zool. Zh. 41: 31-48.

Galkovskaja, G. A. 1963. Utilization of food for growth and conditions for maximum production of the rotifer Brachionus calyciflorus Pallas [in Russian, transl. Fish. Res. Bd. Canada 997]. Zool. Zh. 42: 506-512.

Gilbert, J. J. 1979. Feeding in the rotifer Asplanchna: behavior, cannibalism, selectivity, prey defenses, and impact on rotifer communities. Special Symp. III. Am. Soc. Limnol. Oceanogr. The evolution and ecology of zooplankton communities. W. C. Kerfoot (ed.) in press.

Gilbert, J. J. & Starkweather, P. L. 1977. Feeding in the rotifer Brachionus calyciflorus. I. Regulatory mechanisms. Oecologia 28: 125-131.

Gilbert, J. J. & Starkweather, P. L. 1978. Feeding in the rotifer Brachionus calyciflorus. III. Direct observations of the effects of food type, food density, change in food type, and starvation on the incidence of pseudotrochal screening. Verh. int. Ver. Limnol. 20: 2382-2388.

Gliwicz, Z. M. 1969. Studies on the feeding of pelagic zooplankton in lakes with varying trophy. Ekol. Pol. 17: 663-708.

Halbach, U. & Halbach-Keup, G. 1974. Quantitative Beziehungen zwischen Phytoplankton und der Populationsdynamik des Rotators Brachionus calyciflorus Pallas. Befunde aus Laboratoriumsexperimenten und Freilanduntersuchungen. Arch. Hydrobiol. 73: 272-309.

Haney, J. F. 1971. An in situ method for the measurement of zooplankton grazing rates. Limnol. Oceanogr. 16: 970-977.

Haney, J. F. 1973. An in situ examination of the grazing activities of natural zooplankton communities. Arch. Hydrobiol. 72: 87-132.

Hillbricht-Ilkowska, A. 1972. Morphological variation of Keratella cochlearis (Gosse) (Rotatoria) in several Masurian lakes of different trophic level. Pol. Arch. Hydrobiol. 19: 253-264.

Hirayama, K. & Ogawa, S. 1972. Fundamental studies on physiology of rotifer for its mass culture-I. Filter feeding of rotifer. Bull. Jap. Soc. Sci. Fish. 38: 1207-1214.

Infante, A. de. 1973. Untersuchen über die Ausnutzbarkeit verschiedener Algen durch das Zooplankton. Arch. Hydrobiol. Suppl. 42: 340-405.

Infante, A. de. 1978. Natural food herbivorous zooplankton of Lake Valencia (Venezuela). Arch. Hydrobiol. 82: 347-358.

Ito, T. 1955. Studies on the 'Mizukawari' in eel-culture Ponds. I. The feeding activity of Brachionus plicatilis on phytonannoplankton (as a cause of 'Mizukawari'). Rep. Fac. Fish. Prefect. Univ. Mie. 2: 162-167.

King, C. E. 1967. Food, age and the dynamics of a laboratory population of rotifers. Ecology 48: 111-128.

Lewis, W. M., Jr 1977. Comments on the analysis of grazing in Lake Erken. Limnol. Oceanogr. 22: 966-967.

Naumann, E. 1923. Specielle Untersuchungen über die Ernährungsbiologie und die natürliche Nahrung der Copepoden und der Rotiferan des Limnoplanktons. Lunds Univ. Arsskrift N. F. 2,19: 1-17.

Nauwerck, A. 1959. Zur Bestimmung der Filtrierrate limnischer Planktontiere. Arch. Hydrobiol. Suppl. 25: 83-101.

Nauwerck, A. 1963. Die Beziehungen zwischen Zooplankton und Phytoplankton im See Erken. Symb. Bot. Upsal. 17: 1-63.

Pejler, B. 1957. Taxonomical and ecological studies on planktonic Rotatoria from northern Swedish Lapland. K. Svenska Vetensk. Akad. Handl. 6: 1-68.

Pennington, W. 1941. The control of the numbers of freshwater phytoplankton by small animals. J. Ecol. 29: 204-211.

Pilarska, J. 1977. Eco-physiological studies on Brachionus rubens Ehrbg (Rotatoria) I. Food selectivity and feeding rate. Pol. Arch. Hydrobiol. 24: 319-328.

Pourriot, R. 1965. Recherches sur l'écologie des rotifères. Vie Milieu, Suppl. 21: 1-224.

Pourriot, R. 1977. Food and feeding habits of Rotifera. Arch. Hydrobiol. Beih. 8: 243-260.

Rigler, F. H 1971. Zooplankton. pp. 228-254. In: W. T. Edmondson & Winberg, G. G. (eds.). A Manual on Methods for the Assessment of Secondary Productivity in Fresh Waters. Blackwell Scientific Publications.

Saunders, G. W. 1963. The biological characteristics of freshwaters. Publ. Great Lakes Res. Div., U. Michigan. 10: 245-257.

Scott, J. M. 1977. Rotifer reproduction under controlled experimental conditions. Arch. Hydrobiol. Beih. 8: 169-171.

Starkweather, P. L. 1979. Behavioral determinants of diet quantity and diet quality in Brachionus calyciflorus. Special Symp. III. Am. Soc. Limnol. Oceanogr. The evolution and ecology of zooplankton communities. W. C. Kerfoot (ed.) in press.

Starkweather, P. L. & Bogdan, K. G. 1980. Detrital feeding in natural zooplankton communities: discrimination between live and dead algal foods. In this volume, pp. 83-85.

Starkweather, P. L. & Gilbert, J. J. 1977a. Radiotracer determination of feeding in Brachionus calyciflorus: The importance of gut passage times. Arch. Hydrobiol. Beih. 8: 261-263.

Starkweather, P. L. & Gilbert, J. J. 1977b. Feeding in the rotifer Brachionus calyciflorus. II. Effect of food density on feeding rates using Euglena gracilis and Rhodotorula glutinis. Oecologia 28: 133-139.

Starkweather, P. L. & Gilbert, J. J. 1978. Feeding in the rotifer Brachionus calyciflorus. IV. Selective feeding on tracer particles as a factor in trophic ecology and in situ technique. Verh. int. Ver. Limnol. 20: 2389-2394.

Starkweather, P. L. & Gilbert, J. J. and Frost, T. M. 1979. Bacterial feeding by the rotifer Brachionus calyciflorus: clearance and ingestion rates, behavior and population dynamics. Oecologia 44: 26-30.

Wallace, R. L. 1980. Ecology of sessile rotifers. In this volume, pp. 181-193.

Zimmerman, C. 1974. Die pelagischen Rotatorien des Sempachersees mit spezieller Berücksichtigung der Brachioniden und der Ernahrüngsfrage. Schweiz. Z. Hydrol. 36: 205-300.

IN SITU CLEARANCE RATES OF PLANKTONIC ROTIFERS

Kenneth G. BOGDAN*, John J. GILBERT* & Peter L. STARKWEATHER**

* Department of Biological Sciences, Dartmouth College, Hanover, New Hampshire 03755 U.S.A.
** Department of Biological Sciences, University of Nevada, Las Vegas, Las Vegas, Nevada 89154 U.S.A.

Keywords: rotifer, *in situ* feeding, clearance rate, filtration rate, resource partitioning

Abstract

The *in situ* clearance rates of several rotifer species from a small, temperate eutrophic lake were measured using three radioactive tracer cell-types, a bacterium *(Aerobacter)*, a yeast *(Rhodotorula)*, and alga *(Chlamydomonas)*. Rates were below 10 μl/anim/h but varied significantly among species. *Keratella cochlearis*, *Kellicottia bostoniensis*, and *Conochilus dossuarius* ingested all three tracer cells but rates varied substantially with tracer cell-type. *Polyarthra dolichoptera* and *P. euryptera* ingested only the algal cells. Co-occurring forms of *K. cochlearis* and species of *Polyarthra* differed markedly in size and in tracer cell utilization, indicating niche diversification in food resources.

Introduction

There is a considerable amount of qualitative information on the food selectivity of suspension-feeding rotifers (see reviews by Hutchinson, 1967; Dumont, 1977 and Pourriot, 1977). The feeding rates of natural populations of rotifers, however, except for those of mixed populations studied by Nauwerck (1959) and an anecdotal report of 0.3 μl/anim/h for *Kellicottia* sp. by Haney (1973), are completely unknown. Furthermore, recent laboratory studies by Starkweather & Gilbert (1977) have shown that maximal clearance rates for *Brachionus calyciflorus* can average 45 to 50 μl/anim/h, indicating that rotifers have a potential for exerting an important influence in planktonic communities.

Consequently, we began to investigate the clearance (filtration) rates of rotifers under natural conditions using radioactive tracer cells and a 7.5 liter grazing chamber (Gliwicz, 1968; Haney, 1971). Three types of tracer cells were used, the bacterium *Aerobacter aerogenes* (rod shaped cells 1.8 to 3.1 μm long), the yeast *Rhodotorula glutinis* (ovoid cells 2.7 to 8.7 μm long) and the alga *Chlamydomonas reinhardti* (spherical cells 5 to 8.7 μm diameter). Our immediate goals were (1) to obtain precise and repeatable estimates of *in situ* clearance rates for individual rotifer species and (2) to assess the influence of tracer cell identity on the estimates of such rates.

Materials and methods

All experiments were done between 13.30 and 17.00 h near a single station in Star Lake, a small, eutrophic lake located in Norwich, Vermont, U.S.A. Experiments were generally performed with a single tracer cell-type labeled with [32]phosphorus; however, double-label experiments simultaneously using two different types of tracer cells differentially labeled with ^{33}P and ^{32}P were also performed. Briefly, the technique involves loading the grazing chamber with a concentrated suspension of radioactive cells, lowering it to the desired depth, and then immediately triggering its closure. The closing action simultaneously releases the tracer cells and isolates the captured planktonic community from the lake environment. Thus, the experiment incubates the plankton under ambient biological, chemical, and physical conditions. After a short feeding period, the chamber is raised to the surface, and the experiment is quickly terminated by removing and fixing the zooplankton.

Three types of tracer cells were maintained. *Rhodotorula glutinis* was cultured according to Haney (1973). *Aerobacter aerogenes* was cultured in medium made from a stock solution (Tris base at 6.05 g/l; KCl at 7.35 g/l; MgSO$_4$ · 7H$_2$O at 0.25 g/l; (NH$_4$)$_2$SO$_4$ at 1.00 g/l and

Hydrobiologia 73, 73-77 (1980).   0018-8158/80/0731-0073$01.00.
© *Dr. W. Junk b.v. Publishers, The Hague.*

adjusted to pH 7.4 with HCl) to which sterile solutions of 20% glucose and 5% casamino acids were added at 1.0 ml/100 ml of stock (Frost, 1978). *Chlamydomonas reinhardti* was cultured in a modified Bristol's solution (Starr, 1960) in which only 10% of the prescribed K_2HPO_4 and KH_2PO_4 stock solutions were used to facilitate uptake of the radioisotope. Culture volumes of 50 ml were placed in 125 ml flasks. Isotope addition ($H_3^{32}PO_4$ or $H_3^{33}PO_4$) was generally between 0.5 and 1.5 mCi/50 ml depending on cell-type and the availability of radioisotope. Sterile medium was innoculated with the tracer cell and radio-isotope and shaken continuously for 2-3 days *(Aerobacter and Rhodotorula)* or 7-10 days *(Chlamydomonas)* under constant illumination at room temperature. All cell-types were centrifuged and washed twice with rotifer medium (Gilbert, 1967), diluted with rotifer medium and counted with either a Petroff-Hausser chamber *(Aerobacter)* or an electronic particle counter.

Once at the sampling station, the chamber was loaded, lowered and closed. Fifteen or 20 minutes later, the animals were separated from the food by draining the chamber through a 48 μm Nitex net and were fixed in 10% or 4% formaldehyde. A 100 ml sample of the filtrate was also collected. We did not narcotize the animals before fixation as our preliminary studies with *Brachionus calyciflorus* showed no significant loss of gut-contents with fixation without prior narcotization.

The fixed zooplankton were returned to the laboratory and rinsed with 600 ml of distilled water and then resuspended in small Petri dishes. Individuals from each species were isolated in well-slides, and special care taken to remove all extraneous matter. Known numbers of animals were pipetted onto Whatman GF/C glass fiber filters already inside each scintillation vial. A second vial with filter received a similar volume of liquid from the same well-slide to control for non-animal (residual) activity. The radioactivity of the food suspension in the grazing chamber was assessed from three 0.20 ml samples filtered (under low vacuum) through Whatman GF/C or GF/F filters and washed with 15 ml of distilled water. All filters were thoroughly dried at 65°C. The vials from the double-label experiments were heated, while tightly capped, for an additional 1 1/4 h after receiving 0.4 ml of an oxidant (1 : 1 mixture of 70% perchloric acid and 30% hydrogen peroxide) required for accurate double-label radioisotope counting. Resolution of double-label experiments followed Snipes & Lengemann (1970). Clearance rates (μl/anim/h) were calculated according to Burns & Rigler (1967).

Results and discussion

The *in situ* clearance rates of rotifers during the summer months in Star Lake were always less than 10 μl/anim/h (Table 1). Rates ranged from zero or near zero for *Polyarthra dolichoptera*, *P. euryptera*, *Asplanchna girodi*, and *Sychaeta pectinata* on *Aerobacter* and *Rhodotorula* to 8 μl/anim/h for *Keratella cochlearis* (large form) on *Chlamydomonas*. Rates for the cladoceran, *Daphnia pulex*, ranged from 72 μl/anim/h for individuals (mean length 0.83 mm) feeding on *Aerobacter* to 550 μl/anim/h for animals (mean length 1.63 mm) feeding on *Chlamydomonas*. However, *K. cochlearis* compares quite favorably to *D. pulex* when rates are expressed on per dry weight basis (Bogdan *et al.*, unpubl. data), as the maximum rate of *K. cochlearis* converts to 108 μl/μg dry wt/h while that of the large *D. pulex* on *Chlamydomonas* converts to 32 μl/μg dry wt/h.

Replication, as indicated by the small variation among vials within experiments and by the precision of estimates made on separate days, was considered good. We controlled for non-animal (residual) sources of radioactivity, and the negligible uptake of smaller tracer cells by *S. pectinata* and *A. girodi*, two species considered to be macrophagous, indicates that the adsorption of radioactivity to body surfaces was negligible. We have, therefore, assumed that all the radioactivity in the animal samples represents ingested material.

A summary (Fig. 1) of the results for six rotifer species on the three tracer cell-types illustrates several key points. Clearance rates varied with tracer cell identity in all six species. Species with a ciliated buccal funnel and a malleate mastax, *K. cochlearis*, *Kellicottia bostoniensis* and *Conochilus dossuarius*, ingested all three tracer cell-types at rates substantially above zero. *C. dossuarius* had the smallest range (3.1 to 5.9 μl/anim/h) while *K. cochlearis* (large) had the largest (0.38 to 7.4 μl/anim/h). The two species without a ciliated buccal funnel and with a virgate mastax, *P. dolichoptera* and *P. euryptera* (and also *S. pectinata*) did not ingest the smaller *Aerobacter* or *Rhodotorula*. Thus, a pattern consistent with basic rotifer anatomy and feeding behavior was obtained.

Clearance rates were not correlated with body-length among the rotifers with a malleate mastax as maximum clearance rates varied independently of body-length (i.e. *K. cochlearis*, 80 μm, 2.5 μl/anim/h; *Kellicottia bostoniensis*, 120 μm, 0.85 μl/anim/h; *K. cochlearis*, 120 μm, 7.4 μl/anim/h; and *C. dossuarius* 140 μm, 5.8 μl/anim/h). Nevertheless, body-length was an important characteris-

Table 1. <u>In situ</u> clearance rates of eight rotifer species using three radioactive tracer cell-types and liquid scintillation techniques.

Tracer Cell		Expt. Date (1979)	Water Temp. (°C)	Rotifer Species	Clearance Rate (μl anim^{-1} h^{-1})*	No. of Replicate Vials	Ave. No. of Anim. Per Vial
Type	Conc. (cells μl^{-1})						
Aerobacter	100	5/4	15	Polyarthra dolichoptera	zero[+]	5	71
				Sychaeta pectinata	near zero[++]	1	19
				Asplanchna girodi	zero	4	12
	100	6/6**	25	P. dolichoptera	0.01	1	81
				P. euryptera	zero	2	30
				Keratella cochlearis (small)	0.46	1	75
				K. cochlearis (large)	0.29+0.004	2	56
	10	8/14	21	K. cochlearis (large)	0.47+0.06	3	61
				Conochilus dossuarius	3.07+0.05	2	74
				Kellicottia bostoniensis	0.18+0.02	2	47
Rhodotorula	100	4/26	16	P. dolichoptera	zero	16	56
				S. pectinata	0.08+0.01	10	46
				A. girodi	near zero	6	12
	10	5/31	20	P. dolichoptera	zero	2	50
				P. euryptera	zero	2	27
				K. cochlearis (small)	2.53+0.12	2	54
				K. cochlearis (large)	7.39+1.42	3	28
	10	8/14**	21	C. dossuarius	5.00+0.77	2	52
				Kellicottia bostoniensis	0.77+0.07	2	44
Chlamydomonas	10	5/14	19.5	P. dolichoptera	1.32+0.12	4	71
				K. cochlearis (large)	4.93+0.10	2	40
	10	5/24	19	P. dolichoptera	1.69	1	59
				P. euryptera	0.14+0.04	2	15
				K. cochlearis (small)	0.76+0.10	2	48
				K. cochlearis (large)	6.41+0.11	2	85
	10	5/31	20	P. dolichoptera	1.58+0.19	2	37
				P. euryptera	0.36+0.18	2	14
				K. cochlearis (small)	1.12+0.07	2	40
				K. cochlearis (large)	6.05+0.49	3	26
	10	6/6**	25	P. dolichoptera	2.36	1	81
				P. euryptera	0.30+0.02	2	30
				K. cochlearis (small)	0.75	1	75
				K. cochlearis (large)	8.13+0.33	2	56
	6	8/14**	21	C. dossuarius	5.90+0.20	2	52
				Kellicottia bostoniensis	0.85+0.07	2	44

*   Mean + 1 Standard error of the mean
**   Double-label experiment
\+   Zero = Mean net cpm of replicate vials = 0
\+\+   Near zero = Mean net cpm of replicate vials < 20

tic to consider within genera, as the pattern of tracer cell utilization in *Keratella* and *Polyarthra* varied significantly between taxa. In *Keratella,* for example, the smaller form (80 μm body-length) filtered *Aerobacter* at a slightly higher rate than the larger form (120 μm body-length), but it did not filter as did the larger form, *Chlamydomo-*

nas as rapidly as *Rhodotorula*. In fact, the larger form filtered *Rhodotorula* three times but *Chlamydomonas* seven times faster than the smaller form. The nature of the tracer cell, therefore, can have a strong influence on our perception of the comparative abilities of the two forms. Similarly, the smaller *P. dolichoptera* (130 μm

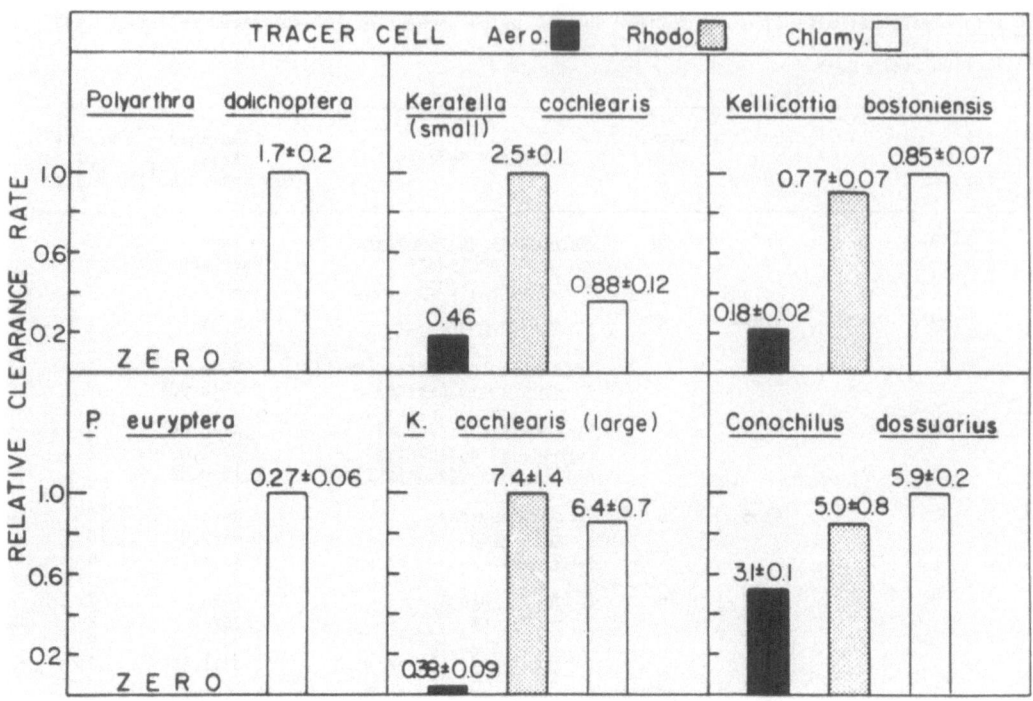

Fig. 1. Patterns of tracer cell utilization in each of six rotifer species. Data from Table 1 were used to calculate grand mean rates (R_{ij}) for species (i) on each tracer cell-type (j) by weighting the mean from each experiment equally. These grand means and ± 1 standard error of the mean are presented above the bars representing the relative clearance rate (R_{ij}/R_{imax}) of each species on each tracer cell-type.

body-length) filtered *Chlamydomonas* six to seven times faster than the larger *P. euryptera* (220 μm body-length). However, we observed numerous *P. euryptera,* but no *P. dolichoptera,* with one to three (or more) *Peridinium* cells in their guts. We hypothesize that *P. euryptera* can consume the large *Peridinium* much more efficiently than *P. dolichoptera* can, and thus, as with *Keratella,* the relative abilities of the two congeneric species are clearly dependent on the choice of tracer cell.

This dependence on tracer cell identity must reflect the differential abilities of each species to respond to each food-type. We hypothesize that the subsequent differences in the patterns of tracer cell utilization among species provides clear evidence for the existence of resource partitioning among rotifers, especially between co-occurring taxa differing only in size, as suggested by Pejler (1957).

We suspect that the response of each rotifer species was dependent on tracer cell identity, size and chemical quality. Size was important to the malleate rotifers. Three of four species showed similar rates on similarly sized food (*Rhodotorula* and *Chlamydomonas)* but much lower

rates on the much smaller food *(Aerobacter).* In the fourth species, *K. cochlearis* (small), the spherical *Chlamydomonas* may have been too wide for it to ingest. However, it is unlikely that both species of *Polyarthra* would have such a dramatic difference between their clearance rates on *Chlamydomonas* and *Rhodotorula* in response to the slight size difference of these food particles. More likely, both species responded to some, larger difference in the chemical quality of the two food particles or perhaps to some morphological difference unrelated to size, such as surface texture or the presence or absence of flagella.

In conclusion, we feel that *in situ* grazing rates of rotifers can be accurately and precisely determined. Special care, however, must be taken regarding the choice of tracer cells and the interpretation of results. Comparisons between species based on one tracer cell-type may be extremely misleading. Our results should yield additional insight into several areas of rotifer ecology, most notably the manner in which the feeding behavior and rates of various species influence competitive interactions and the structure of aquatic communities.

Acknowledgements

We thank Maxine Bean for picking countless rotifers. Support provided by U.S. National Science Foundation research grant DEB 78-2882 to J. J. G. and P. L. S.

References

Burns, C. W. & Rigler, F. H. 1967. Comparison of filtering rates of Daphnia rosea in lakewater and in suspensions of yeast. Limnol. Oceanogr. 14: 492-502.

Dumont, H. J. 1977. Biotic factors in the population dynamics of rotifers. Arch. Hydrobiol. Beih. 8: 98-122.

Frost, T. M. 1978. The ecology of the freshwater sponge Spongilla lacustris. Ph. D. thesis. Dartmouth College, Hanover, New Hampshire, U.S.A.

Gilbert, J. J. 1967. Asplanchna and postero-lateral spine production in Brachionus calyciflorus. Arch. Hydrobiol. 64: 1-62.

Gliwicz, Z. M. 1968. The use of anaesthetizing substance in studies on the food habits of zooplankton communities. Ekol. Polska, Ser. A. 16: 279-295.

Haney, J. F. 1971. An in situ method for the measurement of zooplankton grazing rates. Limnol. Oceanogr. 16: 971-977.

Haney, J. F. 1973. An in situ examination of the grazing activities of natural zooplankton communities. Arch. Hydrobiol. 72: 87-132.

Hutchinson, G. E. 1967. A treatise on limnology. Vol. II. John Wiley & Sons, New York, 1115 pp.

Nauwerck, A. 1959. Zur Bestimmung der Filtrierrate limnischer Planktontiere. Arch. Hydrobiol. Suppl. 25: 83-101.

Pejler, B. 1957. Taxonomical and ecological studies on planktonic Rotatoria from northern Swedish Lapland. K. Svenska. Vetensk. Askd. Handle., Fjärde Ser. Bd. 6, No. 5, 68 pp.

Pourriot, R. 1977. Food and feeding habits of Rotifera. Arch. Hydrobiol. Beih. 8: 243-260.

Snipes, M. B. & Lengemann, F. W. 1970. A practical method for resolution of two β-emittting radionuclides by liquid scintillation counting. Inter. J. of App. Radiation and Isotopes. 22: 513-520.

Starkweather, P. L. & Gilbert, J. J. 1977. Feeding in the rotifer Brachionus calyciflorus II. Effect of food density on feeding rates using Euglena gracilis and Rhodotorula glutinis. Oecologia 28: 133-139.

Starr, R. C. 1960. The culture collection of algae at Indiana University. Amer. J. of Botany. 47: 67-86.

STUDIES ON THE GRAZING RATE OF NOTHOLCA SQUAMULA MÜLLER ON ASTERIONELLA FORMOSA HASS. AT DIFFERENT TEMPERATURES[1]

Linda MAY

Institute of Terrestrial Ecology, 78 Craighall Road, Edinburgh

[1]Part of a dissertation for the Degree of Doctor of Philosophy to the Council for National Academic Awards at Paisley College, Scotland, in conjunction with the Institute of Terrestrial Ecology, Edinburgh, Scotland. Supported by Research Training Grant from the Natural Environment Research Council.

Keywords: Rotifers, diatoms, temperature, grazing

Abstract

The grazing rate of *Notholca squamula* on *Asterionella formosa* has been estimated to be 3.2 cells per female per hour at 6°C and 11.5 cells per female per hour at 10°C.

Introduction

Notholca squamula Müller feeds on *Asterionella formosa* Hass. by breaking open the cell frustules and removing the contents (May, 1980). The broken, empty, frustules, which remain attached to the diatom colony, have been used as an index of the grazing rate of the rotifer since they rarely appear in ungrazed cultures of *A. formosa*.

Material and methods

Approximately 20 adult, female, *Notholca squamula*, from a laboratory culture fed on *Asterionella formosa*, were washed and added to each of thirty 3 cm x 1 cm glass tubes containing 2 ml aliquots of *A. formosa* suspension (1.5×10^4 cells.ml^{-1}). Thirty similar tubes contained control aliquots of *A. formosa*, without rotifers. Both sets of tubes were incubated at 6°C in a shaking waterbath under constant illumination. The experiment was later repeated at 10°C.

At intervals of 0, 3, 9, 21 and 29 hours, 6 tubes were selected at random from both the experimental and control batches and fixed with iodine. The number of broken and undamaged cells in subsamples of the alga were counted in a Lund chamber (Lund, 1959, 1962; Young-man, 1971), and the grazing rate expressed as damaged cells per individual. The dry weight of *A. formosa* was determined by drying to constant weight at 80°C.

Results

The grazing rate of *Notholca squamula* remained constant throughout the experiment at both temperatures (Fig. 1). At 10°C an individual consumed on average 11.5 cells per hour, whereas at 6°C the grazing rate was only 3.2 cells per animal per hour. The dry weight of *Asterionella formosa* was calculated to be 3.8×10^{-4} μg per cell, of which one half can be attributed to the undigestible frustule (Hughes & Lund, 1962). The feeding rate has therefore been estimated as 2.2×10^{-3} μg dry weight per animal per hour at 10°C, and 6.1×10^{-4} μg dry weight per animal per hour at 6°C.

Discussion

The method described, which has not previously been used for rotifers, is very sensitive even when grazing intensities are low. The grazing rate of *Notholca squamula*, expressed as cells per animal per hour, is much lower than those reported for other rotifers (Galkovskaja, 1963; Hirayama & Ogawa, 1972; Halbach & Halbach-Keup, 1974; Pilarska, 1977). This is probably due to the relatively large size of the *A. formosa* cell compared to the smaller food species of the filter feeders, and the difference in the temperatures at which the rates were determined.

Hydrobiologia 73, 79-81 (1980). 0018-8158/80/0731-0079$00.60.
© Dr. W. Junk b.v. Publishers, The Hague.

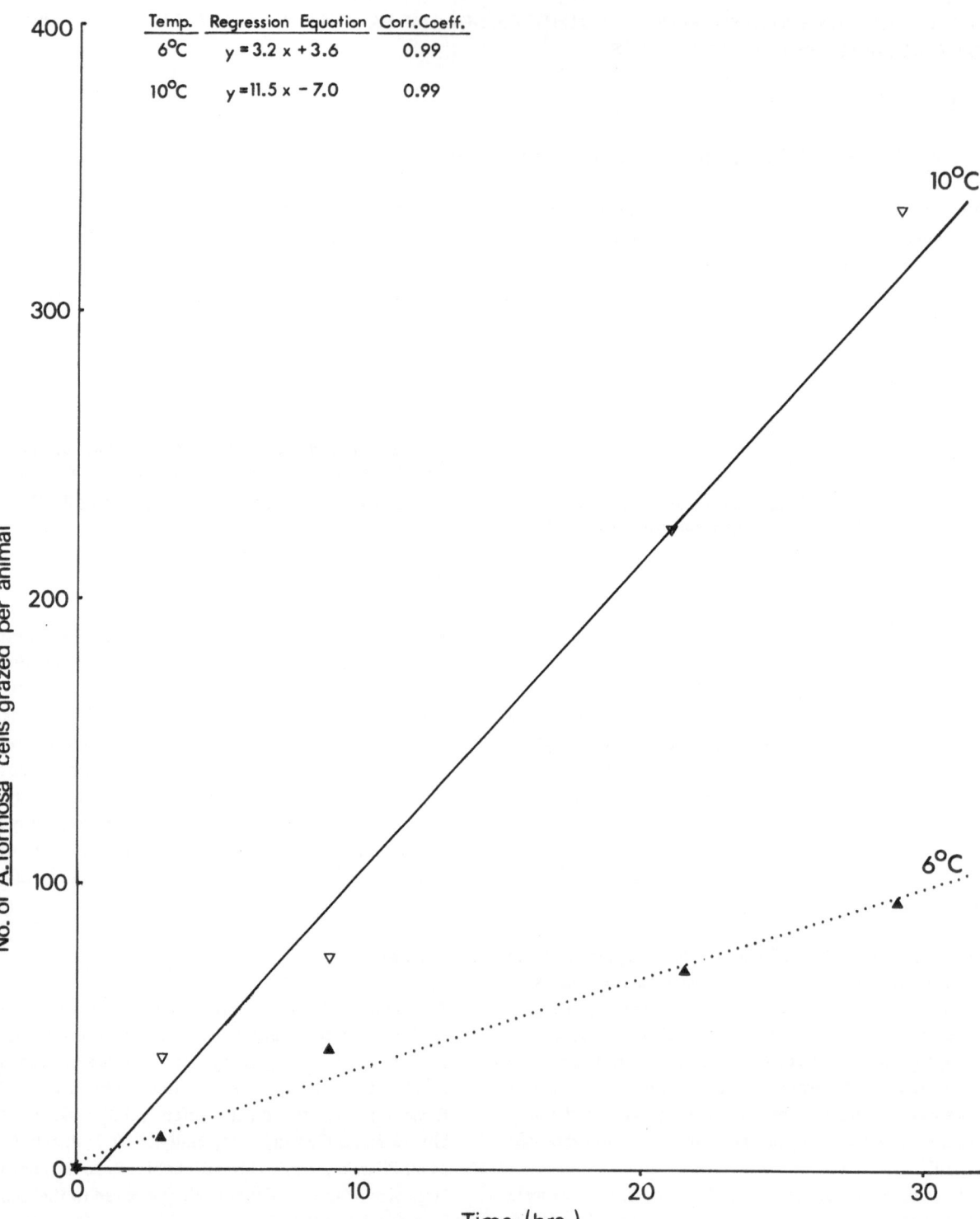

Temp.	Regression Equation	Corr.Coeff.
6°C	$y = 3.2x + 3.6$	0.99
10°C	$y = 11.5x - 7.0$	0.99

Fig. 1. The grazing rate of *Notholca squamula* on *Asterionella formosa* at 6°C and 10°C.

References

Gałkovskaja, G. A. 1963. The nutrition of planktonic Rotifera. Dokl. Akad. Nauk. Bie Bielorissk. SSSR 7: 202-205.

Halbach, U. & Halbach-Keup, G. 1974. Quantitative Beziehungen zwischen Phytoplankton und der Populationsdynamik des Rotators Brachionus calyciflorus Pallas. Befunde aus Laboratoriumsexperimenten und Freilanduntersuchungen. Arch. Hydrobiol. 73: 273-309.

Hirayama, K. & Ogawa, S. 1972. Fundamental studies on the physiology of the rotifer for its mass culture. I Filter feeding of the rotifer. Bull. Jap. Soc. Sci. Fish 38, 11: 1207-1214.

Hughes, J. C. & Lund, J. W. G. 1962. The rate of growth of Asterionella formosa Hass. in relation to its ecology. Arch. Mikrobiol. 42: 117-129.

Lund, J. W. G. 1959. A simple counting chamber of nannoplankton. Limnol. Oceanogr. 4: 57-65.

Lund, J. W. G. 1962. Concerning a counting chamber for nannoplankton described previously. Limnol. Oceanogr. 7: 261-262.

May, L. 1980. On the ecology of Notholca squamula Müller in Loch Leven, Kinross, Scotland. In this volume, pp. 177-180.

Pilarska, J. 1977. Eco-physiological studies on Brachionus rubens Ehrbg. (Rotatoria). I Food selectivity and feeding rate. Pol. Arch. Hydrobiol. 24: 319-328.

Youngman, R. E. 1971. Algal monitoring of water supply reservoirs and rivers. Water Research Association TM63. The Water Research Association, Medmenham.

DETRITAL FEEDING IN NATURAL ZOOPLANKTON COMMUNITIES: DISCRIMINATION BETWEEN LIVE AND DEAD ALGAL FOODS

Peter L. STARKWEATHER & Kenneth G. BOGDAN

Department of Biological Sciences, University of Nevada, Las Vegas, Las Vegas, Nevada 89154 U.S.A.;
Department of Biological Sciences, Dartmouth College, Hanover, New Hampshire 03755 U.S.A.

Keywords: rotifers, detritus, detrital feeding, food selectivity, community structure, niche, microcrustaceans

Abstract

Freshwater zooplankton species differ in their consumption of live and dead algal cells when tested *in situ*. Using isotopically-labeled living and heat-killed *Chlamydomonas reinhardti* as models for phytoplankton and detrital seston, respectively, we tested differential feeding on these foods by 3 rotifers and 2 microcrustaceans. *Keratella cochlearis* selectively feeds on 'detrital' materials while 2 sympatric rotifer species, *Conochilus dossuarius* and *Kellicottia bostoniensis* show no ability to discriminate between the living and dead foods. Both the copepod *Diaptomus spatulocrenatus* and a cladoceran, *Bosmina* sp., differentially consume living cells.

Introduction

The role of particulate detrital material in the diet of suspension-feeding rotifers is poorly understood. Gut content studies have confirmed that detritus is consumed by many species (Erman, 1962; Nauwerck, 1963; Pourriot, 1965, 1977; Dumont, 1977), and laboratory experiments have shown that detritus may be fed upon actively (Naumann, 1923; Spittler, 1969) by rotifers. Suspended detrital material is often abundant in the seston of lakes and ponds, and thus should be considered a potentially important nutritional source for zooplankters. In addition, the differential consumption of detritus versus other particulate foods may serve as a source of niche separation among members of the suspension-feeding community (Edmondson, 1957). This paper reports the results of an experiment performed *in situ* to evaluate the relative consumption of detritus and living phytoplankton by rotifers and microcrustaceans in a small-pond planktonic community.

Materials and methods

We determined clearance rates of rotifers and microcrustaceans *in situ* with the use of a combination sampler-and-incubation chamber modified from that of Haney (1971). The technique is described in detail elsewhere (Bogdan, Gilbert & Starkweather, 1980), but in brief, we introduced previously cultured and isotopically-labeled *Chlamydomonas* cells into a submerged vessel containing zooplankters and the natural seston. After a timed feeding interval (15 min.) we removed the vessel from the water, strained the contents through nitex screening and fixed the retained rotifers and crustaceans in 10% formalin. Quickly transporting the samples (including sample aliquots of the radioactive feeding suspension) to the laboratory, we washed the animals in distilled water and sorted them by species into separate scintillation-counting vials. We included as few as 3 copepods and as many as 50 *Keratella* per vial, preparing between 3 and 6 vials per species.

We grew *Chlamydomonas reinhardti* in Bristol's solution (Starr, 1978) but with only 10% the normal PO_4-phosphorus. This reduction in non-radioactive phosphorus enhanced the uptake of the radioisotope, supplied as either $H_3^{33}PO_4$ for the cultures serving as phytoplankton tracers, or $H_3^{32}PO_4$ for the algae to be heat-killed as laboratory-produced detritus. After incubation for several days under fluorescent illumination at ca. 20-24°C, we heated (1.0°C min^{-1}) the ^{32}P-incubated algae to 65°C and held them at that temperature for 5 min. We then harvested both live and heat-killed cells by centrifugation, washing them twice in rotifer-compatible medium. The

Hydrobiologia 73, 83-85 (1980).   *0018-8158/80/0731-0083$00.60.*
© *Dr. W. Junk b.v. Publishers, The Hague.*

heat treatment causes the *C. reinhardti* to lose their flagella and, over a period of several hours, their green pigmentation. The size-frequency distributions of the live and killed cell suspensions were indistinguishable judging from particle-counter determinations, and visual observation at high magnification revealed no other physical differences between the cell preparations. We diluted the cells so that their final density during the feeding interval would be 2×10^4 *C. reinhardti* ml^{-1} in a 1 : 1 ratio of live : heat-killed cells.

The site for this experiment (Star Lake in Norwich, Vermont, U.S.A.) has been described in other reports of feeding activities of natural rotifer populations (Gilbert & Williamson, 1978; Bogdan *et al.*, 1980). The work was performed during summer stratification in the epilimnetic waters (incubation depth, 1.0 m; temperature, 24°C; dissolved O_2, 7.7 mg l^{-1}).

Results and discussion

Table 1 shows the calculated clearance rates for three rotifer and two microcrustacean species based on their consumption of combinations of live and heat-killed *C. reinhardti*. We have compared the rates obtained with the two food types with one-way analysis of variance and calculated significance probabilities based on the F-distribution.

Keratella cochlearis cochlearis (referred to as the 'large morph' in Bogdan *et al.*, 1980 and Starkweather, 1980) preferentially consumes heat-killed *C. reinhardti* relative to live cells of the same species. This confirms earlier reports (Erman, 1962; Pourriot, 1977) that *Keratella* species are strong consumers of detritus or detritus-bacteria complexes. Sympatric *Conochilus dossuarius,* on the other hand, shows no significant difference in clearance rates

based on the two different tracers. This result places *C. dossuarius* in an intermediate position between its close relative *Conochilus natans*–which may not consume detritus (Pourriot, 1977) and *Conochilus hippocrepis*–an active detrital feeder (Pourriot, 1977). In a similar way, *Kellicottia bostoniensis* demonstrates no preference for either food type, consuming equivalent quantities of both living and heat-killed *C. reinhardti*. These data also point out the differences among rotifer species in terms of absolute feeding rates. Irrespective of tracer type, *K. c. cochlearis* clearance rates exceeded those of *C. dossuarius* which in turn were higher than those of *K. bostoniensis*.

In terms of the microcrustaceans, both *Diaptomus spatulocrenatus* and *Bosmina* sp. had clearance rates on live algae significantly greater than the same parameters based on heat-killed cells. This result differs from that obtained for the three rotifer species which either preferred the detrital food or were indifferent to the 2 tracers provided. It is also useful to mention that while the copepod's clearance rates were at least 3-7 times higher than that of the most rapidly feeding rotifer, the *Bosmina* clearance rates fell in the range of mean values obtained for the rotifers. Considering the size differences among the different species, it is clear that on a per unit weight basis, rotifer feeding may greatly surpass that of the crustacean plankton (see also Bogdan *et al.*, 1980).

Conclusions

It should be kept in mind, that while we have shown significant differences in the consumption of live and heat-killed *C. reinhardti* by natural zooplankton, this experimental procedure provides only a limited view of rotifer or crustacean discriminatory capabilities. We use the two tracer cell types as 'models' of living phytoplankton and fresh detritus derived from phytoplankton. Cer-

Table 1. Clearance rates of zooplankton species (3 rotifers, 2 crustaceans) determined *in situ* with tracer cells of living or heat-killed *Chlamydomonas reinhardti* (rates are mean values $\pm s_{\bar{x}}$).

| Species | $\bar{X}$ clearance rate (μl animal^{-1} h^{-1}) | | F-value | n | P |
	live algae	heat-killed algae			
Keratella cochlearis					
cochlearis	4.70 ± 0.15	6.34 ± 0.35	18.16	4	< 0.01
Conochilus dossuarius	3.62 ± 0.27	3.86 ± 0.41	0.24	4	> 0.05
Kellicottia bostoniensis	0.74 ± 0.08	0.63 ± 0.04	1.45	3	> 0.05
Diaptomus spatulocrenatus	36.59 ± 2.27	20.16 ± 0.78	46.86	6	< 0.01
Bosmina sp.	5.02 ± 0.40	3.14 ± 0.48	9.03	3	< 0.05

tainly, results for other live and dead algal combinations could yield different results, as might selective feeding based on detrital material that had partially decomposed or been colonized by bacteria. Both Naumann (1923) and Nauwerck (1963) have made a distinction between fresh and aged detritus relative to suspension-feeder diets.

In a similar way, the differential feeding which we have observed here may be unique to the particular conditions of the experiment, relative to the animal species studied and to the array of natural seston available. Any generalization for either species or species groups would be premature and should await more comprehensive experimentation in this and other systems.

The absolute rates of feeding by the animals in this study are generally in agreement with other results. The clearance rate values for the rotifers are within the range of values found earlier (Bogdan *et al.,* 1980) and the copepod values are considerably higher than those of the rotifers, as expected. The clearance rates for *Bosmina* sp. are lower than anticipated for a cladoceran of this size and might indicate the inappropriateness of *C. reinhardti* as a tracer for the population or a methodological inadequacy in terms of cladoceran fixation.

An additional factor which could influence the outcome of an experiment of this type is a lack of homogeneity in the distribution of tracer cells and animals within the feeding chamber. Live *C. reinhardti* are phototactic and the interior of the feeding chamber is probably not absolutely consistent in illumination due to both interior and exterior hardware. This condition might have resulted in a complex distribution of live algal cells dissimilar to that of the 'detrital' material. If zooplankton distributions were also heterogeneous within the chamber, a situation which we do, however, consider unlikely, errors in differential feeding of unknown magnitude could result. To avoid this potential difficulty, subsequent experiments should incorporate the use of non-motile or non-phototactic algal types in making comparisons between live and heat-killed cells.

In summary, we feel that these data provide an indication that certain rotifer and crustacean suspension-feeders are able to distinguish between living algal and dead sestonic material (particulate detritus). Once more, it is clear that the distinction is made on the basis of food characteristics other than size. We as yet have no clue regarding the nature of the factor or factors allowing such discrimination, but suggest that surface chemistry, surface texture or cell malleability may be involved.

Acknowledgements

The research was supported by N.S.F. research grant DEB 78-02882 (to Dartmouth College; J. J. Gilbert and P. L. Starkweather, Principal Investigators). We thank J. J. Gilbert for valuable advice and Mrs. Maxine Bean for technical assistance during all phases of this work. Ms. Dorothy N. Edelman prepared the final manuscript.

References

Bogdan, K. G., Gilbert, J. J. & Starkweather, P. L. 1980. In situ clearance rates of planktonic rotifers. In this volume, pp. 73-77.

Dumont, H. J. 1977. Biotic factors in the population dynamics of rotifers. Arch. Hydrobiol. Beih. 8: 98-122.

Edmondson, W. T. 1957. Trophic relations of the zooplankton. Trans. Am. microsc. Soc. 76: 225-245.

Erman, L. A. 1962. On the utilization of the reservoirs trophic resources by plankton rotifers. [in Russian]. Byull. Mosk. Obshch. Ispyt. Prir. 67: 32-47.

Gilbert, J. J. & Williamson, C. E. 1978. Predator-prey behavior and its effect on rotifer survival in associations of Mesocyclops edax, Asplanchna girodi, Polyarthra vulgaris, and Keratella cochlearis. Oecologia 37: 13-22.

Haney, J. F. 1971. An in situ method for the measurement of zooplankton grazing rates. Limnol. Oceanogr. 16: 970-977.

Naumann, E. 1923. Specielle Untersuchungen über die Ernährungsbiologie des tierischen Limnoplanktons. II. Über den Nahrungsbiologie und die natürliche Nahrung der Copepoden und der Rotiferan des Limnoplanktons. Lunds Univ. Arsskr. N.F. 2, 19: 1-17.

Nauwerck, A. 1963. Die Beziehungen zwischen Zooplankton und Phytoplankton im See Erken. Symb. Bot. Upsal. 17: 1-163.

Pourriot, R. 1965. Recherches sur l'écologie des rotifères. Vie Milieu, Suppl. 21: 1-224.

Pourriot, R. 1977. Food and feeding habits of Rotifera. Arch. Hydrobiol. Beih. 8: 243-260.

Starkweather, P. L. 1980. Aspects of the feeding behavior and trophic ecology of suspension-feeding rotifers. In this volume, pp. 63-72.

Starr, R. C. 1978. The culture collection of algae at the University of Texas at Austin. J. Phycol. 14 Suppl.: 47-100.

Spittler, P. 1969. Zucht- und Fütterungsversuche an einigen Rädertierarten der Hiddenseer Gewässer. Limnologica 7: 207-211.

OBSERVATIONS ON THE SUSCEPTIBILITY OF SOME PROTISTS AND ROTIFERS TO PREDATION BY ASPLANCHNA GIRODI

John J. GILBERT

Department of Biological Sciences, Dartmouth College, Hanover, New Hampshire 03755 U.S.A.

Keywords: *Asplanchna,* predation, rotifers, behavior

Abstract

Direct observations of behavioral interactions show that the predator *A. girodi:* 1) easily ingested *Synchaeta pectinata,* two forms of *Keratella cochlearis cochlearis,* and individuals of *Conochilus unicornis* and *C. dossuarius* enzymatically dissociated from their matrix; 2) rarely, if ever, captured *Kellicottia bostoniensis* and intact *Conochilus;* and 3) generally rejected the peritrich ciliate *Rhabdostyla* sp. and the dinoflagellate *Peridinium* sp. Coloniality and secretion of a gelatinous matrix in *Conochilus* can be viewed as adaptations to limit mortality from invertebrate predation. Intraspecific variability in the feeding responses of *A. girodi* is considered.

Introduction

Since *Asplanchna* can exhibit both functional and numerical responses to prey density, it may have an important regulatory effect in freshwater ecosystems. Some field correlations between the population density of *A. priodonta* and the death rate of *Keratella cochlearis* (Edmondson, 1960; Zimmerman, 1974) provide indirect evidence for such control.

We are initiating a study to determine the extent to which predation by *A. girodi* controls the population dynamics of a variety of potential prey in a small, eutrophic lake. We plan to look for correlations between the abundance and birth rate of *A. girodi* and· the population dynamics of its prey and to conduct *in situ* experiments in which *A. girodi* is enclosed with or excluded from the zooplankton community. Before undertaking such studies, however, it seemed desirable to first identify those prey which were most susceptible to, and hence most likely to be regulated by, *A. girodi* predation. Attention

could then be focused on interactions among these organisms. One such study has been completed (Gilbert & Williamson, 1978) and showed that *Keratella cochlearis* was readily captured and eaten while *Polyarthra vulgaris* almost always escaped capture. The purpose of this study was to observe behavioral interactions between *A. girodi* and other, previously untested, potential prey from its environment.

Material and methods

A. girodi were collected from Star Lake, Norwich, Vermont, U.S.A. or taken from a laboratory population, clone 5A1, originating from Tampa, Florida, U.S.A. and cultured on *Paramecium aurelia* as described elsewhere (Gilbert & Litton, 1978).

Procedures for analyzing behavioral interactions between *A. girodi* and its prey were similar to those described previously (Gilbert & Williamson, 1978). Predators were starved for 1-8 hours and then generally observed singly in the presence of first one and then another type of prey in lake water, the least preferred type usually being tested first so that the predator would be likely to encounter both types before eating a food item. Predators were usually discarded after ingesting a prey. The behavior of *A. girodi* was observed after it physically contacted a prey with the center of its corona, and the numbers of prey so encountered, attacked, captured, and ingested were recorded.

All of the following prey organisms were, at least at times, very common in the plankton of Star Lake and were collected from there during the spring and summer of 1979: the dinoflagellate *Peridinium* sp., a peritrich

87

Hydrobiologia 73, 87-91 (1980). 0018-8158/80/0731-0087$01.00.
© Dr. W. Junk b.v. Publishers, The Hague.

ciliate provisionally identified as swarmers of *Rhabdo-styla* sp., and the rotifers *Conochilus unicornis, C. dos-suarius, Kellicottia bostoniensis, Synchaeta pectinata,* and a large and small form of *Keratella cochlearis cochlearis.*

Conochilus individuals from both species were dissociated from their gelatinous matrix by incubation in filtered lake water containing 0.01% of the proteolytic enzyme Pronase (B grade, Calbiochem, La Jolla, California, U.S.A.) at room temperature for about 15 minutes. The separated rotifers were then washed and tested as prey.

Results and discussion

A. Susceptibility of Kellicottia, Keratella, Rhabdostyla, and Peridinium to predation by A. girodi, clone 5A1

Predator-prey interactions between *A. girodi* and several types of potential prey were observed and quantified in some detail. All of these prey were of a size that could be captured and ingested, approximate body lengths or diameters being 60 μm for *Peridinium*, 80 μm for the small form of *K. cochlearis*, 120 μm for both *K. bostoniensis* and the large form of *K. cochlearis*, and 140 μm for *Rhabdostyla*. The results are shown in Table 1. Although there was considerable variability among replicates, perhaps partly due to variation in starvation times, a number of points can be made.

K. bostoniensis was rarely attacked and, once attacked, was rarely captured and never ingested. The basis for the failure of this prey to elicit attacks is not known. However, attacked individuals were clearly well protected from capture by their long anterior and posterior spines. Some specimens of *K. bostoniensis*, however, are occasionally found in the stomachs of *A. girodi* collected from Star Lake (R. E. Magnien, unpublished). Thus, this species is not entirely unavailable to *A. girodi*. It is also interesting to note that specimens of *K. longispina* have been found in the stomachs of *A. herricki* and *A. priodonta* (Ghilarov, 1977).

Both large and small forms of *K. cochlearis* were readily attacked, and about 20% of those attacked were subsequently eaten. Experiment 3 showed that there was little difference between the two forms in their susceptibilities to attack, capture, and ingestion. These results are similar to those obtained previously for the large form and are consistent with the common occurrence of *K. cochlearis* in the stomachs of individuals from Star Lake and other natural systems (Gilbert & Williamson, 1978; Gilbert,

unpublished; R. E. Magnien, unpublished).

The peritrich *Rhabdostyla* was rarely attacked and, once attacked, was very rarely captured. It was unquestionably distasteful to *A. girodi*, since every individual that was captured was subsequently rejected.

The dinoflagellate *Peridinium* was also distasteful to *A. girodi*. The attack probability was low, compared to that for *K. cochlearis*, and, although every individual attacked was captured, every captured individual was rejected.

B. Susceptibility of intact and dissociated Conochilus to predation by A. girodi, clone 5A1

When adult *A. girodi*, starved for 6 hours, were placed together with colonies of *C. unicornis*, they would frequently attack those individuals of the colony that they contacted with the centers of their coronae. During these attacks, the predator would open its mouth and attempt to engulf the anterior part of the individual extending from the colony matrix. Often the predator succeeded, but almost invariably it soon released its prey, apparently unharmed. In only two instances was an individual ever dislodged from the matrix and eaten. Sometimes individuals would slowly contract into the colony matrix when attacked.

C. unicornis was definitely well protected against attacking *A. girodi*. The colony itself was much too large (about 900 μm in diameter) to be captured, and the individuals within it were too firmly embedded in the matrix to be extracted.

Adult individuals in *C. unicornis* colonies were about 140 μm in length and, when enzymatically dissociated from the colony matrix, were extremely vulnerable to *A. girodi* predators which had been starved for only 1-2 hours. Eight predators attacked and then ate the first specimen encountered. A ninth predator attacked the first four specimens encountered and ate three of these.

Slightly different results were obtained with *C. dossuarius*, a solitary species in which individuals and their developing young are surrounded by a gelatinous matrix. Extended individuals were attacked much less commonly than in *C. unicornis*. Attacks never led to captures, because *C. dossuarius*, with its matrix, was too large (about 600 μm in length) and because individuals could not be dislodged from their matrix.

As with *C. unicornis*, individuals dissociated from their matrix were vulnerable to *A. girodi*. All of the 14 young individuals encountered were attacked and ingested. Of the 11 adult individuals contacted, 7 (64%) were attacked

Table I. Feeding responses of adult *Asplanchna girodi,* clone 5AI, to the rotifers *Kellicottia bostoniensis* and *Keratella cochlearis cochlearis,* large (lg) and small (sm) forms, to the peritrich ciliate *Rhabdostyla,* and to the dinoflagellate *Peridinium.* The abbreviations Enc., Att., Cap., and Ing. refer to numbers of encounters, attacks, captures, and ingestions, respectively. *A. girodi* starved for 4-8, 2-8, 2-5, and 1.5-2.5 hours in experiments I, 2, 3 and 4, respectively.

Experiment and replicate	Prey	Number of predators tested	Number of contacts with prey recorded	Feeding responses			
				Att. Enc.	Cap. Att.	Ing. Att.	Ing. Cap.
1a	*Kellicottia*	7	154	.058	.111	0	0
1b		7	147	.061	0	0	0
mean				.060	.056	0	0
1a	*Keratella* (lg+sm)	7	49	.592	.276	.241	.875
1b		7	62	.694	.209	.163	.778
mean				.643	.243	.202	.827
2a	*Rhabdostyla*	8	135	.089	0	0	0
2b		6	92	.109	0	0	0
2c		8	249	–	.076	0	0
2d		10	318	–	.138	0	0
mean				.099	.054	0	0
2a	*Keratella* (lg)	8	56	.732	.341	.195	.571
2b		6	45	.867	.487	.154	.316
2c		8	40	1	.375	.225	.600
2d		10	106	1	.274	.094	.345
mean				.900	.369	.167	.458
3a	*Keratella* (lg)	11	43	.721	.548	.355	.647
3b		18	97	.629	.525	.279	.531
3c		13	100	.490	.367	.143	.389
mean				.613	.480	.259	.522
3a	*Keratella* (sm)	6	21	.333	.857	.429	.500
3b		9	59	.424	.320	.120	.375
3c		11	95	.537	.392	.118	.300
mean				.431	.523	.222	.392
4a	*Peridinium*	6	62	.210	1	0	0
4a	*Keratella* (sm)	6	13	.923	.583	.500	.857

and 5 (45%) were captured and ingested. Adult individuals of this species were about 300 μm in length and thus were somewhat more difficult to capture when dissociated than those of *C. unicornis*.

It is interesting to note that *Conochilus* spp. have been recorded from the stomachs of *A. girodi* (Guiset, 1977) and that *C. unicornis* has been recorded from those of *A. herricki* and *A. priodonta* (Ghilarov, 1977). Such ingested individuals may have been pulled from their matrix by the *Asplanchna* or might have somehow become dissociated from their matrix and then encountered and ingested by the *Asplanchna*.

The great vulnerability of *Conochilus* individuals dissociated from their matrix suggests that secretion of a gelatinous covering and coloniality in this genus are extremely adaptive in limiting mortality from invertebrate predators such as *Asplanchna*. These features were probably inherited from sessile ancestors and may have pre-adapted such forms for invasion of open-water habitats.

C. Susceptibility of Synchaeta and Rhabdostyla to A. girodi from Star Lake

A. girodi, starved for 6-7 hours, were tested with *S. pectinata* (about 400 μm in length), or *Rhabdostyla*, or both. Five predators exposed only to *S. pectinata* encountered a total of 31 individuals and, of these, attacked 14 (45%) and ingested 8 (26%). Three predators exposed only to *Rhabdostyla* ingested the first 6 individuals encountered. Four predators were tested first with *Rhabdostyla* and then *Synchaeta*. Each of two of these immediately ate a small peritrich, contacted 8 more without responding, and then attacked 3 out of the 4-5 *Synchaeta* encountered, ingesting the last of these. Another predator contacted 10 peritrichs; it attacked, captured, and rejected 2 of these and then ate the first *Synchaeta* encountered. The other predator attacked the first 4 peritrichs contacted without capturing any of them and then attacked and ingested the fifth *Synchaeta* it met.

These observations clearly show that *S. pectinata* was readily attacked, quite easily captured after an attack, and always ingested after capture. This is consistent with the common occurrence of *S. pectinata* and other species of this genus in the stomachs of *A. girodi* from natural populations (Guiset, 1977; R. W. Magnien, unpublished; Salt et al., 1978).

Interpretation of the observations on *Rhabdostyla* is more difficult. The peritrichs were frequently attacked and ingested. However, some predators, which had eaten them, found them distasteful and subsequently ignored them, even though they were still responsive to *Synchaeta*. To this extent, the Star Lake *A. girodi* reacted to *Rhabdostyla* very much like the 5A1 clone of *A. girodi*. The former, however, were more apt to attack, capture, and ingest the peritrich. It is possible that the Star Lake predators had not been feeding pior to collection and, therefore, were more hungry than the laboratory clone. It is also possible that the two populations of predators were somewhat different with regard to their behavioral responses.

Other differences have been noted between the Star Lake *A. girodi* and the 5A1 clone *A. girodi*. For example, the former have been observed to capture *Kellicottia bostoniensis* and appear to feed more efficiently on the small form of *Keratella cochlearis cochlearis* (Gilbert, unpublished). The possible existence of such differences suggests some caution in generalizing from one population of predators to another. The preferences of a predator towards various prey and the ability of a predator to handle different prey could vary with genotype, environmental condition, and historical factors.

Summary and general conclusions

Observations were made on interactions between the predator *A. girodi* and various potential prey. Of the prey tested, only *Synchaeta pectinata* and two forms of *Keratella cochlearis cochlearis* were readily attacked and could be easily captured and ingested. Both the peritrich ciliate *Rhabdostyla* and the dinoflagellate *Peridinium* were distasteful to *A. girodi*, being very rarely attacked and regularly rejected if captured. *Kellicottia bostoniensis*, with its long spines, was very well defended against capture and was also attacked much less readily than *Keratella*. Extended individuals of *Conochilus dossuarius* and especially *C. unicornis* were frequently attacked, but, except for rare instances in the latter species, they could not be dislodged from their matrix and eaten. In both species, individuals enzymatically dissociated from their matrix were readily attacked and ingested. Coloniality and the secretion of matrix in *Conochilus* can be viewed as adaptations to limit mortality from invertebrate predation.

These and other data suggest that predation by *A. girodi* in Star Lake probably can exert an important controlling effect on only a few species, such as *S. pectinata* and *K. cochlearis*. Other common, potentially available prey are either extremely difficult to capture, such as *Kellicottia, Conochilus,* and *Polyarthra*, or are distasteful, such as *Rhabdostyla* and *Peridinium*.

Some differences in behavior between *A. girodi* from Star Lake and a laboratory clone indicate that the degree and basis of such intraspecific variability should be examined before attempting generalizations on the prey preferences and handling capabilities of this predator.

Acknowledgements

I thank Maxine M. Bean for expert technical assistance and R. E. Magnien and C. E. Williamson for help with field collections. This project was supported by National Science Foundation research grant DEB-782882.

References

Edmondson, W. T. 1960. Reproductive rates of rotifers in natural populations. Mem. Ist. Ital. Idrobiol. 12: 21-77.

Ghilarov, A. M. 1977. Observations on food composition in rotifers of the genus Asplanchna (in Russian). Zool. Zh. 56: 1874-1876.

Gilbert, J. J. & Litton, J. R., Jr. 1978. Sexual reproduction in the rotifer Asplanchna: effects of tocopherol and population density. J. expl. Zool. 204: 113-122.

Gilbert, J. J. & Williamson, C. E. 1978. Predator-prey behavior and its effect on rotifer survival in associations of Mesocyclops edax, Asplanchna girodi, Polyarthra vulgaris, and Keratella cochlearis. Oecologia 37: 13-22.

Guiset, A. 1977. Stomach contents in Asplanchna and Ploesoma. Arch. Hydrobiol. Beih. 8: 126-129.

Salt, G. W., Sabbadini, G. F. & Commins, M. L. 1978. Trophi morphology relative to food habits in six species of rotifers (Asplanchnidae). Trans. am. microsc. Soc. 97: 469-485.

Zimmerman, C. 1974. Die pelagischen Rotatorien des Sempachersees mit spezieller Berücksichtigung der Brachioniden und der Ernährungsfrage. Schweiz. Z. Hydrol. 36: 205-300.

PHYLOGENETIC RELATIONSHIPS OF ROTIFERS, AS DERIVED FROM PHOTORECEPTOR MORPHOLOGY AND OTHER ULTRASTRUCTURAL ANALYSES

Pierre CLÉMENT

Laboratoire d'Histologie et de Biologie tissulaire, Université Lyon I (Claude Bernard) 69622, Villeurbanne, France

Keywords: Phylogeny, evolution, adaptations, rotifers, ultrastructures, eyes, ocelli, photosensitivity, pigments, ovogenesis, muscles, protonephridia, integument, pseudocoel, collagen, glia

For the Summary to this contribution see pp. 113-114.

I. Introduction

1963: disagreements on lower metazoan phylogeny

'Are phylogenetic theories subjective views? Can any man propose his own phylogeny or can we get definite scientific solutions?' asked Remane (1963) in criticism of Hadzi's (1944, 1953), Steinböck's (1952, 1958, 1963) and Hanson's theories (1958, 1963), stating that acoels turbellarians are derived from plasmodial ciliates and are the most primitive metazoa.

The first trap in constructing any phylogeny is to mistake a convergent similarity for an homology. To recognize homologies, Remane (1955) defined precise criteria. However, even with these criteria (that I shall criticize in the next chapter), it is possible to see two ways in a phyletic line. Thus for Remane (1955, 1958, 1960, 1963), Marcus (1958) and Jägersten (1955, 1959), Platyhelminths and Nemathelminths stem from coelomates by regression.

According to these two groups of theories (lower metazoa come from acoels or from coelomates), Cnidaria would be less primitive than Platyhelminths: their radial symmetry would derive from the bilateral symmetry of the Anthozoa. 'Emotion, too, sometimes, seems to substitute to reason' says Hand (1963). Hyman (1959) and Hand (1959, 1963) brought some classical theories back into fashion: Cnidaria would be primitive and the first bilateral symmetry of the Anthozoa would be primitive too. They assert that 'the early worm was a planula or a planuloïd organism and the planula did not come from early worms'.

In 1963, the battle raged. In the book published by Dougherty, Ax regarded the Cnidaria as the most primitive metazoa, but he suggested that the other metazoa were derived from a primitive coelomate; Beklemishev favoured a polyphyletic origin for the metazoa from different Coelenterate ancestors; and Remane repeated his ideas about the coelomate ancestor for pseudocoelomates. In the same book, Ruttner-Kolisko prudently ends her paper on the origin of rotifers by proposing two possibili-

ties: 1) from the Turbellaria, 2) from 'forms that might be traced back to the Turbellaria', like *Diurodrilus* (Dinophilidae).

Today a relationship between Rotifera and Turbellaria is generally favoured. The anatomy and the embryology (De Beauchamp, 1907, 1909, 1965; Nachtwey, 1925) suggest the origin of rotifers 'from some low grade creeping bilateral type such as a primitive flatworm' (Hyman, 1951). 'There is no fact indicating a case of reduction from more highly developed, coelomate worms (no rudimentary coelom or mesoderm)' (Ruttner-Kolisko, 1974).

Nevertheless, Koste (1978) still favours the hypothesis of a coelomate ancestor of rotifers, thus supporting Remane *et al.,* 1972, 1976.

Evolution of basis for arguments on phylogeny

From the first observations of rotifers by Leeuwenhoek to the recent treatises on rotifers (De Beauchamp, 1965; Ruttner-Kolisko, 1972; Koste, 1978), the principal source of information has been morphological or histological observation under optical microscopy. The problem of homology is the main one. On this basis, different relationships were successively suggested for rotifers: Infusoria, Polypa, Crustacea, Annelida, Molluska, Turbellaria... (Hyman, 1951; De Beauchamp, 1965). This agitated history originates from:

1. Imprecision of the observations: rotifers were classified as infusoria when their nuclei and cells were not yet observed; when Huxley (1853) saw their protonephridia, he put rotifers in Vermes.

2. Confusion between specialized rotifers *(Hexarthra* or *Trochosphaera)* and archetypes of the group.

Recent ultrastructural studies on rotifers (review in Clément, 1977b) are of much better quality, and throw a new light on the problem of homologies. However, ultrastructures of only a few species of rotifers are actually described, so that danger of confusion between specialized structures and archetypes remains present. In this work, I shall dis-

93

cuss the phylogeny of rotifers with the help of ultrastructural results on seven genera of rotifers: *Trichocerca, Notommata, Brachionus, Rhinoglena, Asplanchna, Philodina, Habrotrocha*. Obviously, the size and the phylogenetic distribution of this sample must be taken into account when making generalizations. For some organs, we have ultrastructural information on only one or two species (an exception is the integument, known for the seven genera, and also in *Mytilina, Keratella* and *Synchaeta*).

Animal behaviour is very important to study evolution and to try to understand trans-specific evolution (Mayr, 1974). Unfortunately ethological work on rotifers is scarce, and only just beginning to grow.

Finally, recent progress in genetics and ecology of rotifers (review in King, 1977) will help to understand speciation and evolution in this group. King (1977) detected by electrophoresis a great variation in different clones of the same species. Nevertheless, we have no information about correlations between genetic variation and structural or behavioral variation. So, when I speak in this text about hypothetical 'chromosomic segments', characteristic of a precise ultrastructure in rotifers or other animal groups, this will be speculation. I know the danger of such speculations when some biologists write that individuals are nothing else than their gene pools (Wilson, 1979). We begin to know the origin of the variability of the responses of single rotifers which have the same genome: the reasons are in their own history, and in the history of their parents, grand-parents and other ascendants (Clément, 1977a; Clément & Pourriot, 1979 and 1980). So, even in rotifers, it is impossible to reduce an individual to its (until now unknown) gene pool.

II. Photoreceptors and photosensitivities in rotifers

Are photoreceptors good indicators of phylogeny?
Eakin (1965) proposed 'a speculation as a catalyst of research': some zoological groups, leading to deuterostomia, would have only ciliary-type photoreceptors; others groups (acoelomates, pseudocoelomates and protosomia) would have only rhabdomeric-type photoreceptors. In the first case, the photoreceptoral organelles are derived from the ciliary membrane; in the second one, from the distal cell membrane.

Eakin himself (1968, 1972) proposed a number of exceptions to his diphyletic theory. We now know that both types of photoreceptors are present in most zoological groups.

Vanfleteren & Coomans (1975) summarized these exceptions and made a new, monophyletic theory: the photoreceptoral organelles would always be induced 'by a ciliary information which, after initiating membrane proliferation, may become more or less abortive (rhabdomeric type) or may develop further into a ciliary organelle (ciliary type)'. They concluded that the photoreceptor structure is not useful to distinguish large phyla like protostomia and deuterostomia, but only to study closer phylogenetic relationships.

In a more recent synthesis, Salvini-Plawen & Mayr (1977) proposed a different idea about the photoreceptor types: they described at least 40 (possibly up to 65 or more) independant phyletic lines, which can be grouped in a ganglionic or an epidermal category by their localization and embryology. The ganglionic (diverticular) type would be rhabdomeric and there would be three epidermal types: ciliary (enlargement of the ciliary membrane), rhabdomeric (enlargement of the distal cell portion) and unpleated (surface enlargement through increase in cilia number). For these authors, 'similar photoreceptor types differentiated convergently several times' and 'their distribution in various phyla of animals cannot safely be used as the basis for the construction of phylogenesis'.

These three theories ('diphyletism' of Eakin, 'monophyletism' of Vanfleteren & Coomans and 'aphyletism' of Salvini-Plawen & Mayr) use mainly morphological observations, often neglecting biochemical, physiological and ethological aspects. Moreover the scarcity of ultrastructural descriptions enhances a danger that I presented above: confusion between a specialized structure and an archetype of the zoological group. For instance, until now, the cerebral eye of *Asplanchna* (Eakin & Westfall, 1965) was considered to be the unique photoreceptor type of rotifers. Salvini-Plawen & Mayr (1977) considered a second 'epidermal' type with the anterior ocelli. The first ultrastructural observations of these ocelli and others photoreceptors in rotifers will allow me to discuss this precise point but also to propose a new ('polyphyletic') theory about the evolution of photoreceptors and photosensitivities. As much as possible I shall try not to limit my descriptions and conclusions to morphological features.

Photosensitivities and photopigments in rotifers
The vision of rotifers is very primitive. No female, even if carnivorous, seems to see her food. No male seems to see the female he tries to fecundate. Instead, these meetings occur by random encounters facilitated by taxes.

Among these taxes, phototaxis was studied first (Jennings, 1901; Viaud, 1940-1943; Menzel & Roth, 1972; Preissler, 1977; Clément, 1977a-c; Hertel, 1979). It is a resultant of two components: phototaxis, *sensu strictu,* providing the orientation of the animal, and photokinesis, directing its movements.

Three cases are possible:

1. Regular phototaxis: in planktonic species, in particular those only moving by swimming. A variability in this behaviour has been noted but not studied in different animals or clones.

2. Irregular phototaxis: in particular in rotifers which often settle or creep. In *Notommata copeus,* a species reputedly not phototactic, the phototaxis is inhibited by contact with a filament and others factors (Clément, 1977a).

3. Apparently non existant phototaxis: *Synchaeta pectinata* (Menzel & Roth, 1972); *Resticula gelida* (Viaud, 1943); perhaps many bdelloïds. More precise studies on these species are needed: are there particular inhibitions as for *Notommata copeus* or is there complete inhibition?

Phototaxis is characterized by a peak about 540 nm (Viaud, 1940-43; Menzel & Roth, 1972; Clément, 1977a-c), except in *Filinia longiseta* whose peak is about 450 nm (Menzel & Roth, 1972).

Photokinesis is characterized by a regular increase of speed *(Brachionus calyciflorus)* or of proportion of swimming animals *(Notommata copeus)* as the light changes from blue to red (Clément, 1977a-c).

A third photosensitivity was discovered by Pourriot (1963) and studied by Pourriot & Clément (review in Clément, 1977a and in Pourriot *et al.,* in press): in three species *(N. copeus, N. codonella* and *Trichocerca rattus)* photoperiod controls the production of mictic females, and therefore the production of males and resting eggs. The action spectrum of *N. copeus* is different from that of the other two species; it has peaks at approximately 310, 360, and 450 nm, and there is no response to red light. So, the same animal shows three different photosensitivities.

It is possible to hypothesize that β-carotene or riboflavine or pterine is responsible for photoperiod influence, rhodopsin or porphyrin is responsible for phototaxis, and phytochrome is responsible for photokinesis (Clément, 1977a).

Note that Champ (1976) and Wallace (1980) have pointed out a photoperiod influence on the hatching rhythm in *Sinantherina socialis,* and Pourriot & Rougier (1980) have demonstrated a light effect on the hatching of resting eggs in *Brachionus rubens.* These two effects of light have not been studied in detail and we do not know

whether they use one of the three preceding pigments.

Lastly, we do not know if rotifers have a shadow response. However, there have been studies of the influence of light intensity on photoperiod, phototaxis and photokinesis (Clément, 1977a); this influence can explain the avoidance of shores by some planktonic rotifers (Preissler, 1977).

Photoreceptors of rotifers

The synthesis of Remane (1929-32) takes into account only the pigment cups. In the absence of pigment, it was thought that there were no eyes (the 'blind' *Asplanchna* of Viaud, 1940-43). However, with the electron microscope, it is possible to demonstrate that the red pigment is only an accessory epithelial cell associated with nervous structures (Eakin & Westfall, 1965; Clément, 1975). Other presumed photoreceptive structures can exist without the presence of a pigment cup, and probably correspond to the 'dermatoptic sensibility' of Viaud (1940-43).

The following is an annotated list of photoreceptors that have been described in rotifers:

a/ *Trichocerca rattus:*

As in *Notommata copeus,* there are three photosensitivities. Three presumed photoreceptors have been described:

– The cerebral eye (Fig. 1 to 4) (Clément, 1975): the red cup is located in a single epithelial cell. The lamellar photoreceptive neurites are piled up and embedded in the cytoplasm of the sensory neuron.

– Paired cerebral receptors (Fig. 16) (Clément, 1977a): located on both sides of the brain. Some neurons bear microvilli.

– Anterior ocelli (Fig. 13, 14, 15) (Clément *et al.,* 1980): these two complex apical sensory organs have specialized short, ampulla-shaped cilia containing electrondense material.

b/ *Asplanchna brightwelli:*

Only the cerebral eye has been described (Eakin & Westfall, 1965). The pigment of the cup is flatter than in *Trichocerca rattus,* and is arranged in several superposed layers. The sensory neurons bear lamellar photoreceptive rhabdomeres that pile up like onion leaves.

c/ *Brachionus calyciflorus:*

Only the cerebral eye has been described (Clément *et al.,* in press) (Fig. 5, 6, 7). Two pigment cells form the red cup. The pigment resembles that of the cerebral eye of *Trichocerca rattus.* Two neurons form the sensory part. Cylindri-

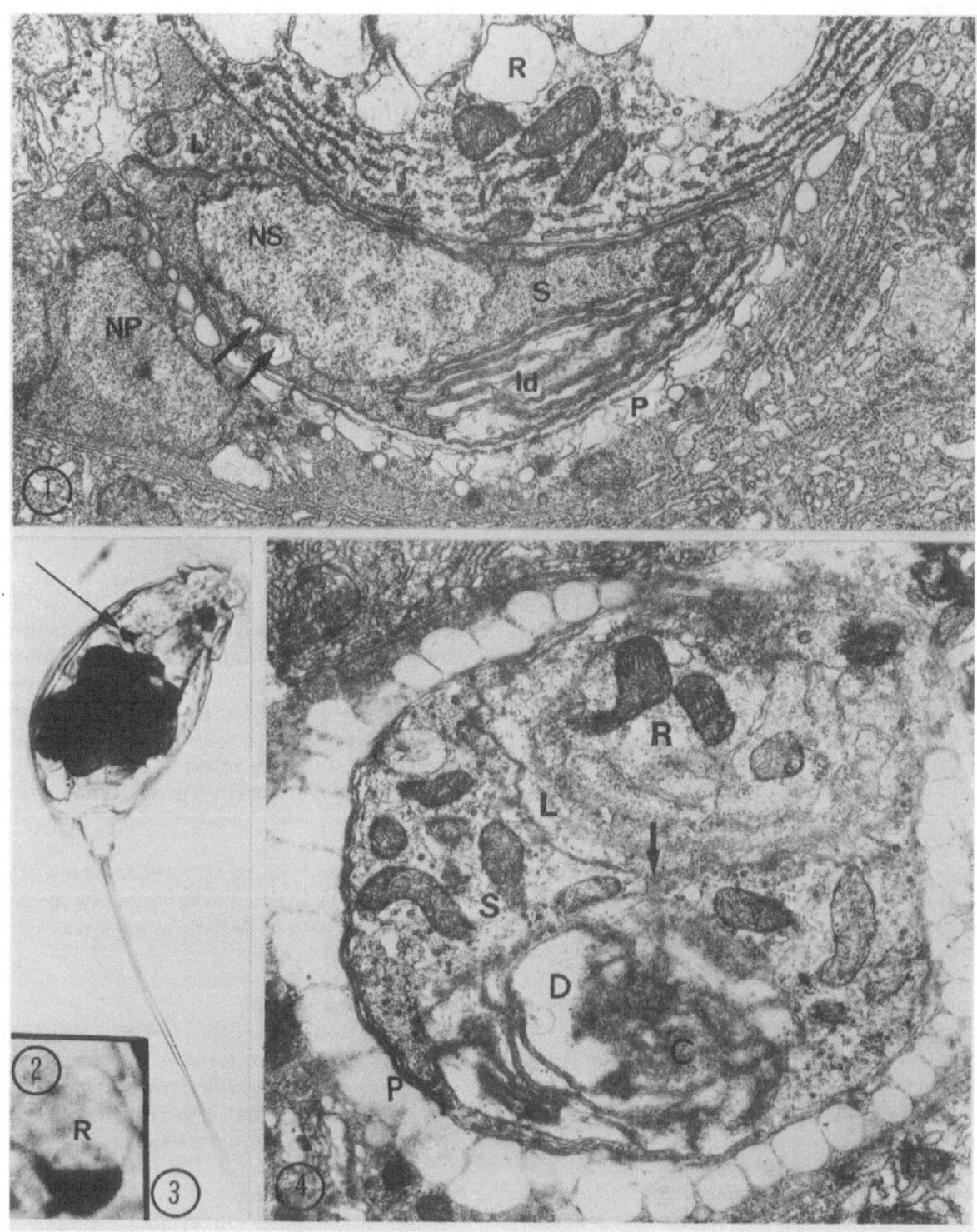

cal neurites of the first neuron penetrate the cytoplasm of the second neuron.

d/ *Rhinoglena frontalis:*

Only the anterior ocelli have been observed (Clément *et al.,* in press) (Fig. 11 and 12). The pigment cup is intraepithelial. The sensory structures are piled dendritic lamellae, coming from cerebral neuron processes. They are everse ocelli (the preceding cerebral eyes were inverse).

e/ *Philodina roseola:*

– The cerebral eyes (Fig. 8, 9, and 10) (Clément *et al.,* in press) are located on each side of the brain. The pigment is different from the pigment of the cup's eyes of the Monogononta described above. In each eye, the photosensory strutures are ampullae-shaped cilia containing electron-dense material.

– An anterior receptor (Fig. 17, 18, and 19) (Clément *et al.,* in press): it is a median apical receptor, located in the pseudocoel. Beneath an epithelial anterior cell, a nerve process contains a spherical cavity filled with numerous flattened and piled lamellar cilia. Each cilium bears lateral lamellar expansions that are also piled up.

The polyphyletic origin of the photoreceptors of rotifers

Ampullae-shaped cilia containing electron-dense material are characteristic of a first phyletic line. In rotifers, I found these cilia in the cerebral eyes of *Philodina roseola* as well as in the anterior ocelli of *Trichocerca rattus.* Elsewhere, to my knowledge, this kind if cilia has only been described from the stigma of some phytoflagellates (Fauré-Fremiet & Rouillet, 1957; Fauré-Fremiet, 1961). They look like the parabasal apparatus of the phytoflagellates (Wolken, 1971).

The presumed anterior ocellus of *Philodina roseola* represents a second phyletic line. We find exactly the same structures and organization in the cercaria of *Schistosoma mansoni* (Short & Cagné, 1975). Very similar organs are found in:

– other Platyhelminths, in which an intraneural spherical cavity contains some cilia with slightly modified axonems, but with piled lamellar expansions (Wilson, 1970;

Brooker, 1972; Lyons, 1972).

– some Annelida (Clark, 1967; Röhlich *et al.,* 1970) and Pogonophora (Norrevang, 1974) in which an intraneural spherical cavity, called the 'phaosome', contains piled lamellar expansions, with sometimes regressed cilia or only ciliary rootlets. In no case is this very special organ associated with a pigmented epithelial structure. The function of these organs is always presumed to be photoreception, except by Wilson (1970) who proposed that it functions in gravity reception (Vanfleteren & Coomans, 1975, disagree with Wilson).

The impaired cerebral eyes of monogononts represent at least one more phyletic line: cylindrical or lamellar rhabdomeres juxtaposed to a pigmented epithelial cup. Clément *et al.* (in press) detail the comparison of these eyes: primitive characeristics are noted in *B. calyciflorus* and specializations in *Asplanchna brightwelli.* Rhabdomeric structures, also issued from a cerebral neuron, and juxtaposed to a pigmented epithelial cup, are found in the anterior ocelli of *Rhinoglena frontalis.* As in the first phyletic line (ampullae-shaped cilia), we find here anterior ocelli as well as cerebral eyes in the same phyletic line. This phyletic line represents the ganglionic diverticular type of Salvini-Plawen & Mayr (1977), in which the rhabdomeric structure seems to differentiate without cilia from a ganglionic cell juxtaposed with a pigment cell. This photoreceptor type is present in Platyhelminthes (see also Fournier & Combes, 1978, and a review in Fournier, in press), Aschelminths, Polychaeta and some Arthropoda.

In conclusion, we know of at least three phyletic lines of photoreceptor types in lower metazoa. All three lines are present in rotifers and homologies can be established with photoreceptors of others zoological groups. On this basis, I propose a polyphyletic origin of rotifer photoreceptors.

Salvini-Plawen & Mayr (1977) suppose no ciliary induction in the photoreceptors of their ganglionic type. This hypothesis, which criticises the monophyletic theory of Vanfleteren & Coomans (1975), could be tested by an ultrastructural embryological study of some of these receptors.

Our observations on rotifers are in contradiction with the classification of photoreceptor types proposed by Sal-

Figs. 1-4. Cerebral eye of *Trichocerca rattus.*
Fig. 1. x 21000. Axial section. The eye caps a retrocerebral gland (R). It is made of a dendritic blade (L), a sensory neuron (S and its nucleus NS), and a pigmented cell (NP: its nucleus, P: pigments of the cup). The sensory neurocytoplasm (S) contains dendritic lamellar expansions (the piled dendritic lamellae: ld) and some cylindrical expansions (arrows); Fig. 2. x 2500. The pigmented

cup caps a retrocerebral gland (R); Fig. 3. x 500. The arrow points to the eye; Fig. 4. x 18500. Transversal section of the eye. The pigmented cup (P) surrounds the sensory neurone (S) and a part of the retrocerebral gland (R). The arrow shows the communication between the dendritic blade (L) and a dendritic lamella (D). This lamella (D) is sectionned tangentially, as is the part of cytoplasm (C) located between two piled lamellae.

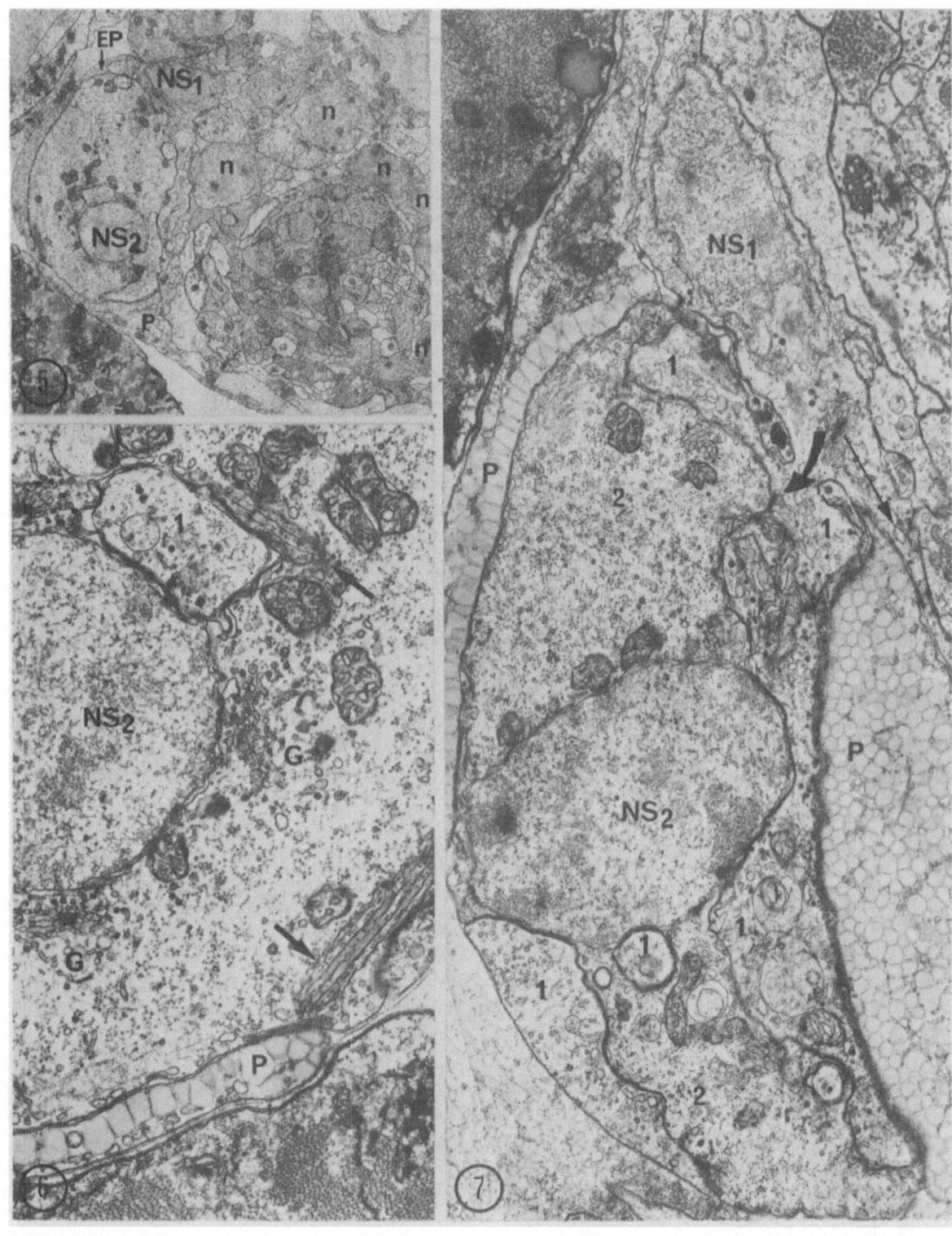

vini-Plawen & Mayr (1977). Their fundamental distinction between ganglionic and epidermal photoreceptors is not supported by our observations: the previously presumed 'epidermal' anterior ocelli are in fact feedings of cerebral neurons (see above). I think that it is not pertinent to propose photoreceptor types based on epidermal or nervous origin, because both have the same neuro-ectodermal origin. I therefore prefer more precise definitions of photoreceptor types. I have three other points of disagreement with Salvini-Plawen & Mayr:

1. In rotifers, some cerebral neurons bear photoreceptor cilia (eyes of *Philodina roseola,* ocelli of *Trichocerca rattus*); other cerebral neurons associated with eyes or ocelli bear rhabdomers. Because of this, I disagree with the rhabdomeric 'ganglionic type' proposed by Salvini-Plawen & Mayr. As discussed earlier, the third phyletic line of rotifer photoreceptors has three characteristics: ganglionic, rhabdomeric, and associated with an epithelial pigment cup.

2. Another point concerns the ampulla-shaped cilia (see above: first phyletic line). These cilia have no place in the receptor types defined by Salvini-Plawen & Mayr. In this phyletic line, there is no enlargement of membranes but accumulation of electron dense material inside the short cilia. The presence of the same material, closely juxtaposed to the pigment stigma in the parabasal apparatus of the phytoflagellates, suggests the photosensitivity of the material.

3. Our second phyletic line ('ocellus' of *Philodina roseola*) is defined to be preecise structure named 'phaosome' in Annelids. In this line, there is a progressive evolution from ciliary to rhabdomeric types.

Why, then, did Salvini-Plawen & Mayr find no phylogenetic significance in the distribution of their photoreceptor types? The reason is perhaps an insufficient precision in their definitions of these types.

I have tried to formulate precise definitions for the photoreceptors that can be observed in rotifers. However, completely satisfactory definitions must take into account both ultrastructural features and biochemical and physiological aspects.

Phylogeny and evolution of photoreceptors and photosensitivities

I have described (paragraph 2) at least three photosensitivities, and three photopigments, in rotifers. Rhodopsin is the only pigment which can be implied in phototaxis. Wolken (1970) states that rhodopsin is the only visual pigment found in invertebrates, but presents no data on lower metazoa. It is possible that in rotifers and in other lower metazoa, one of the photoreceptors is associated with rhodopsin and represents a primitive form of future visual organs. It would be interesting to construct a phylogeny of the animal kingdom by comparing all photoreceptors associated with a rhodopsin.

The other pigments of rotifers are involved in photokinesis and in determinism of mixis by photoperiod. Their localization is unknown. Yet, we can postulate that it is extraocular, in those presumed photoreceptors that do not possess a pigmented epithelial cup. It would be interesting to establish a correlation between a pigment, a photoreceptor, and a behaviour. If such a relationship could be found, its comparative evolution in the animal kingdom would have a real phylogenetic interest. Unfortunately, even in rotifers, we have no knowledge of these correlations.

For instance, in all animal groups, the receptor involved in the sensitivity to photoperiod is unknown. The action spectra found in other groups are often, but not always, the same as those in rotifers (review in Pourriot & Clément, 1973). Different mechanisms were proposed in Arthropoda for the influence of photoperiod (review in Saunders, 1976). The mechanism present in *Notommata copeus* is a primitive one, without endogenous rhythm (Pourriot *et al.,* in press). I am sure that the comparison of the photoreceptors and nervous and endocrine structures involved in these influences of photoperiod on animals, would have a phylogenetic sense.

In summary, I think that a phylogeny must simultaneously consider the evolution of structures, pigments and functions of photoreceptors. In this sense, rotifers are primitive metazoa, having primitive responses to light, no

Figs. 5-7. Cerebral eye of *Brachionus calyciflorus.*
Fig. 5. x 3600. Axial section of the brain. The neuropile is surrounded by small neurons (n: their nuclei). Towards the back of the brain, the two sensory neurons (NS1 and NS2 their nuclei) occupy a large volume. The biggest (NS2), is capped by two pigmented epithelial cells (EP and P) which contain the pigmented cup of the eye; Fig. 6. x 22000. Detail of the large sensory cytoplasm; NS2: its nucleus, P: pigmented cup, (1) dendritic

expansions originate in the sensory neuron NS1. G: golgi apparatus. The arrows indicate a very peculiar piled cytoplasmic structure; Fig. 7. x 14000. The sensory neuron NS1 gives expansions (large arrow) between the pigmented cup (P) and the cytoplasm of the second sensory neuron (NS2); some of these dendritic expansions (1) go down in to the cytoplasm (2) of the second sensory neuron. Note that another dendritic expansion (fine arrow) of the sensory neuron NS1 goes towards the cerebral neuropile.

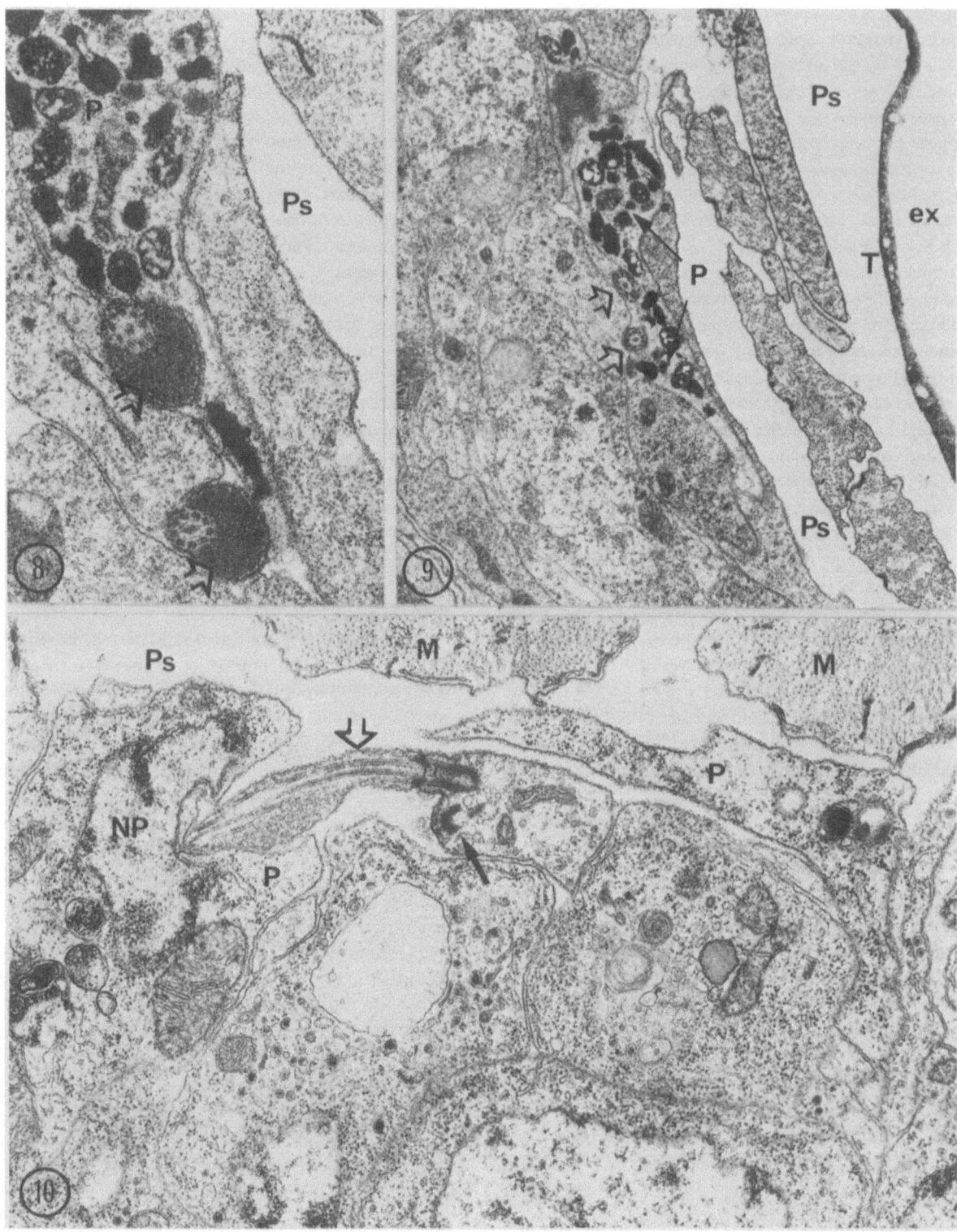

real vision, and photoreceptors with only one or two neurons.

One hypotheis is that the most primitive metazoa, like rotifers, evolved different photosensitivities, with a large number of pigments and different but simple structures of photoreceptors. With subsequent evolution, one of these adaptations was successful. The result was the use of one pigment (rhodopsin) for vision; however some variation was retained between different phyletic lines for the different kinds of rhodopsins, for photoreceptor structures, and for organisation of the eyes (multiplication of sensory cells and of accessory structures). Extraocular sensitivities within different phyletic lines can persist or disappear.

A remark about the phylogeny of lower metazoa
We know that the genome of metazoa is considerably richer than suggested by the limited set of different phenotypes present at any one time. Since an important part of this genetic potential does not express itself.

It is well-known that the same genome can be expressed, for example, in a miracidium, in a cercaria, in a metacercaria, or in an adult form of a parasitic Platyhelminth. Each stage has unique structures and functions not expressed in other stages.

The classical criteria of homology (Remane, 1955) seem to be too rigid and not always justified when juxtaposed to this view of the genome. For instance, with Remane's criteria, the cerebral eyes of Monogononta are homologous, and can constitute a phyletic line; but this is not true of the ampullae-shaped cilia in the ocelli of *Trichocerca rattus* and in the eyes of *Philodina roseola*. These cilia are not found in *Brachionus* or in *Rhinoglena*. So, the classical point of view says that they represent a convergent analogy, as does the same cilium in a phytoflagellate. I do not agree with these conclusions. In my opinion, the hypothetical, chromosome segment involved in the differentiation of the ampullae-shaped photoreceptor cilia in some phytoflagellates, is transmitted and is present in the primitive lower metazoa but expresses itself only in some of them.

The same argument is possible for the receptors with phaosome. The corresponding part of the genome is perhaps present in platyhelminths, rotifers and lower coel-omates; it expresses itself only in some cases such as in cercaria of some parasitic platyhelminths, *P. roseola,* some annelids and pogonophora, etc.

I do not mean to imply that each similarity can be an homology. The chromosome segments involved in the construction of the eyes of peridinians and cephalopods are surely different. But when the anterior ocelli of *Rhinoglena* have the same structures (rhabdomers borne by a cerebral neuron) and the same function (phototaxis) as the cerebral eye of a monogonont, are differences (such as localization, and perhaps one pigment more in the cerebral eyes) important enough to say that the same structures come from convergent independent mutations? I do not think so. Instead, it seems likely that the similarities between these two photoreceptors come from the same genome, and the differences from additional genetic information.

Finally, I think that there are two possibilities for approaching the phylogeny of a zoological group. The first one is to understand their richness in behaviour and related structures. I began here with photoreception. The next chapter summarizes other possible approaches. One is to study the structures and functions which are constant within a group, and then to compare them with other groups. This I shall try in the last chapter.

III. Adaptations and evolution

The success of rotifers is probably due to their rapid parthenogenetic reproduction. Rapid reproduction is only possible if the animal can get enough food. Therefore I first discuss moving and feeding mechanisms. Next, I consider those adaptations that foster survival in unstable biotopes often colonized by rotifers (ponds, mosses, lichens.). After briefly discussing the cycle of reproduction, I end this chapter with some hypotheses about the mechanisms of evolution in rotifers.

Moving and feeding behaviour
Recent studies on rotifers have substantially increased our knowledge of moving and feeding behaviours.

Figs. 8, 9, 10. Cerebral eyes of *Philodina roseola.*
Fig. 8. x 30000. Tranversal section of the two photoreceptive cilia (arrows): the electron dense substance is lateral to the cilium axonema. P: pigments in an epithelial pigmented cell. Ps: pseudocoel; Fig. 9. x 13000. Tranversal section of the basis of two receptive cilia (arrows). The pigmented cell (P) is located at the periphery of the brain (left). Ps: pseudocoel, T: integument, ex: external medium; Fig. 10. x 26000. Axial section of one of the two photoreceptive cilia (wide arrow). Its extremity goes down into the pigmented cell (P) whose nucleus (NP) is visible. The insertion of the second cilium near the base of the first is indicated by a black arrow. The lower part of the picture is occupied by peripheral cerebral neurons. Ps: pseudocoel, M: muscles.

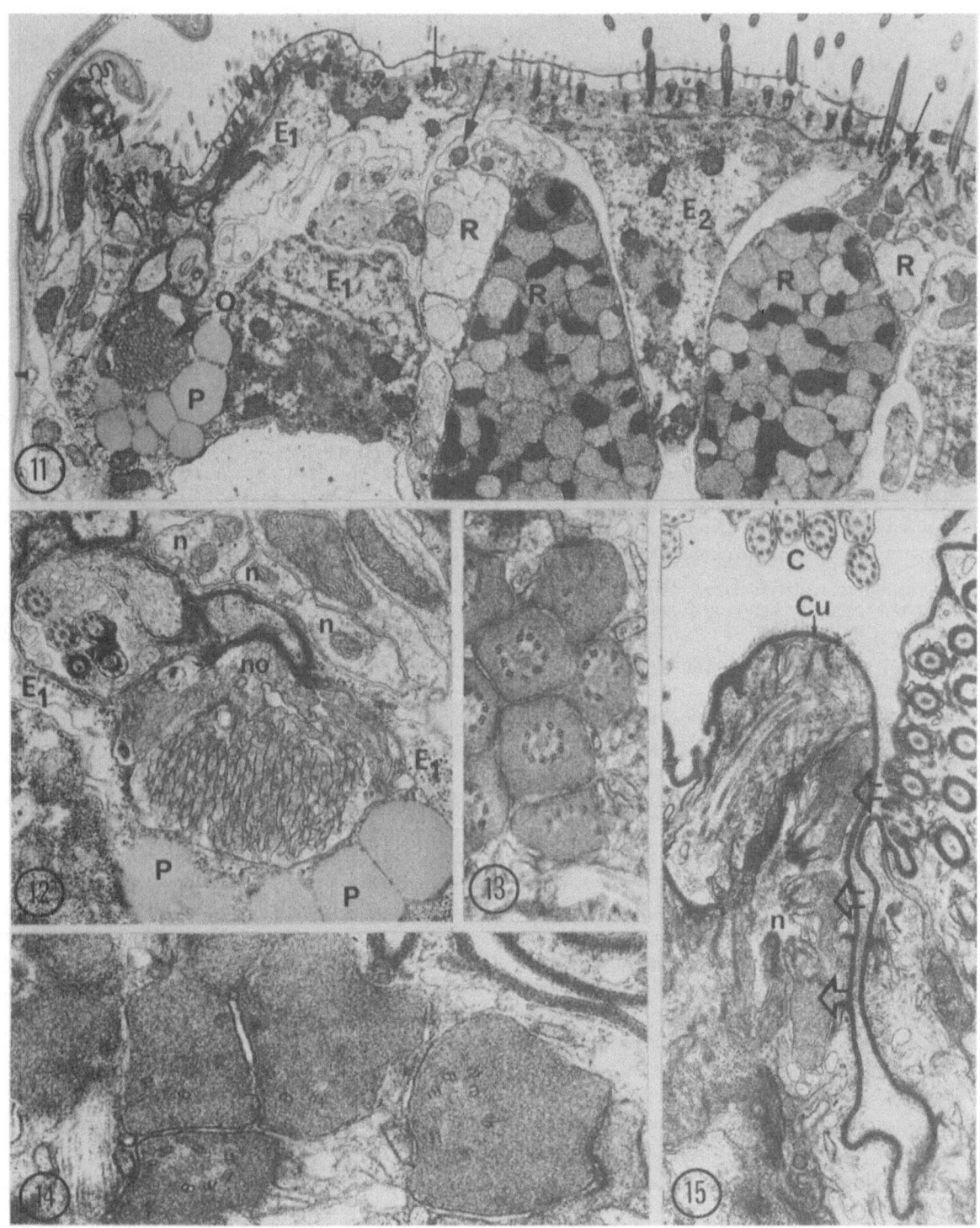

The classical work of De Beauchamp (1907, 1909) on the modifications and lack of foot, pedal glands, retrocerebral apparatus and different parts of the rotatory apparatus, is currently being expanded by two types of approaches.

First, the ultrastructural approach is used to study the different categories of cilia, muscles and sensory receptors involved in these behaviours (Clément, 1977a and b; Amsellem & Clément, 1977; Clément *et al.* a, b, c in this volume). Second, behaviour is being studied directly by a variety of experimental and observational approaches: see Gilbert (1977a, b), Gilbert & Starkweather (1977) and Starkweather (in this volume) for studies on feeding; Wallace (1980) for studies on sessile rotifers; and the preceding chapter for studies on phototaxis.

These results are too new and too voluminous to review them here. I only tried with photosensitivities (Chapter II) and the ultrastructural approach to feeding behaviour (this volume). About this last point, the classical work of De Beauchamp (1909) on the digestive tract of rotifers begins to be completed by electron microscopy, from which new questions arize: for instance, why are the pharyngeal cilia, which are the only cilia until now known to contain striated material, not exactly the same in *Philodina* and *Brachionus* (the striated material is immediately under the cytomembrane in *Brachionus* and inside the axonema in *Philodina*)? These cilia probably have the same function, as the malleate and ramate mastax seem to have the same function.

The dietary specialization of each rotifer (review in Pourriot, 1965, 1977 and Starkweather in this volume) is a complex problem: it is dependent as much on the type of mastax and digestive tract as on the different specializations of the sensory receptors and behaviours of the species. Photoreceptors and photosensitivities are only one part of these multifaceted problems.

Ovogenesis and cycle of reproduction

The reproduction of rotifers is more specialized than that of Platyhelmintha for two reasons: parthenogenesis and lack of scissiparity, and power of regeneration. The first forms of parthenogenesis appear in Platyhelmintha. In different primitive zoological groups, parthenogenesis exists in some individuals. With the exception of one genus *(Seison),* all rotifers can reproduce by parthenogenesis. For this reason, parthenogenesis appears to be a primitive trait of the entire group. Variations of the rate of reproduction with temperature, feeding and other factors, probably express adaptations of the parthenogenetic reproduction to precise biotopes.

The number of ovocytes of rotifers is determined prior to birth. These ovocytes are situated in the follicular epithelium (Fig. 20, Bentfeld, 1971a, b; Clément, 1977a, b) which sometimes surrounds the whole female genital apparatus (*Philodina roseola,* Fig. 20) or sometimes surrounds only the ovocytes (*Trichocerca rattus,* Clément, 1977a, b). An ovocyte grows with substances which come from both the vitellarium and pseudocoel (sometimes via the follicular epithelium: Fig. 20). Then the ovule secretes its shell and is layed.

The different egg deposition behaviours also express adaptations to precise biotopes. The eggs of some planktonic species can float; others are carried by the mother, either internally (in which case the female is ovoviviparous) or externally after laying. In some periphytic species, such as *Notommata copeus,* the mother turns for ten minutes around the egg she has layed, and thus fastens it to a filamentous alga. This alga is one of the food species of *N. copeus.*

The capacity for anhydrobiosis of Bdelloïdea is an adaptation to environments that frequently dry up (e.g. mosses and lichens). This ability perhaps explains the complete lack of males in this group.

Figs. 11 and 12. Anterior ocelli of *Rhinoglena frontalis* (in lateral sensory organs).
Fig. 11. x 8000. Localization of one of the two ocelli. The pigments (P) of the red cup are located in the epithelial cell (E₁) under the photoreceptive part (O) Left, the syncitial integument. Right, the four sections of symmetrical ducts of retrocerebral organ (R). Two epithelial ciliary cells (E1 and E2). In the cell E1, are embedded the neurites of the lateral sensory complex which contains the ocellus. The arrows indicate the apical sensory receptor neurites which are also symmetrical: they are located more centrally near the openings of the retrocerebral organ; Fig. 12. x 20000. Detail of an ocellus. Many sensory neurites (n) form the lateral anterior sensory receptor. They are surrounded by only one epithelial cell (E₁). This cell (E₁) contains pigments (P) which form the pigment-cup of the ocellus. The photoreceptive parts are the piled branches originating from the neurite (no).
Figs. 13-14-15. Anterior ocelli of *Trichocerca rattus* (in apical anterior sensory organs). Fig. 13. x 28000 and Fig. 14. x 54000. Transversal sections of dense ampullae shaped cilia; Fig. 15. x 26000. Axial section of one of the two anterior sensory organs of *T. rattus*. The ampullar-shaped cilia whose content is dense (arrows) are cut axially or obliquely; they are inserted on a neurite (n). This neurite is situated in an epithelial cell of the pseudotrochus (E). (C) pseudotrochus cilia (in the external medium). (Cu) anterior fine cuticle.

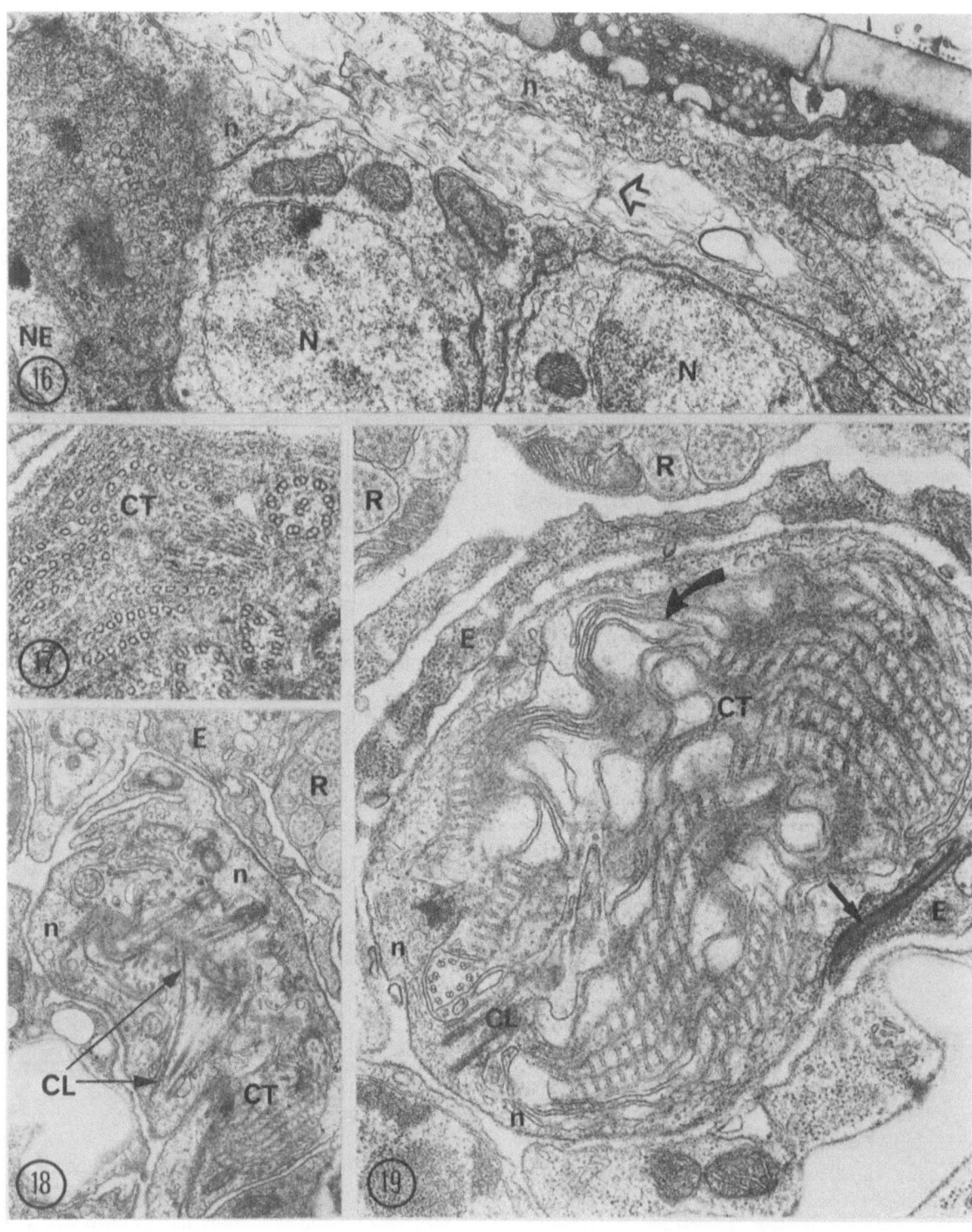

The function of males and of sexual reproduction is indeed, in Monogononta, to produce resting eggs. These eggs retain viability after being frozen or dessicated. Males seem to be absent in some clones of Monogononta which live in stable environments, such as big lakes (Ruttner-Koklisko, 1974).

The factors controling mixis can sometimes, but not always, be understood. In some cases, there is continuous production of mictic females (Pourriot & Rougier, personnal communication); in other cases, there are alternative phases of parthenogenesis and sexual reproduction. In the latter cases, mixis is produced by a precise factor: photoperiod in *Notommata copeus* or α-tocopherol in *Asplanchna* (review in Clément, 1977a and Gilbert, 1977c). In all cases, exogenous and endogenous factors control the percentage of mictic females (Clément *et al.,* 1976; Clément, 1977a).

The first appearance of such an heterogonic cycle and formation of resting eggs is in the Platyhelminthes. The influence of population density and photoperiod can already be noticed. In more advanced animal groups, very similar cycles controlled by the same factors can also be observed (Cladocera).

Hypotheses on the mechanisms of evolution in rotifers
The general scheme could be the following one:
– initially rich genome with both multiple and primitive potentialities;
– acquisition of the unchanging characteristics of the group (see chapter IV, and above about parthenogenesis);
– diversification of the forms which keep these characteristics, and the initial rich genome, but express a diversity of more or less different phenotypes.

The specialization of rotifers and of the main lines of their classification can probably be explained by modifications of their genome during sexual reproduction: crossing-over, mutation... (King, 1977). Yet, other specializations of rotifers can be due to peculiar mechanisms.

The first of these mechanisms is mutation occurring during mitoses of the parthenogenesis. This parthenogenesis is probably mitotic (King, 1977) and not endomeïo-tic. It seems to be primitive in rotifers (see above), and rapid reproduction increases the probability of mutations. The speciation of Bdelloïda, and possibly of a lot of Monogononta, is perhaps due to his mechanism (Pourriot & Clément, in press).

The second mechanism may be maternal effects. I am using the term to include all reversible maternal influences expressed over some generations: Lansing effect (Lansing, 1947, 1954); influence of the age of the grand-mothers (Pourriot & Rougier, 1977; Clément & Pourriot, 1979); influence of substances sent out by *Asplanchna* on the appearance of tegumentary spines in *Brachionus, Filina, Asplanchna* (Gilbert, 1967, 1977; Pourriot, 1974); influence of substances related to crowding in the induction of mixis (Clément & Pourriot, in this volume). In the last case, mixis can disappear when females of *N. copeus* are in a crowded situation, but this is reversible. Nevertheless, could be apparently complete disappearance of mixis and males in planktonic rotifers of big lakes (Ruttner-Kolisko, 1972) have a similar origin?

IV. Ultrastructures and phylogeny

In the second and the third chapter, I discussed the diversity of the phenotypes of rotifers: I criticized the often too arbitrary rigidity of Remane's criteria for homology, and I made some hypothesis about the possible use of this diversity for phylogeny.

In this fourth chapter, I am considering some structures that are constant in all rotifers, in spite of adaptative modifications from an animal to an other. I have choosen five examples that seem to be of interest for comparing rotifers to other zoological groups. These are the integument, the flame-cells, the body cavity, the thick myofilaments, and the nervous system.

The syncytial integument
The skeleton of rotifers is peripheral but not extracellular: the extracellular cuticle is always gelatinous and never skeletal.

Fig. 16. x 23000. Cerebral paired receptors of *Trichocerca rattus.* Detail of one of them. The sensory neurite (n) bears thin microvilli (arrows); (N) nuclei of cerebral neurons; (NE) nucleus of an epithelial cell located at the brain periphery.
Fig. 17. x 68000; 18. x 20000; 19. x 35000. Anterior unpaired receptor of *Philodina roseola.* The sensory neurite (n) forms a sort of sphere in which lamellar cilia are piled. These cilia are inserted on both sides of the neurite (Fig. 18). The cilia base shows a classical axonema (Fig. 17). The two central tubules then disappear while the other tubules form the parallel ribs in the ciliar lamella (CT); (CL) an axial section of the base of a lamellar cilia; (E) anterior epithelial cell on which the neurite is fixed by a desmosome (arrow, Fig. 19); (R) retrocerebral organ. The large arrow (Fig. 19) indicates that the membrane of cilia extends in flattened villi.

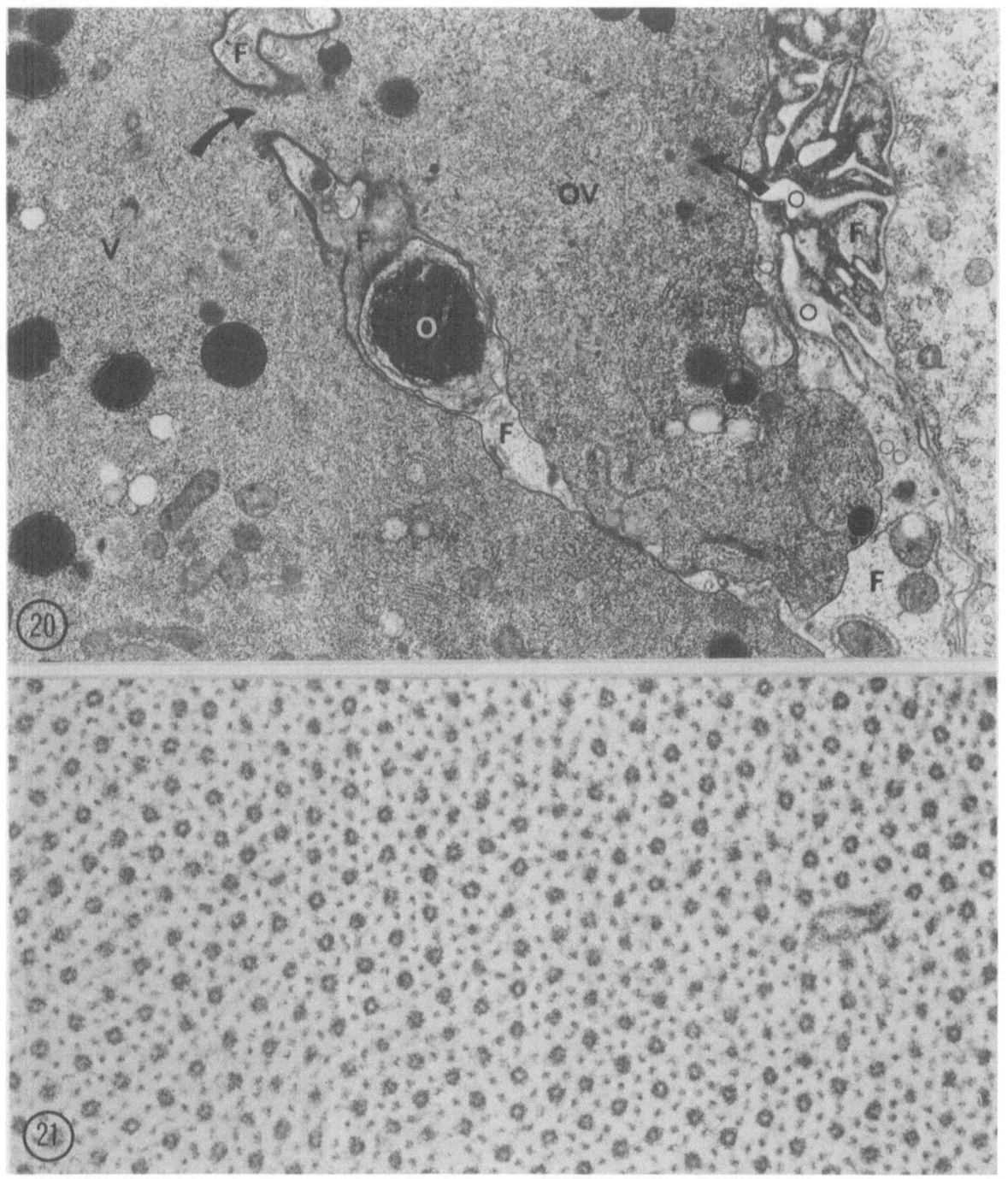

Fig. 20. Ovogenesis in *Philodina roseola*. x 16400. The follicular epithelium (F) surrounds all the vitellarium (V) in which the ovocytes (O and OV) are found. The ovocyte which grows to form an ovule (OV) communicates with the vitellarium by a cytoplasmic bridge (left arrow). At the level of this ovocyte (OV), the follicular cell shows many infoldings: this indicates a probable entrance of substances from the pseudocoel (right arrow); Fig. 21. Musculuture-Tranversal section of the mastax striated muscle of *Trichocerca rattus* (x 200000). The ratio between thick and thin myofilaments is 1/3. See comments in chapter IV, par. 4.

The peripheral skeleton, which muscles attach to, is a dense intracytoplasmic lamina located inside the syncytial integument. This skeletal lamina is either thick and rigid, at the level of the trunk (Fig. 22, 23, 24, 25, 27) or supple, at the level of the articulations or at the front of the animal (Fig. 26).

Three kinds of skeletal lamina have been observed (Clément, 1969), However, four categories can be distinguished now:

1. The *Philodina* type (Fig. 27, and Schramm, 1978b for *Habrotrocha*), in which only the internal layer of the skeletal lamina is thickened. In the three other types, only the external layer of the skeletal lamina is thickened.

2. The *Trichocerca* type (Fig. 24), in which the external layer of the lamina is uniformly dense (see also *Keratella* in Koelher, 1966 and in Hendelberg *et al.,* 1979).

3. The *Brachionus* type (Fig. 22, 23; Clément, 1969, 1977b; Storch & Welsh, 1969), in which the external layer of the lamina is made of juxtaposed vertical tubules (see also *Mytilina* in Clément, 1969).

4. The *Notommata* type (Fig. 25; Clément, 1969) in which the external layer of the lamina is made of stacked lamellae (see also *Asplanchna* in Koelher, 1965, and *Synchata* in Clément, 1969).

In the four cases, the function of this skeletal lamina remains the same. Variations of structure are therefore good indicators of the phylogeny among rotifers.

The internal layer of the skeletal lamina is thicker than the external layer in the bdelloïds; but this is also the case in some young monogononts (*Brachionus,* Clément, 1977b).

An intracytoplasmic peripheral skeleton seems to exist in only one other zoological group, the Acanthocephala. In other Aschelminths, in Annelida, Mollusca, and Arthropoda, the external skeleton is cuticular, i.e. extracellular.

The phyletic origin of the skeleton of rotifers is to be sought in animals with soft integument: for example in the terminal webb or in the infraciliature of a ciliary integument. A soft non-cuticular and ciliary integument can be observed in Platyhelmints as well as in Cnidaria and Ctenaria.

The flame-cells

In 1853, Huxley emphasized the phylogenetic importance of the flame-cell of rotifers. Its study by electron microscopy enabled both a better definition of its characteristics and variations in rotifers, and the postulation of possible relationships with other animal groups.

In rotifers (Fig. 28 to 32) the flame cell has been studied in some Monogononta: *Asplanchna* (Pontin, 1964, 1966; Braun *et al.,* 1966; Warner, 1969), *Notommata* (Fig. 28, Clément, 1967, 1968, 1969, 1977a), *Trichocerca* (Fig. 29, Clément, 1977a-b), *Rhinoglena* (Fig. 31), and in some Bdelloïda: *Philodina* (Fig. 30, Mattern & Daniel, 1966) and *Habrotrocha* (Schramm, 1978a).

In all cases, it is a hollow, cylindrical or ampulla-shaped cell with a non-apical nucleus. The apical cap contains the bases of the cilia of the flame. The cavity of the cell communicates with the lumen of the protonephridial tube. The membranes of the filtering wall are situated between thin parallel cytoplasmic columns. This first grid subtended inside a second grid that has a more skeletal function. The second grid is formed by cytoplasmic extensions called pillars.

There are three types of variations:

1. Size and number of flame-cells. The volume of the animal and the surface of the filtering wall are correlated (Clément, 1977a). The number of cilia of the flame and the size of the cell are also related. In the large species of rotifers, there are two ways possible to increase the surface area of the filtering membranes: (1) to increase the size of the flame-cells (*N. copeus* Fig. 28), (2) to increase their number (*Asplanchna* Pontin, 1966).

2. Structure of the filtering-wall. In *Trichocerca* (Fig. 29) and in bdelloïds (Fig. 30) pillars and columns are often bound together. In the other Monogononta, pillars and columns are two distinct parallel grids.

3. Structure of the dense material of the pillars. It is cross-wise striated, as a ciliary root in bdelloïds, but not in monogononts (Fig. 31 and 32).

The flame-cell of Platyhelminths is very similar to that of rotifers (Kümmel & Brandenburg, 1961; McKanna, 1968; Swiderski *et al.,* 1975): flame with many cilia, and filtering-wall with pillars and columns. Nevertheless in Platyhelminths the nucleus is always apical, and the filtering-wall is a grid in which a column and a pillar containing electron dense material alternate regularly. The filtering membrane is located between each column and pillar.

In nematods and nematophores, flame-cells do not exist.

In Priapulids, the protonephridial apparatus has groups of typical solenocytes, quite similar to those of some Annelida: apical nucleus, one cilium only, filtering wall with one grid only (Kümmel & Brandenburg, 1961).

In Gastrotricha (Brandenburg, 1962; Teuchert, 1973), the flame-cells are solenocytes too, going by pair or grouped.

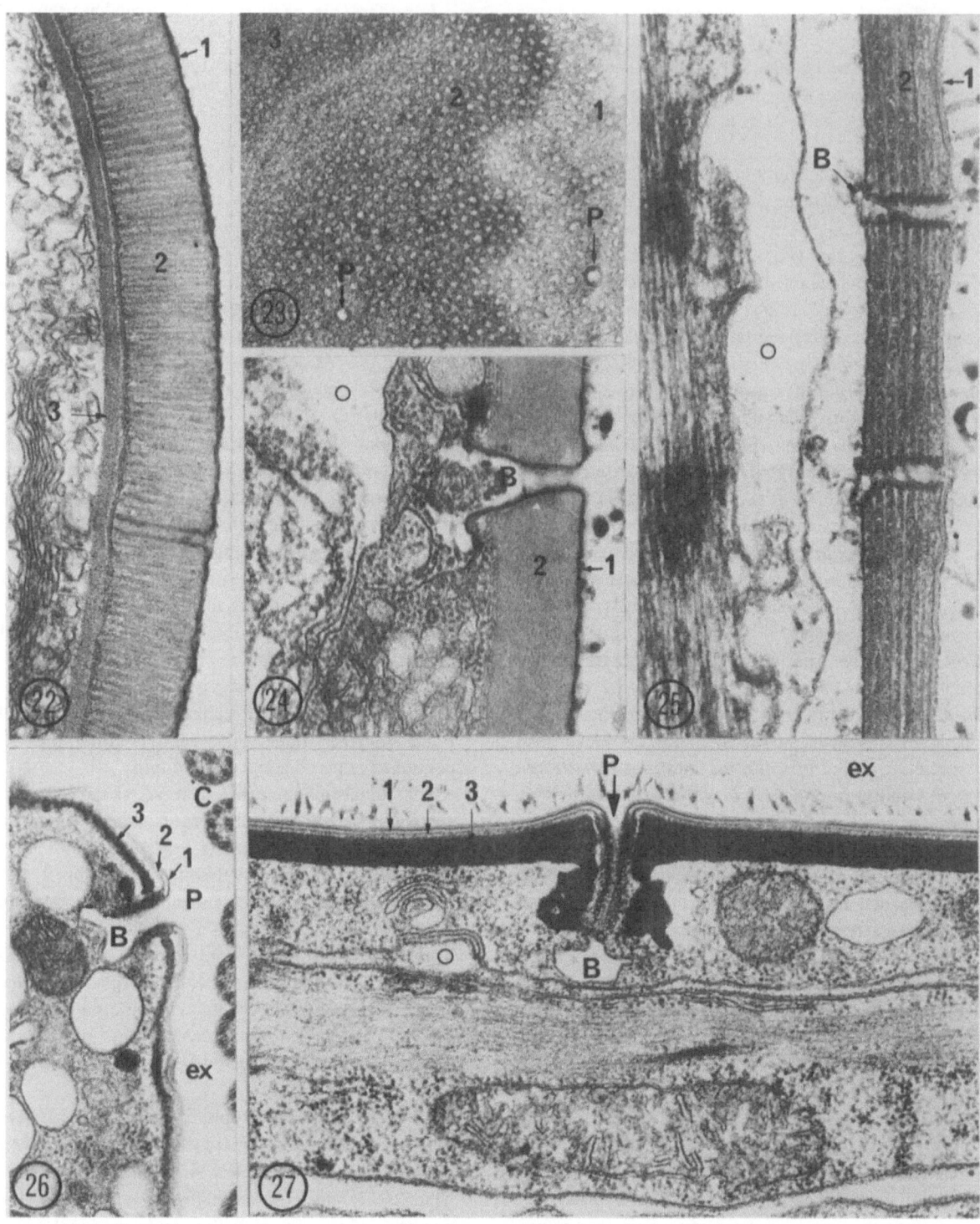

108

In Kinorhincha, the flame-cell seems to have several nuclei and each flame is made of one or two cilia (Hyman, 1951). The ancestor of the plurinucleated flame-cell could be the ciliary rosette of the Ctenaria (Franc, 1972).

In Cnidaria, no structure looks like a flame-cell. Some cells are more like the choanocytes of sponges whose function is more digestive than excretive, but which are fairly similar to that of the solenocytes.

These facts suggest some hypotheses about the phylogeny of lower metazoa:

1. A close relationship between rotifers and Platyhelminths; but the non-apical flame-cell nuclei distinguish rotifers from Platyhelminths.

2. An early separation between bdelloïds (striated pillars) and monogononts (non-striated pillars).

3. A great homogeneity of bdelloïds, perhaps reflecting a rapid disappearance of sexual reproduction in this order. In contrast, a relative heterogeneity in monogononts, and in Platyheminths, in which these variations of the flame-cell are good indicators for the phylogeny.

4. A possible relationship between gastrotrichs, priapulids, and annelids, which all possess solenocytes and no flame-cells.

5. A more speculative relationship between Ctenaria, Kinorhincha, and Nemertina, with only the basis of their pluricellular flame-cells.

The body cavity

Pseudocoel and mesenchyme

In rotifers, electron microscopy has demonstrated that the pseudocoel is directly limited by the integument on the outside and by the digestive epithelium on the inside. No thin membrane looking like a regressed coelomic wall has been observed, contrarily to what Remane suggested (1963). The presence of basal lamellae between a cell and the pseudocoel is very variable (compare Figs. 28, 29 and 30 for a basal lamina around the flame-cell).

Free mesenchymatic cells do not seem to exist in the pseudocoel of rotifers. The starry cells of the pseudocoel described by Nachtwey (1925) and Remane (1929-32) are scarce and do not seem to be free. In electron microscopy, some very thin cellular expansions are sometimes observed in the pseudocoel, but they often seem to be expansions of muscular cells.

Many moving cells, associated with fibrous structures, are observed in the pseudocoel of Kinorhincha, Priapulida and Nematomorpha (Hyman, 1951). So, in these animals, the pseudocoel seems to be a mesenchyme less compact than that of Platyhelminths. In nematods, there is an intermediate situation. The pseudocoelocytes are fixed, they are neither phagocytic nor amoeboïd and will not take up vital dyes (Hyman, 1951). Are they real mesenchymatic cells? Lastly, in Gastrotricha, there are no free amoeboïd cells in the pseudocoel. (Hyman, 1951).

No ultrastructural morphological argument allows us to suggest that the pseudocoel of rotifers is a regressed coelom.

On the other hand, the presence of mesenchymatic cells in the pseudocoel of most of the Aschelminths raises another question: is this 'cavity' a mesenchymatic or a classical conjunctive space? The answer is perhaps yes for Priapulida, Nematomorpha, and Kinorhincha. In nematods, gastrotrichs, and rotifers, the question remains without answer. We are attempting to solve it by another criterium: the intercellular collagen which, in all the animal kingdom, characterizes the conjunctive spaces. First, however, I wish to stress a correlation between eutely and lack of characteristic mesenchyme with active and free amoeboïd cells.

Eutely

The number of nuclei is remarkably constant in rotifers: from birth to death and from one animal to another in the same species. Exceptions to this last point are rare and concern individual variations of the number of nuclei in polyploïd syncytial organs: vitellarium and gastric glands (Birky & Field, 1966).

This perfect eutely in rotifers explains the absence of regeneration in these animals. So, it is usually admitted that on the genealogical tree of the animal kingdom, the rotifers are apparently out on a limb from which there is

Figs. 22-27. x 50000. The skeletal lamina of the integument. Fig. 22. *Brachionus calyciflorus;* Fig. 23. *Brachionus calyciflorus:* tangential section; Fig. 24. *Trichocerca rattus;* Fig. 25. *Notommata copeus;* Fig. 26. *Philodinia roseola:* anterior supple integument (C: pseudotrochus cilia); Fig. 27. *Philodina roseola:* trunk integument

ex	: external medium
o	: pseudocoel
B	: bulb
P	: pore
I	: apical cytoplasmic membrane
2 and 3	: the two layers of the skeletal lamina.

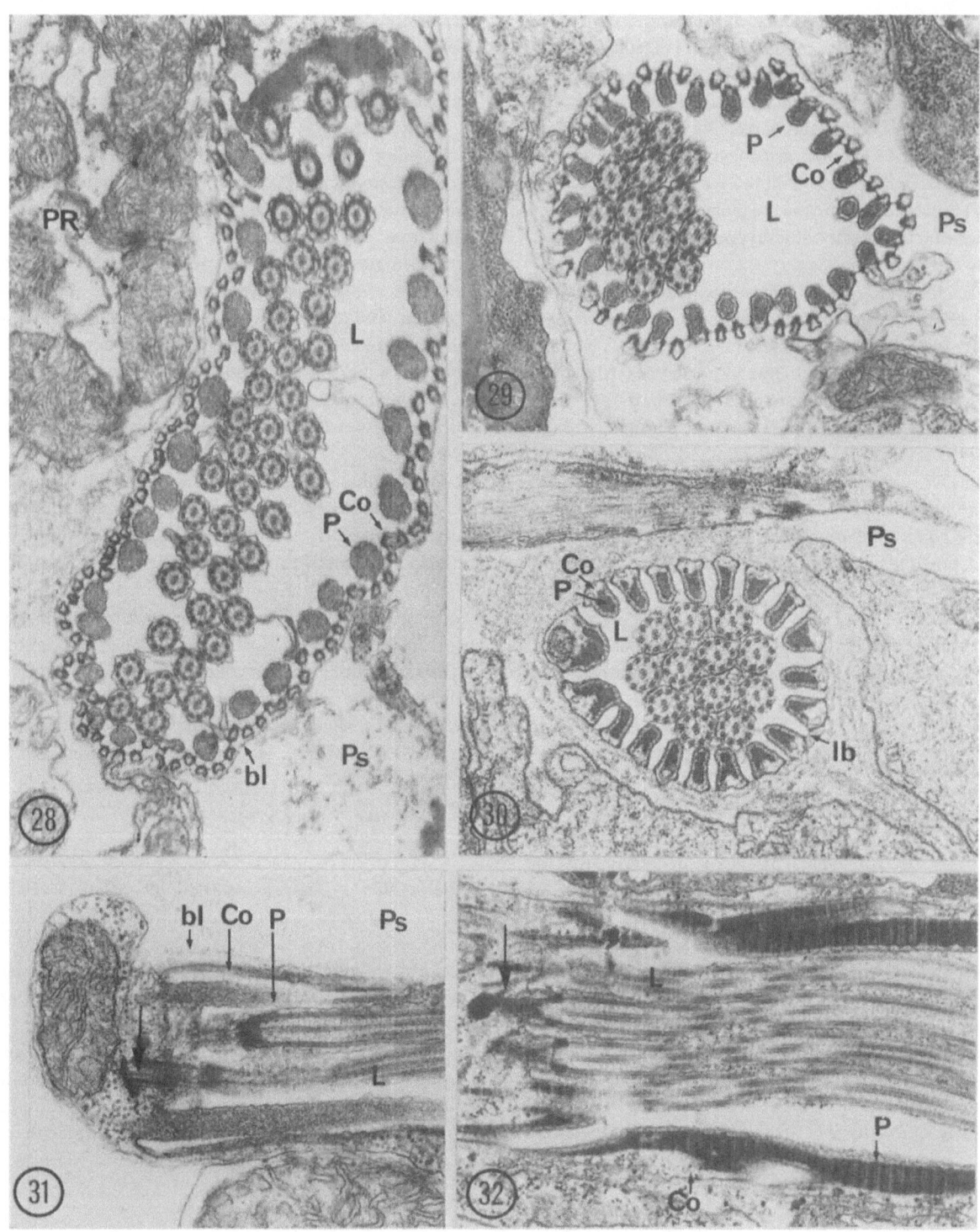

nowhere to go (Ruttner-Kolisko, 1963). Hyman (1951) propounds that entoprocts come from rotifers. Most certainly the similarities with the Collothecacea are impressive. But the entoprocts regenerate and multiply themselves by asexual reproduction: these characters observable in Turbellaria also are primitive and do not exist in Collothecacea. They exclude any possibility of direct relationships between the specialized sessile rotifers and the Entoprocta.

Such a cell constancy is observed neither in Acoelomates nor in other Pseudocoelomates, except for nematods (Hyman, 1951). In nematods, however, eutely is less perfect; it does not apply to gonads and variations for other organs have also been observed in young stages. In particular, in Aschelminths, the Gastrotrichs, Kinorchinchs and Priapulids have no cell constancy (Lang, 1963).

Collagen and collagen fibrils

Collagen has been found in all pluricellular animals where it has been looked for (Adams, 1978), mostly as transversely striated fibrils. This is probably a primary animal characteristic (Pikkarainen & Kulonen, 1969).

However, some variation exists in primitive Invertebrates (reviews in Bairati, 1972; Garonne, 1975; Adams, 1978). In platyhelminths and nematods, the fibrils seem less structured than in Porifera and in Anthozoa (Cnidaria). In Ctenaria, there are no structured fibrils but a network of microfilaments which contain some hydroxyprolin (Franc *et al.,* 1976). After extraction, this collagen precipitates in the shape of distinct fibrils. The collagen of a parasitic platyhelminth does the same: the fibrils are more distinct in vitro than in vivo. Garonne (1975) suggests the presence in vivo of a factor limiting the organization of the fibrils of collagen in Platyhelminths and in Ctenaria.

In rotifers, we see neither microfilaments nor fibrils of collagen in the pseudocoel. The basal lamellae that seem to have a type IV collagen in vertebrates are not always present according to the cells and the species of rotifer. Lastly, the gelatinous cuticle of rotifers does not show the same aspect as that of nematods, nematomorphs or annelids which contain massive collagen (reviews in Bairati, 1972 and in Garonne, 1975; see also Eakin & Brandenburger, 1974 for the nematomorphs).

A biochemical assay of hydroxyprolin, aminoacid characteristic of collagen, would allow to see if non-fibrillar collagen exists in rotifers. Nevertheless a large number of rotifers would be necessary for such an assay especially if their located is only in the basal lamellae.

Lastly, in comparison with Ctenaria, the presence of non-fibrillar collagen will represent either an homology or an analogy. More advanced biochemical analyses of the different chains of collagen, comparable with those being carried out in vertebrates, would be necessary in invertebrates for the presence and the shape of collagen to become good phylogenetic indicators.

The body cavity and the relationships of rotifers

Rotifers are very different in their eutely and their lack of mesenchyma from Platyhelminths and most of the other Aschelminths apart from nematods. Nematods and rotifers seem to represent endings of phyla having lost all plasticity while becoming more specialized.

From which type of organization the parallel and irreversible acquisition of (1) eutely, (2) factors preventing the apparition of (visible) collagen, and (3) factors preventing the existence of little differenciated and labile cells, has been made?

Thick myofilaments
All muscles of rotifers, slow or fast, smooth or striated, have two kinds of myofilaments characterized by their diameter (Fig. 21). The thin myofilaments have the same morphological characteristic than classical myofilaments of actin. The thick myofilaments look like the myofilaments made of myosin of the arthropods, in particular of the Crustacea (Atwood, 1972; Pringle, 1972).

Rotifers are different from the zoological groups which have very thick myofilaments of paramyosin. These myofilaments are characterized by their very large diameter

Figs. 28 to 32. x 30000. The flame-cell.
Fig. 28, 29 and 30. Tranversal sections in the filtering wall;
Fig. 28. *Notommata copeus;* Fig. 29. *Trichocerca rattus;* Fig. 30. *Philodina roseola.*
Fig. 31 and 32. Axial sections of the cap and the filtering wall.
Fig. 31. *Rhinoglena frontalis;* Fig. 32. *Philodina roseola.*
The filtering wall is made of the pillars (P), the columns (C), the filter membrane between, the columns, and sometimes the basal lamina (bl). The vibratile flame cilia are located in the flame-cell lumen (L) and inserted on the cap (arrows). The pillars show a transversal striation in *P. roseola* (Fig. 32) but not in the Monogononts (Fig. 31). In *P. roseola,* the pillars and small columns are nearly always fused (Fig. 30). They are sometimes fused in *Trichocerca rattus* (Fig. 29) and practically never in *Notommata copeus* (Fig. 28). (Ps) pseudocoel, (PR) protonephridial tubule.

and by their striation. Myofilaments of paramyosin are well known in Annelida and Mollusca but have also been observed in Platyhelminths (Reger, 1976; Kryvi, 1973; Fournier, pers. comm.), in nematomorphs (Eakin & Brandenburger, 1974; Lanzavecchia *et al.*, 1979) etc...

In other respects, the thick myofilaments of the Ctenaria and Cnidaria have the same diameter as those of rotifers (Hernandez-Nicaise & Amsellem, in press).

Nervous system and endocrine secretions

A glia-free nervous system

Rotifers do not seem to have glial cells: it is epithelial cells and sometimes muscular ones which surround some nerves or ganglia and an imporant part of the brain.

Such a lack of glia is only seen in the most primitive invertebrates. Horridge (1968) reports that sea anemones, jelly fishes, siphonophores and ctenophores do not have glia. But contrary to these animals, rotifers do not possess nervous nets; the nervous system is very concentrated: brain, two main nerves, a few ganglia.

In Platyhelminths, groups of nervous cells are surrounded by thin glial sheaths (Morita & Best, 1966 and mainly Koopowitz & Chien, 1974), as in nematomorphs (Eakin & Brandenburger, 1974).

In nematods, Ward *et al.* (1975), Ware *et al.* (1975) and Wergin & Endo (1976) describe a sheath structure around each anterior sensorial neurite. This sheath cell is likely to be epithelial. Such a disposition is also seen in most anterior sensorial receptors of rotifers (Clément, 1977a). But Ware and coll. (1975) also report glial cells in the nerve ring of *Caenorhabditis elegans*. They locate the cells in a precise tridimensional reconstruction. They notice that the aggregation of cell bodies of neurons 'are occasionally separated from the surrounding hypodermal cells by thin glial processes but more often are in direct contact with them'.

Neurosecretions and hormones

The neurons of rotifers show a variety of vesicles containing neurotransmitters of at least four types (Villeneuve-Amsellem & Clément, 1971; Clément, 1977a). This variety is already seen in the most primitive invertebrates, in particular in Platyhelminths (Lentz, 1968).

Among the vesicles with dense center, the biggest have the morphological features of neurosecretions (Clément, 1977a, p. 193-198). Neurosecretions have been described in most primitive animal groups. A neurohormone has been isolated in *Hydra* (Schaller, 1976): it acts either on mitoses or by induction of the transformation of indifferentiated cells into neurons. The neurosecretion also plays a role in the regeneration and scissiparity of planarians (Lender, 1976). As there is not differentiation or cellular multiplication in rotifers, neurosecretions could control ovogenesis, in particular production of mictic females (Clément, 1977).

I have suggested (Clément, 1977a and b) the existence of an endocrine integumental secretion in *T. rattus*. However, ultrastructural features on which this hypothesis was based have not been found in other rotifers. This problem remains without answer.

The endocrine system of rotifers is little developed; this feature is probably related to eutely, to lack of moultings and plasticity of the structures.

V. Conclusions

Phylogeny in rotifers

Inside the class of rotifers, I have argued for a very early separation between Bdelloïds and Monogononta. Beside the classical differences there are ultrastructural differences in the skeletal lamina of the integument, pillars of the flame-cells, eyes and ocelli, pharyngeal cilia, stomach, etc...

I have also tried to show that:

– the study of sensory organs and, at the same time, the study of behaviour, is necessary to understand the evolution of rotifers. For instance, what are the relations between the photosensitivities and the biotope of a specific rotifer (cf. chapter II)?

– the mechanisms of evolution rotifers are not limited to classical meiotic mutations; there are also mutations during parthenogenesis and maternal effects to be considered. The implications of these on phylogeny merit further study (cf. chapter III).

– Some structures are constant in all rotifers, and have the same functions. In these cases small variations are good indicators of the phylogeny of rotifers. Such structures for instance, are the skeletal lamina of the integument and the flame-cells (cf. chapter IV). Nevertheless, it is difficult to use these new criteria for classification and phylogeny of rotifers because, to date, the number of ultrastructural studies is still very inadequate.

Relationships of rotifers

Comparing the embryogenesis of different groups of pseudocoelomates, Joffe (1979) distinguished three types: 1/ Priapulida and Nematomorpha; 2/ Rotifera and Acanthocephala and 3/ Nematoda and Gastrotricha. He also emphasized the similarities between the embryology of Rotifera (and Acantocephala) and Turbellaria.

Our ultrastructural study of the integument of rotifers also shows a possible relationship between Rotifera, Acantocephala and Platyhelminths: the other pseudocoelomates and the lower coelomates have a cuticular external skeleton. Other ultrastructural features seem to join rotifers and Platyhelminths: the flame cells, the cerebral eyes, the ocellus which is a ciliary phaosome, etc....

Moreover, the pseudocoel of rotifers is not a regressed coelom. Neither embryological nor morphological evidence supports Remane's theory of the coelomate origin of rotifers.

Nevertheless, I have shown in this paper some major differences between Rotifera and Platyhelmints. The most important of these is that rotifers have eutely and lack mesenchyme. These caracters are related to the absence of regeneration. Beside well-known differences (parthenogenesis, cloaca). I have pointed out new ones concerning for instance the thick myofilaments, the glia etc... These differences indicate that Rotifera do not directly stem from Platyhelminths. However, the two groups probably had a common ancestor.

The success of rotifers is based on their mechanisms of reproduction, well adapted to their aquatic biotope; at the same time, their cells are few, and constant in number, and very specialized. So, even with a great richness of receptors and with a centralized nervous system, their behaviour remains simple (taxis) and they lack capacity to learn.

Conversely, and although their ancestor was the same, the Platyhelminths remained a labile and totipotent group. The same species can be successively adapted, by different morphological forms, to different free or parasitic biotopes. The structure of the sensory receptors and of the nervous system are not very different from the Rotifera; a lot of sensitivities and behaviours which have been studied in the two groups, seem to be the same in Platyhelminths and Rotifera. However, some behaviours seem to be more evolved in Platyhelminths, for instance, the possibility of learning, and are perhaps related to the presence of glia or of some pluricellular sensory receptors.

Finally, this work suggest an indirect relationship between the Rotifera and Platyhelminths and a phyletic heterogeneity of the Aschelminths. This last point must be further clarified by syntheses on each pseudocoelomate group.

Another exciting point is the origin of the lower metazoa. In the second chapter, I suggested that Rotifera possess at least a part of the genome of the Phytoflagellates. If this relationship is not direct and comes through the acoelomate groups, we should observe in some species of Porifera, Cnidaria or Ctenaria, some characteristics of Phytoflagellates (for instance the photoreceptor ampulla-shaped cilia). Future ultrastructural work will perhaps give an answer to this question.

Summary

The first chapter summarizes the state of the disagreements about the phylogeny of rotifers and lower metazoa in 1963. The only arguments were morphological, and the only problem was the definition of homologies. There are today more diversified approaches of the evolution: electron microscopy, ethology, genetics and ecology.

The second chapter shows, using an example, that phylogeny is very complex. A synthesis is made on the photosensitivities and the photoreceptors of rotifers, with several original ultrastructural descriptions (ocelli of *Rhinoglena frontalis* and *Philodina roseola;* cerebral eyes of *Brachionus calyciflorus* and *P. roseola*). After a criticism of several theories on the use of photoreceptors in phylogeny, a new polyphyletic theory is proposed and the classical criteria of homology (Remane, 1955) are discussed.

The third chapter considers two major evolutionary features of rotifers: parthenogenetic reproduction, which is correlated with feeding, and special adaptations promoting survivorship in harsh environments (anhydrobiosis in Bdelloïdea, resting eggs production in Monogononta). In addition to classical meiotic recombination, evolutionary mechanisms in the Rotatoria include mutation during parthenogenesis and maternal effects.

The forth chapter describes some constant ultrastructural features in rotifers, and compares them to homologous structures in related groups: skeletal integument, flame-cells, pseudocoel, thick myofilaments and a glia-free nervous system. Since some of these structures (integument and flame-cell) have the same fonctions in all rotifers, their variations are good indicators of phylogeny.

In conclusion (V), not one argument corroborates Remane's hypothesis of the coelomate origin of rotifers. The hypothesis of Josse (1979), founded on embryological works, is corroborated by several ultrastructural features discussed herein, although rotifers have been placed in the phylum Aschelminthes, several aspects of their ultrastructural morphology suggest more relationships to the Acanthocephala and Platyhelminths than to the other classes of Aschelminths. Other ultrastructural observations show that this relationship Rotatoria-Platyhelmintes is not direct: they have a common ancestor. The relationship Rotifera-Phytoflagellates is also discussed. Finally it is necessary to carry on other ultrastructural, ethological and genetic work on both rotifers and related groups.

Acknowledgements

Jacqueline Amsellem provided essential aid with the electron microscopy, preparation of the micrographs, and the typing of this text. Charles E. King and Roger Pourriot discussed this work with fruitfull criticism, as did John Stewart for the genetic part. Roger Pourriot and Claudia Ricci gave us the clones of rotifers that Anne Luciani and Annie Cornillac maintain in culture in Lyon. E. A. Kutikova and K. Plasota helped with the russian bibliography. Fiona Hemig, Pascal Rousset and Charles E. King participated in the translation of this work from french to english. Financial support came from the Laboratoire d'Histologie et de Biologie tissulaire, L.A. C.N.R.S. 244 (Université Lyon I); electron microscopy was conducted in the C.M.E.A.B.G. (Université Lyon I).

References

Adams, E. 1978. Invertebrate Collagens. Marked differences from vertebrate collagens appear in only a few invertebrate groups. Science, 302: 591-598.

Amsellem, J. & Clément, P. 1977. Correlation between ultrastructural features and contraction rates in rotiferan muscles. I. Preliminary observations on longitudinal retractor muscles in Trichocerca rattus. Cell. Tiss. Res., 181: 81-90.

Atwood, H. L. 1972. Crustacean muscle. in 'The structure and function of muscle' 2nd ed., C. H. Bourne ed., Acad. Press, I: 422-490.

Ax, P. 1963. Relationships and phylogeny of the Turbellaria. in: 'The Lower Metazoa', E. C. Dougherty ed., Univ. Calif. Press, p. 191-224.

Bairati, A. 1972. Collagen: an analysis of phylogenetic aspects. Boll. Zool., 39: 205-248.

Beauchamp, P. de 1907. Morphologie et variations de l'appareil rotateur des Rotifères. Arch. Zool. exp. gén., 6: 1-29.

Beauchamp, P. de 1909. Recherches sur les Rotifères: les formations tégumentaires et l'appareil digestif. Arch. Zool. exp. gén., 10: 1-410.

Beauchamp, P. de 1965. Classe des Rotifères. in 'Traité de Zoologie, Anatomie, Systématique, Biologie' P. P. Grasse ed., IV, 3: 1225-1379.

Beklemishev, V. N. 1963. On the relationship of the Turbellaria to other groups of the animal kingdom. in: 'The Lower Metazoa', E. C. Dougherty ed., Univ. Calif. Press, p. 234-246.

Bentfeld, M. E. 1971a. Studies of oogenis in the rotifer Asplanchna. I. Fine structure of the female reproductive system. Z. Zellforsch., 115: 165-183.

Bentfeld, M. E. 1971b. Studies of oogenis in the rotifer Asplanchna. II. Oocyte growth and development. Z. Zellforsch., 115: 184-195.

Birky, C. W. & Field, B. 1966. Nuclear number in the rotifer Asplanchna: intraclonal variation and environmental control. Science, 151: 585-587.

Brandenburg, J. 1962. Elektronmikroskopische Untersuchung des Terminalapparatus von Chaetonotus sp. (Gastrotrichen) als ersten Beispiels einer Cyrtocyte bei Askelminthen. Z. Zellforsch., 57: 136-144.

Braun, G., Kummel, G. & Mangos, J. A. 1966. Studies on the ultrastructure and function of a primitive excretory organ, the protonephridium on the rotifer Asplanchna priodonta. Pflügers Archiv., 289: 141-154.

Brooker, B. E. 1972. The sense organs of trematode miracidia. in: 'Behavioural aspects of parasite transmission', Canning, E. U. & Wright, C. A. eds., Acad. Press, London, p. 171-180.

Champ, P. 1976. Etude des populations d'un Rotifère épiphyte dans la Loire. Thèse doctorat 3ème cycle, Univ. Paris VI, 80 p.

Clark, A. W. 1967. The fine structure of the eye of the leech, Helobdella stagnalis. J. Cell Sci., 2: 314-348.

Clément, P. 1967. Ultrastructure du système osmorégulateur d'un Rotifère Notommata copeus. Conclusions physiologiques et phylogénétiques. Thèse doctorat 3ème cycle, Univ. Lyon I, 248, 116 p.

Clément, P. 1968. Ultrastructures d'un Rotifère, Notommata copeus. I. La cellule-flamme. Hypothèses physiologiques. Z. Zellforsch., 89: 478-498.

Clément, P. 1969a. Ultrastructures d'un Rotifère Notommata copeus. II. Le tube protonéphridien. Z. Zellforsch., 94: 103-117.

Clément, P. 1969b. Premières observations sur l'ultrastructure comparée des téguments de Rotifères. Vie Milieu A, 20: 461-482.

Clément, P. 1975. Ultrastructure de l'oeil cérébral d'un Rotifère, Trichocerca rattus. J. Microsc. Biol. cell., 22: 69-86.

Clément, P. 1977a. Introduction à la photobiologie des Rotifères dont le cycle reproducteur est contrôlé par la photopériode. Approches ultrastructurale et expérimentale. Thèse doctorat Etat, Univ. Lyon I, 7716, 262 p.

Clément, P. 1977b. Ultrastructural research on rotifers. Arch. Hydrobiol. Beih. 8: 270-297.

Clément, P. 1977c. Phototaxis in rotifers (action spectra). Arch. Hydrobiol. Beih., 8: 67-70.

Clément, P., Amsellem, J., Luciani, A. & Cornillac, A. 1980. Ultrastructure des yeux cérébraux des Rotifères. Colloque 'La vision chez les Invertébrés', C.N.R.S. Paris sept. 1979, in press.

Clément, P., Amsellem, J., Cornillac, A. & Luciani, A. 1980b. A la recherche (ultrastructurale) des photorécepteurs extraoculaires chez les Rotifères. Ibid., in press.

Clément, P., Amsellem, J., Cornillac, A. M., Luciani, A. & Ricci, C. 1980c. idem. I. The buccal velum. In this volume, pp. 127-131.

Clément, P., Amsellem, J., Cornillac, A. M., Luciani, A. & Ricci,

C. 1980d. An ultrastructural approach to feeding behaviour in Philodina roseola and Brachionus calyciflorus. II. The oesophagus. In this volume, pp. 133-136.

Clément, P., Amsellem, J., Cornillac, A. M., Luciani, A. & Ricci, C. 1980e. Idem. III. Cilia and muscles. Conclusions. In this volume, pp. 137-141.

Clément, P. & Pourriot, R. 1979. Influence de l'âge des grands-parents sur l'apparition des mâles chez le Rotifère Notommata copeus. Int. J. Invert. Repr. I: 89-98.

Clément, P. & Pourriot, R. 1980. About a transmissible influence through several generations in a clone of the Rotifer Notommata copeus Ehr. In this volume, pp. 27-31.

Clément, P., Rougier, C. & Pourriot, R. 1976. Les facteurs exogènes et endogènes qui contrôlent l'apparition des mâles chez les Rotifères. Bull. Soc. Zool. Fr., 101, suppl. 4: 86-95.

Eakin, R. M. 1965. Evolution of photoreceptors. Cold Spring Harb. Symp. quant. Biol., 30: 363-370.

Eakin, R. M. 1968. Evolution of photoreceptors. In: 'Evolutionary biology', Dobzhansky, T., Hecht, M. K. & Steere, W. C. eds, New York, p. 194-242.

Eakin, R. M. 1972. Structure of invertebrate photoreceptors. In: 'Handbook of sensory physiology', Springer-Verlag, 7: 625-684.

Eakin, R. M. & Brandenburger, J. L. 1974. Ultrastructural features of a Gordian Worm (Nematomorpha). J. Ultrastr. Res., 46: 351-374.

Eakin, R. M. & Westfall, J. A. 1965. Ultrastructure of the eye of the rotifer Asplanchna brightwelli. J. Ultrastr. Res. 12: 46-62.

Fauré-Fremiet, E. 1961. Cils vibratiles et flagelles. Biol. Rev., 36: 464-536.

Fauré-Fremiet, E. & Rouiller, C. 1957. Le flagelle interne d'une Chrysomonadale: Chromulina psammobia. C. R. Acad. Sci. Fr., 244: 2655-2657.

Fournier, A. 1980. Les photorécepteurs des Plathelminthes parasites. Colloque 'La vision chez les invertébrés', C.N.R.S., Paris sept. 1979, in press.

Fournier, A. & Combes, F. 1978. Structure of photoreceptors of Polystoma integerrimum (Platyhelminths, Monogena). Zoomorphologie, 91: 147-155.

Franc, J. M. 1972. Activités des rosettes ciliées et leurs supports ultrastructuraux chez les Cténaires. Z. Zellforsch, 130: 527-544.

Franc, S., Franc, J. M. & Garrone, R. 1976. Fine structure and cellular origin of collagenous matrices in primitive animals: Porifera, Cnidaria and Ctenophora. Front. Matrix. Biol., 3: 143-156.

Garrone, R. 1975. Nature, génèse et fonctions des formations conjonctives chez les Spongiaires. Thèse Doctorat Etat. Univ. Lyon I, Fr., 302 p.

Gilbert, J. J. 1967. Asplanchna and postero-lateral spine production in Brachionus calyciflorus. Arch. Hydrobiol. 64: 1-62.

Gilbert, J. J. 1977a. Selective cannibalism in the rotifer Asplanchna sieboldi. Arch. Hydrobiol. Beih. 8: 267-269.

Gilbert, J. J. 1977b. Mictic-female production in monogonont rotifers. Arch. Hydrobiol. Beih. 8: 142-155.

Gilbert, J. J. & Starkweather, P. L. 1977. Feeding in the rotifer Brachionus calyciflorus. I. Regulatory mechanisms. Oecologia, 28: 125-131.

Hadzi, J. 1944. Turbelaryska Teorija Knidarijev. (Turbellarien-Theorie der Knidarier). Slov. Akad. Znan. Um., Ljubljana (in Slovenian with German summary).

Hadzi, J. 1953. An attempt to reconstruct the system of animal classification. Syst. Zool., 2: 145-154.

Hand, C. 1959. On the origin and phylogeny of the coelenterates. Syst. Zool., 8: 191-202.

Hand, C. 1963. The Early Worm: a Planula. in: 'The Lower Metazoa', E. C. Dougherty ed; Univ. Calif. Press, p. 33-39.

Hanson, E. D. 1958. On the origin of the Eumetazoa. Syst. Zool. 7: 16-47.

Hanson, E. D. 1963. Homologies and the ciliate origin of the Eumetazoa. in 'The lower Metazoa', E. C. Dougherty ed., Univ. Calif. Press, p. 7-22.

Hendelberg, M., Morling, G. & Pejler, B. 1979. The ultrastructure of the lorica of the rotifer Keratella serrulata (Ehrbg). Zoon, 7: 49-54.

Hernandez-Nicaise, M. L. & Amsellem, J. 1980. A giant multinucleated smooth muscle cell: the muscle fiber of Beroe (Ctenophora). J. Cell Sci., in press.

Hertel, H. 1979. Phototactic reations of Aslanchna priodonta to monochromatic light. Z. Naturforsch., 34: 1-2.

Horridge, G. A. 1968. Interneurons. Their origin, action, specificity, growth and plasticity. W. H. Freeman and al. ed., London and San Francisco, p. 1-83.

Huxley, 1853. cited by Hyman, 1951.

Hyman, L. H. 1951. Class Rotifers. in: 'The Invertebrates', Mc. Graw-Hill Book Company Inc., 3, p. 59-151.

Jägersten, G. 1955. On the early phylogeny of the Metazoa. Zool. Bidr. Uppsala, 30: 321-354.

Jägersten, G. 1959. Further remarks on the early phylogeny of the Metazoa. Zool. Bidr. Uppsala, 33: 79-108.

Jennings, H. S. 1901. On the significance of spiral swimming of organisms. Amer. Nat., 35: 369-378.

Joffe, B. I. 1979. (The comparative embryological analysis of the development of Nemathelminthes). (in russian)., Proc. Zool. Inst. Akad. Sci. URSS, 84: 39-62.

King, C. E. 1977. Genetics of reproduction, variation and adaptation in rotifers. Arch. Hydrobiol. Beih., 8: 187-201.

Koehler, J. K. 1965. A fine study of the rotifer integument. J. Ultrastr. Res., 12: 113-134.

Koehler, J. K. 1966. Some comparative fine structure relationships of the rotifer integument. J. Exp. Zool., 162: 231-243.

Koopowitz, H. & Chien, P. 1974. Ultrastructure of the nerve plexus in flatworms. I. Peripheral organization. Cell Tissue Res., 155: 337-351.

Koste, W. 1978. Rotatoria. Borntraeger, Berlin, 2 vols., 673 pp. 234 pls.

Kryvi, H. 1973. Ultrastructural studies of the sucker cells in Hemiurus communis (Trematoda). Norm. J. Zool., 21: 273-280.

Kümmel, G. & Brandenburg, J. 1961. Die Reusengeisselzellen (Cyrtocyten). Z. Naturforsch., 16b: 692-697.

Lansing, A. I. 1947. A transmissible, cumulative, and reversible factor in aging. J. Gerontology, 2: 228-239.

Lansing, A. I. 1954. A non-genic factor in the longevity of rotifers. Ann. N.Y. Acad. Sci., 57: 455-464.

Lanzavecchia, G., Valvassori, R., Eguileor, M. de, & Lanzavecchia, P. 1979. Three dimensional reconstruction of the contractile system of the Nematophore muscle fiber. J. Ultrastr. Res., 66: 201-223.

Lender, Th. 1976. Rôle de la neurosécrétion au cours de la régénération et de la scissiparité des Planaires d'eau douce. in: 'Actualités sur les hormones d'Invertébrés', Colloques internationaux du C.N.R.S., 251: 39-48.

Lentz, T. L. 1968. Primitive nervous systems. Yale Univ. Press, New Haven and London. 141 p.

Lyons, K. M. 1972. Sense organs of monogeneans. in: 'Behavioural aspects of parasite transmission', Canning, E. U. & Wright, C. A. eds., London: Acad. Press, p. 181-199.

McKanna, J. A. 1968. Fine structure of the protonephridial system in Planaria. I. Flame cells. Zellforsch., 92: 509-523.

Marcus, E. 1958. On the evolution of the animal phyla. Quart. Rev. Biol. 33: 24-58.

Mattern, C. F. T. & Daniel, W. A. 1966. The flame-cell of rotifer. Electron microscope observations of supporting rootlets structures. J. Cell Biol., 29: 547-551.

Mayr, E. 1974. Populations, espèces et évolution. Herman ed., Paris, 496 p.

Menzel, R. & Roth, F. 1972. Spektrale Phototaxis von Planktonrotatorien. Experientia, 28: 356-357.

Morita, M. & Best, J. B. 1966. Electron microscopic studies of planaria. III. Some observations on the fine structure of planarian nervous tissue. J. Exp. Zool., 161: 391-411.

Nachtwey, R. 1925. Untersuchungen über die Keimbahn Organogenese und Anatomie von Asplanchna priodonta Gosse. Z. wiss. Zool., 126: 239-492.

Nørrevang, A. 1974. Photoreceptors of the phaosome (hirudinean) type in a pogonophore. Zool. Anz., 193: 297-304.

Pikkarainen, J. & Kulonen, E. 1969. Collagen. Some generalizations on comparative chemistry. Nature, Lond., 223: 839-841.

Pontin, R. M. 1964. A comparative account of the protonephridia of Asplanchna (Rotifera) with special reference to the flame bulbs. Proc. Zool. Soc. Lond., 143: 511-525.

Pontin, R. M. 1966. The osmoregulatory function of the vibratile flames and the contractile vesicle of Asplanchna (Rotifera). Comp. Biochem. Physiol., 17: 1111-1126.

Pourriot, R. 1963. Influence du rythme nycthéméral sur le cycle sexuel de quelques Rotifères. C.R. Acad. Sci., 256: 5216-5219.

Pourriot, R. 1965. Recherches sur l'écologie des Rotifères. Thèse Doct. d'état, Vie Milieu, Suppl. 21, 224 p.

Pourriot, R. 1974. Relations prédateur-proie chez les Rotifères. Influence du prédateur (Asplanchna brightwelli) sur la morphologie de la proie (Brachionus bidentata) Ann. Hydrobiol., 5: 43-55.

Pourriot, R. 1977. Food and feeding habits of Rotifera. Arch. Hydrobiol. Beih., 8: 243-260.

Pourriot, R. & Clément, P. 1973. Photopériodisme et cycle hétérogonique chez Notommata copeus (Rotifère Monogononte). II. Influence de la qualité de la lumière. Spectres d'action. Arch. Zool. exp. gén., 114: 277-300.

Pourriot, R., Clément, P. & Luciani, A. 1980. Influence de la photopériode sur un Rotifère. Hypothèses sur les mécanismes. in: 'Colloque sur la vision chez les invertébrés', C.N.R.S., Paris Sept. 79, in press.

Pourriot, R. & Rougier, C. 1976. Influence de l'âge des parents sur la production de femelles mictiques chez Brachionus calyciflorus Pallas et B. rubens Ehr. (Rotifères). C.R. Acad. Sci. Paris, 283: 1497-1500.

Pourriot, R., Rougier, C. & Benest, D. 1980. Hatching of Brachionus rubens O. F. Muller resting eggs (Rotifers). In this volume, pp. 51-54.

Preissler, K. 1977. Horizontal distribution and 'avoidance of shore' by rotifers. Arch. Hydrobiol. Beih., 8: 43-46.

Pringle, J. W. S. 1972. Arthropod muscles. in: 'The structure and function of muscle', 2nd ed., G. H. Bourne ed., Acad. Press, I: 491-542.

Reger, J. F. 1976. Studies on the fine structure of cercarial tail muscle of Schistosoma sp. (Trematoda). J. Ultrastr. Res. 57: 77-86.

Remane, A. 1929-1933. Rotatoria. in: Dr. H. G. Bronn's Klassen und Ordnungen des Tier-Reichs', IV (Vermes), 2 (Aschelminthes), I (Rotatorien, Gastrotrichen und Kinorhynchen), 3: 1-448.

Remane, A. 1955. Morphologie als Homologienforschung. Verh. dtsch. Zool. Ges., Tübingen (1954). Also in: Zool. Anz., Suppl. 18: 159-183.

Remane, A. 1958. Zur Verwandtschaft und Ableitung der niederen Metazoen. Zool. Anz., Suppl. 21: 179-195.

Remane, A. 1960. Die Beziehungen Zwischen Phylogenie und Ontogenie. Zool. Anz., 164: 306-337.

Remane, A. 1963. The systematic position and phylogeny of the pseudocoelomates. in: 'The lower Metazoa', E. C. Dougherty ed., Unif. Calif. Press, p. 247-255.

Remane, A., Storch, V. & Welsch, V. 1972. Kurzes Lehrbuch der Zoologie. I-X, Stuttgart, G. Fischer ed., p. 1-493.

Remane, A., Storch, V. & Welsch, U. 1976. Systematische Zoologie. Stuttgart. p. 1-678.

Röhlich, P., Aros, B. & Viragh, S. 1970. Fine structure of photoreceptor cells in the earthworm Lumbricus terrestris. Z. Zellforsch., 104: 345-357.

Ruttner-Kolisko, A. 1963. The interrelationships of the Rotatoria. in: 'The Lower Metazoa', E. C. Dougherty ed., Univ. Calif. Press, p. 263-272.

Ruttner-Kolisko, A. 1972. Plankton Rotifers. Biology and Taxonomy. Binnengewässer, 26: 99-234.

Salvini-Plawen, L. V. & Mayr, E. 1977. On the evolution of photoreceptors and eyes. Evolutionary Biol., 10: 207-263.

Saunders, D. S. 1976. Insect Clocks. Pergamon Press. 280 p.

Schaller, H. C. 1976. Action of a neurohormone from Hydra. in: 'Actualités sur les hormones d'invertébrés', Colloques Internat. C.N.R.S., 251: 33-38.

Schramm, U. 1978a. On the excretory system of the rotifer Habrotrocha rosa Donner. Cell tiss. Res., 189: 515-524.

Schramm, U. 1978b. Studies on the ultrastucture of the integument of the Rotifer Habrotrocha rosa Donner (Aschelminthes). Cell tiss. Res., 189: 167-177.

Short, R. B. & Gagne, H. T. 1975. Fine structure of a possible photoreceptor in cercariae of Schistosoma mansoni. J. Parsitol., 61: 69-74.

Starkweather, P. L. 1980. Aspects of the feeding behavior and trophic ecology of suspensionfeeding rotifers. In this volume, pp. 63-72.

Steinböck, O. 1952. Keimblätterlehre und Gastraea-Theorie. Pyramide, 2: 13-15; 26-31.

Steinböck, O. 1958. Zur Phylogenie der Gastrotrichen. Zool. Anz., Suppl. 21: 128-169; Schlusswort zur Diskussion Remane-Steinböck. Ibid p. 196-218.

Steinböck, O. 1963. Origin and affinities of the Lower Metazoa: the 'aceloid' ancestry of the Eumetazoa. In: 'The Lower Metazoa', Dougherty, E. C. ed., Univ. Calif. Press, p. 40-54.

Storch, V. & Welsch, U. 1969. Über den Aufbau des Rotatorienintegumentes. Z. Zellforsch., 95: 405-414.

Swiderski, Z., Euzet, L. & Schonenberger, N. 1975. Ultrastructures du système néphridien des Cestodes Cyclophyllides. Cellule, 71: 7-18.

Teuchert, G. 1973. Die Feinstruktur des Protonephridialsystems von Turbanella cornuta Remane, einem marinen Gastrotisch der Ordnung Macrodasyoidea. Z. Zellforsch., 136: 277-289.

Vanfleteren, J. R. & Coomans, A. 1975. Photoreceptor evolution and phylogeny. Z. Zool. Syst. Evolut. -forsch, 14: 157-169.

Viaud, G. 1940-1943: Recherches expérimentales sur le phototropisme des Rotiferes: Bull. Biol. Fr. Belg., 74: 249-308 (1940); 77: 68-93 (1943); 77: 224-242 (1943).

Villeneuve-Amsellem, J. & Clement, P. 1971. Le neuropile du cerveau de Rotifère:observations ultrastructurales préliminaires. J. Microscopie Fr., 11: 108.

Wallace, R. 1980. Ecology of sessile rotifers. In this volume, pp. 181-193.

Ward, S., Thomson, N., White, J. G. & Brenner, S. 1975. Electron microscopical reconstruction of the anterior sensory anatomy of the nematode Caenorhabditis elegans. J. Comp. Neurology, 160: 313-337.

Ware, R. W., Clark, D., Crossland, K. & Russell, R. L. 1975. The nerve ring of the nematode Caenorhabditis elegans: sensory input and motor output. J. Comp. Neurology, 162: 71-110.

Warner, F. D. 1969. The fine structure of the protonephridia in the Rotifer Asplanchna. J. Ultrastructure Res., 29: 499-524.

Wergin, W. P. & Endo, B. Y. 1976. Ultrastructure of a neuro-sensory organ in a root-knot nematode. J. Ultrastructure Res., 56: 258-276.

Wilson, E. O. 1979. L'humaine nature. Essai de sociobiologie. ed. Stock, Paris, 318 p.

Wilson, R. A. 1970. Fine structure of the nervous system and specialized nerve endings in the miracidium of Fasciola hepatica. Parasitol., p. 399-410.

Wolken, J. J. 1971. Invertebrate photoreceptors: a comparative analysis. Acad. Press. ed., New York, London.

A SIMPLIFIED METHOD FOR THE PREPARATION OF ROTIFERS FOR TRANSMISSION AND SCANNING ELECTRON MICROSCOPY

J. AMSELLEM & P. CLÉMENT

Laboratoire d'Histologie et de Biologie tissulaire, CNRS: L.A. 244, et CMEABG, Université Claude Bernard (Lyon I), Villeurbanne, France

Keywords: Scanning e.m., transmission e.m., technique, rotifers

Abstract

Tor TEM and SEM prepatations, rotifers are placed in little panels made of plastic Beem capsules normally used for embedding, with their conical parts cut off and closed by plankton filter cloth. Thus, the risk of losing animals is considerably reduced.

Methods

The main difficulty in preparing Rotifers for TEM and SEM is linked to their small size. In the two processes, if the animals are transferred from one bath to another or if the solutions are changed in the same evaporating dish, the disadvantages are the same: dilution of the solutions, risk of loosing animals, and the considerable amount of time involved.

For the SEM process, in particular, the use of the critical point technique requires an adaptation of the commercial accessories provided with the apparatus to the small size of the animals. So, for all these purposes, we use plastic Beem capsules for embedding (Polaron, M 206). We cut off the conical part of the capsule. Two separate caps are placed over the cylindrical body. A hole approximately 5 x 5 mm is cut in each cap. Finaly a fifty mesh plankton filter cloth is placed between the body and each cap. (Fig. I).

For TEM, we use a cylinder with only one cap in an evaporating dish. The anaesthetized animals are placed in the chamber and the fixatives and dehydrating solutions are added sequentially. Because of the mesh, the solutions can be removed from the evaporating dish whilst the following solution is added to the chamber. This process is faster and easier and there is no risk of loosing animals (Fig. 2).

For SEM, we use a cylinder with only one cap for fixing and dehydrating; the second cap is then added before placing the chamber into the half cylinder support of the critical point apparatus. As these chambers are completely closed, they do not need to be fixed in place within the half cylinder. After the critical point process, the filter cloth is removed under a binocular microscope. The dried

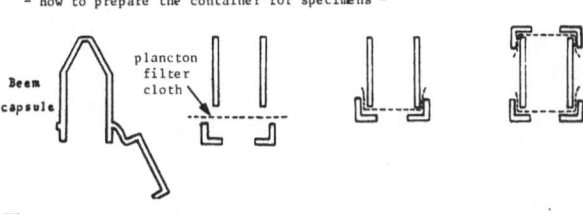

- How to prepare the container for specimens -

Beem capsule · plancton filter cloth

Fig. I.

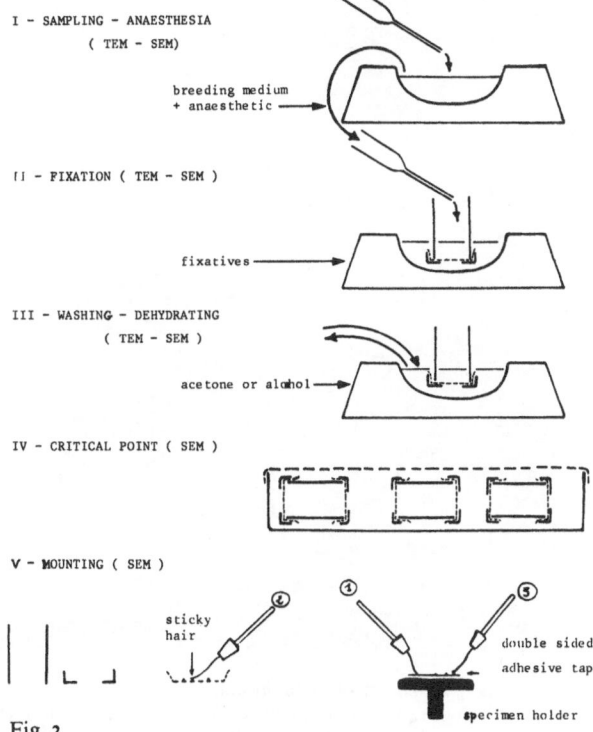

I - SAMPLING - ANAESTHESIA (TEM - SEM)

breeding medium + anaesthetic

II - FIXATION (TEM - SEM)

fixatives

III - WASHING - DEHYDRATING (TEM - SEM)

acetone or alcohol

IV - CRITICAL POINT (SEM)

V - MOUNTING (SEM)

sticky hair

double sided adhesive tape

specimen holder

Fig. 2.

119

Hydrobiologia 73, 119-122 (1980). 0018-8158/80/0732-0119$00.80.

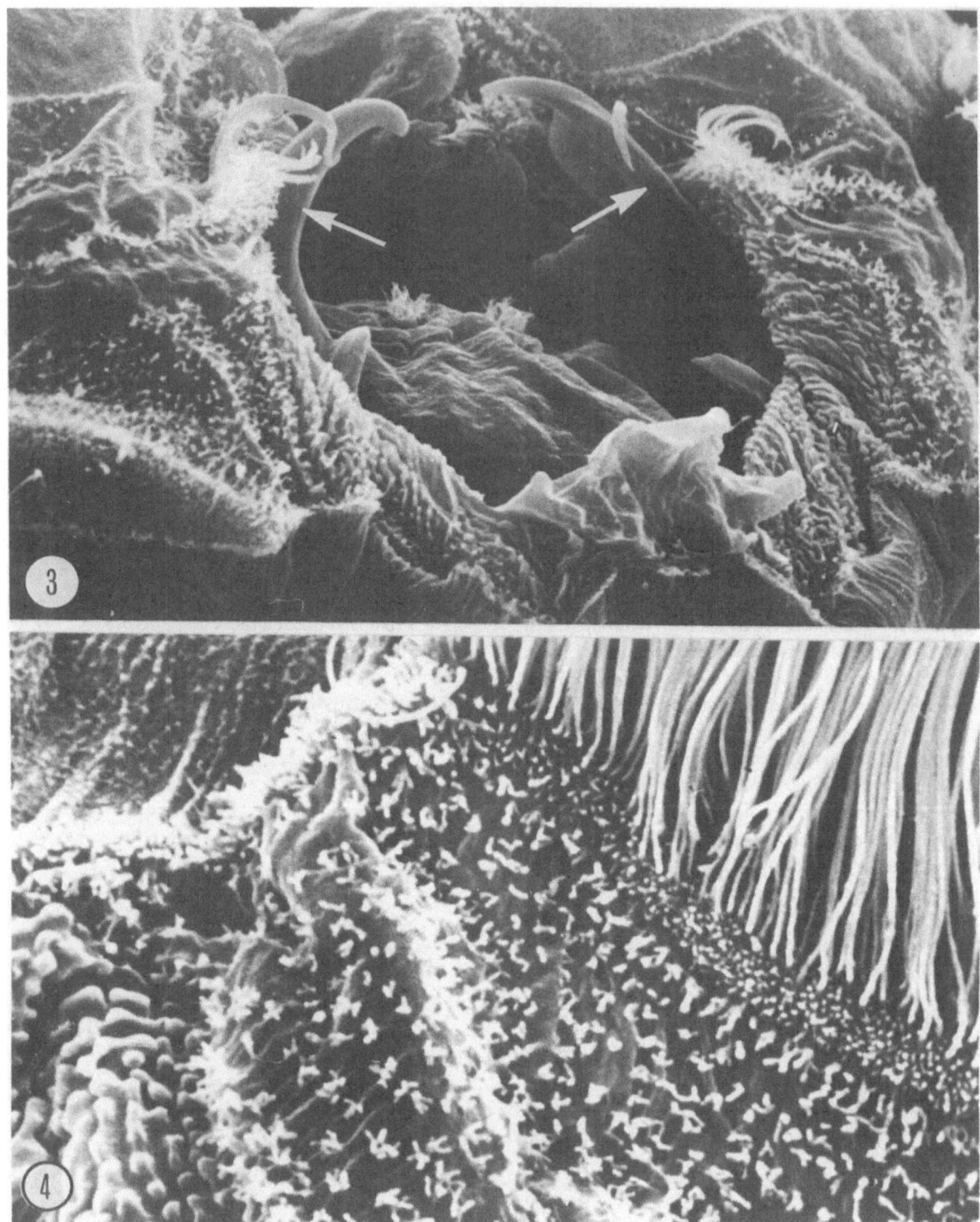

Fig. 3-4. *Asplanchna brigthwelli.*
Fig. 3. x 1400. Mouth and jaws. Ventral view. The mouth is wide open. Under the jaws (arrows) the floor of the mastax with two bunches of sensory cilia (sensory receptors of the mastax). On both sides of the mouth, near the jaws, microvilli and a bunch of sensory cilia. The tegument of the trunk is visible on the lower part of the figure. Fig. 4. x 4000. Detail of the apical part of the rotatory appparatus. To the right: base of the locomotory cilia. Different zones of the apical tegument (with or without microvilli) can be seen in the center and to the left of the photo respectively.

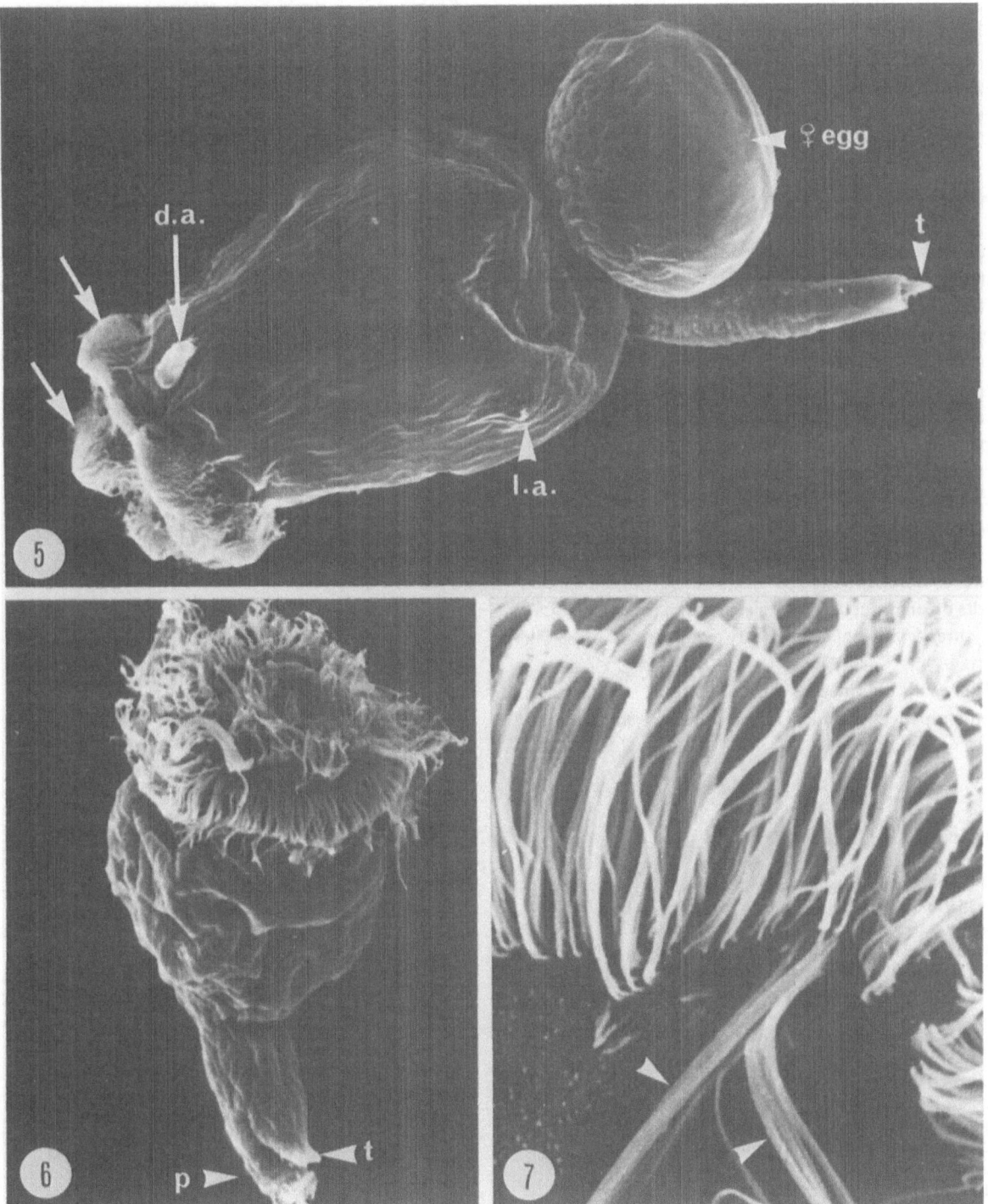

Fig. 5, 6, 7. *Brachionus calyciflorus.* (var. *sericus* in Fig. 5).
Fig. 5. x 430. amicitic female bearing an egg (♀ egg). arrow: cilia of the rotatory apparatus, d. a.: the dorsal antenna,
l. a.: a lateral antenna, t: toes; Fig. 6. x 800. male, t: toes; p: ciliated penis, Fig. 7. x 3850. detail of the rotatory
apparatus of the male; the sensory cilia (arrows) are longer than the epithelial locomotory cilia.

rotifers are collected using a hair which has been made sticky by touching cellotape; they are then attached to the adhesive tape on the stub and metallized (Fig. 2). An illustration of some results obtained with SEM is given in Figs. 3 to 7.

Results

Until now, the only results obtained with rotifers in scanning e.m. deal with loricate species (Maehler & Dewall, 1978; Hendelberg *et al,* 1979), with resting eggs (Wurdak *et al.,* 1977) or with the trophi of the mastax (Koehler & Hayes, 1969a and b; Salt *et all.,* 1978). Whith this material, it is possible to use the air drying technique because damaging effects due to surface tension are not important. Nevertheless, it was impossible to see the soft parts of the loricate rotifers and to observe the illoricate ones because the air drying technique distorts and obscures the details of the soft parts.

Using the critical point technique and with an appropriate osmolarity (150-200 mosM), we obtained good results on anaesthetized illoricate rotifers; even the fragile structures (epithelial and sensory cilia) are well preserved (Figs. 3-7).

The results obtained in transmission e.m. are illustrated in other papers included in this book (Clément, 1980, Clément *et al.,* 1980a, b, c).

References

Clément, P. 1980. Phylogenetic relationships of rotifers, as derived from photoreceptor morphology and other ultrastructural analyses. In this volume, pp. 93-117.

Clément, P., Amsellem, J., Cornillac, AM., Luciani, A. & Ricci, C. 1980a. An ultrastructural approach to the feeding behaviour of Philodina roseola and Brachionus calyciflorus (Rotifers). I. The buccal velum. In this volume, pp. 127-131.

Clément, P., Amsellem, J., Cornillac, AM., Luciani, A. & Ricci, C. 1980b. An ultrastructural approach to the feeding behaviour of Philodina roseola and Brachionus calyciflorus (Rotifers). II. The oesophagus. In this volume, pp. 133-136.

Clément, P., Amsellem, J., Cornillac, AM., Luciani, A. & Ricci, C. 1980c. An ultrastructural approach to the feeding behaviour of Philodina roseola and Brachionus calyciflorus (Rotifers) III. Cilia, muscles and conclusions. In this volume, pp. 137-141.

Hendelberg, M., Morling, G. & Pejler, B. 1979. The ultrastructure of the lorica of the rotifer Keratella serrulata (Ehrbg). Zoon, 7: 49-54.

Koehler, J. K. & Hayes, Th. L. 1969a. The rotifer jaw: a scanning and transmission electron microscope study. I. The trophi of Philodina acuticornis odiosa. J. Ultrastr. Res., 27: 402-418.

Koehler, J. K. & Hayes, Th. L. 1969b. The rotifer jaw: a scanning and transmission electron microscope study. II. The trophi of Asplanchna sieboldi. J. Ultrastr. Res., 27: 419-434.

Maehler, A. & Dewall, V. 1978. In: W. Koste (ed) Rotatoria, p. 45, Borntraeger, Berlin, 2 vols., 673 pp. 234 pls.

Salt, G. W., Sabbadini, G. F. & Commins, M. L. 1978. Trophi morphology relative to food habits in six species of rotifers (Asplanchnidae). Trans. amer. microsc. Soc., 97: 469-485.

Wurdak, E., Gilbert, J. J. & Jagels, R. 1977. Resting egg ultrastructure and formation of the shell in Asplanchna sieboldi and Brachionus calyciflorus. Arch. Hydrobiol. Beih., 8: 298-302.

ULTRASTRUCTURE AND HISTOCHEMISTRY OF THE RUDIMENTARY GUT OF MALE ASPLANCHNA SIEBOLDI (ROTIFERA)

E. S. WURDAK & J. J. GILBERT*

* Dartmouth College, Hanover, New Hampshire, 03755, U.S.A.

Keywords: rudimentary gut, autophagic vacuole, beta-glycogen

Abstract

The fine structure of the rudimentary gut of male *Asplanchna sieboldi* in late stage embryos and at 0, 12 and 24 hours after birth is described. The results of histochemical tests for acid phosphatase and glycogen indicate that glycogen, mitochondria, endoplasmic reticulum, and nuclei are subjected to autolysosomal breakdown, while glycogen remains as the major component of the gut in old males.

Introduction

At birth, the rudimentary gut of male *A. sieboldi* consists of 4-5 fused solid spheres with no connection with the oral region. It has been repeatedly observed that the rudimentary gut decreases in size with age (Lange, 1911; Buchner *et al.*, 1970). This observation led several investigators to conclude that certain constituents of the gut were being utilized to sustain the male during his 2-3 day lifespan (Hudson, 1883; Hudson & Gosse, 1886; Lange, 1911; Powers, 1912; Waniczek, 1930). The chief reserve substance in the gut has been tentatively identified as glycogen (Buchner *et al.*, 1970).

The present report follows the fate of the rudimentary gut in successively older males at the electron microscope level and correlates the morphological findings with histochemical tests for glycogen and acid phosphatase, a marker enzyme for lysosomes.

Materials and methods

Male rotifers used in this study were obtained from cultures of *A. sieboldi* clone 10C6, maintained as described by Gilbert (1968). Mictic female production was induced by the addition of 5 x 10^{-7} M alpha-tocopherol to the culture medium (Gilbert & Birky, 1971). Gravid mictic females carrying late-stage male embryos and adult males 0-5 hours, 12-16 hours and 24 hours of age were placed in 2.5-3.5% glutaraldehyde in .1 M phosphate buffer, pH 7.3-7.5 for 1-4 hours at 4°C. They were postfixed in 1% phosphate buffered OsO₄ for 1 hour at 4°C, dehydrated in ethanol and embedded in Epon 812 (Luft, 1961) or in the Spurr low-viscosity epoxy mixture (Spurr, 1969).

The presence of glycogen in the rudimentary guts of late stage embryos, neonates and 24-hour old individuals was confirmed at the level of the light microscope by the PAS technique (Humason, 1962). Ultrathin sections of 0-5 hour and 12-16 hour old males were treated by the PATO method (Vye & Fishman, 1971) to achieve selective staining of glycogen at the electron microscope level.

Acid phosphatase was localized in the rudimentary gut of newborn males at the electron microscope level by using a modified lead phosphate method (Miller & Palade, 1964).

Results and discussion

The rudimentary gut of a late-stage embryo is a syncytial mass enclosed by the plasma membrane. The nuclei are usually spherical or oval and have large granular nucleoli. The rough endoplasmic reticulum (RER) is moderately well developed; mitochondria are numerous. Doughnut, sickle-shaped and other unusual mitochondrial configurations are common. The Golgi apparatus shows pronounced activity. The ground cytoplasm is filled with 200-300 A° particles. Several islets of cytoplasm are enclosed by concentric arrays of rough and smooth-sur-

123

Hydrobiologia 73, 123-126 (1980).   0018-8158/80/0732-0123$08.00.
© Dr. W. Junk b.v. Publishers, The Hague.

faced membranes. These formations are regarded as early autophagosomes.

In the stomach rudiment of 0-5 hour old males the RER is more fragmented, and the association of the endoplasmic reticulum and the Golgi apparatus is less extensive than in the embryo (Fig. 1). Definitive autophagic vacuoles are present, containing ER segments, mitochondria, glycogen and small, smooth-surfaced vesicles (Fig. 3). They are bounded by one, two or several concentric membranes. Complex bodies consisting of concentric layers of smooth-surfaced membranes and glycogen are also present (Fig. 6). These formations have been referred to as 'glycogen bodies' (Schiaffino & Hanzlikova, 1972). The ground cytoplasm is rich in particles identified as beta glycogen by the PAS and PATO techniques (Fig. 6a). A positive reaction for acid phosphatase is present over large heterogeneous, autolytic vacuoles, delimited by a single membrane and showing extensive degradation of contents (Fig. 4).

By 12-16 hours there is a marked concentration of beta glycogen in the interior of the gut, while the nuclei, mitochondria, dictyosomes and ER vesicles are located exclusively at the periphery (Figs. 5a, 6b). Glycogen bodies of large dimensions (Fig. 5b) and dense bodies containing membraneous and lipid deposits (Fig. 5a) are prominent.

Marked nuclear changes take place in the interval between 12 and 24 hours. Nuclei become wholly or partially encircled by isolation membranes (Fig. 2). The nuclear membrane proper is thrown into extensive folds. In the surrounding cytoplasm there are dense patches of granulo-fibrillar material which resemble the nuclear contents (Fig. 2). These areas may derive from the nuclei by the fragmentation of the nuclear membrane and the ensuing release of the contents into the cytoplasm. Dense bodies reach their greatest number in 24 hour old males.

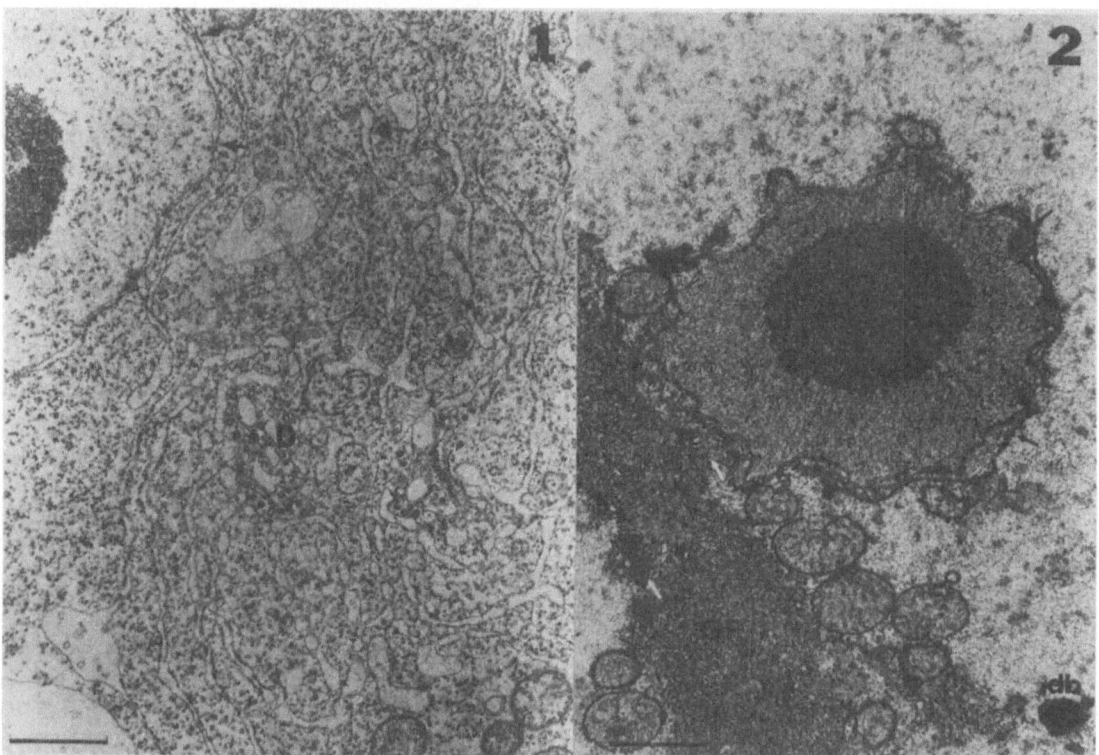

Fig. 1. Periphery of the rudimentary gut of a newborn male. Note ribosome-covered segments of the outer nuclear membrane (arrows). Dictyosomes (D) are active and vesicles and cisternae of granular endoplasmic reticulum are abundant. Scalar indicates 1 micron.

Fig. 2. Rudimentary gut of a 24 hour old male. An apparently normal nucleus is partly surrounded by densely-staining membranes (black arrows). Nearby granulo-fibrillar material (GF) closely resembles the nuclear contents. It is flanked by dictyosomes (white arrows). Dense body (db) is occluded with an electron-dense deposit. Scalar indicates 1 micron.

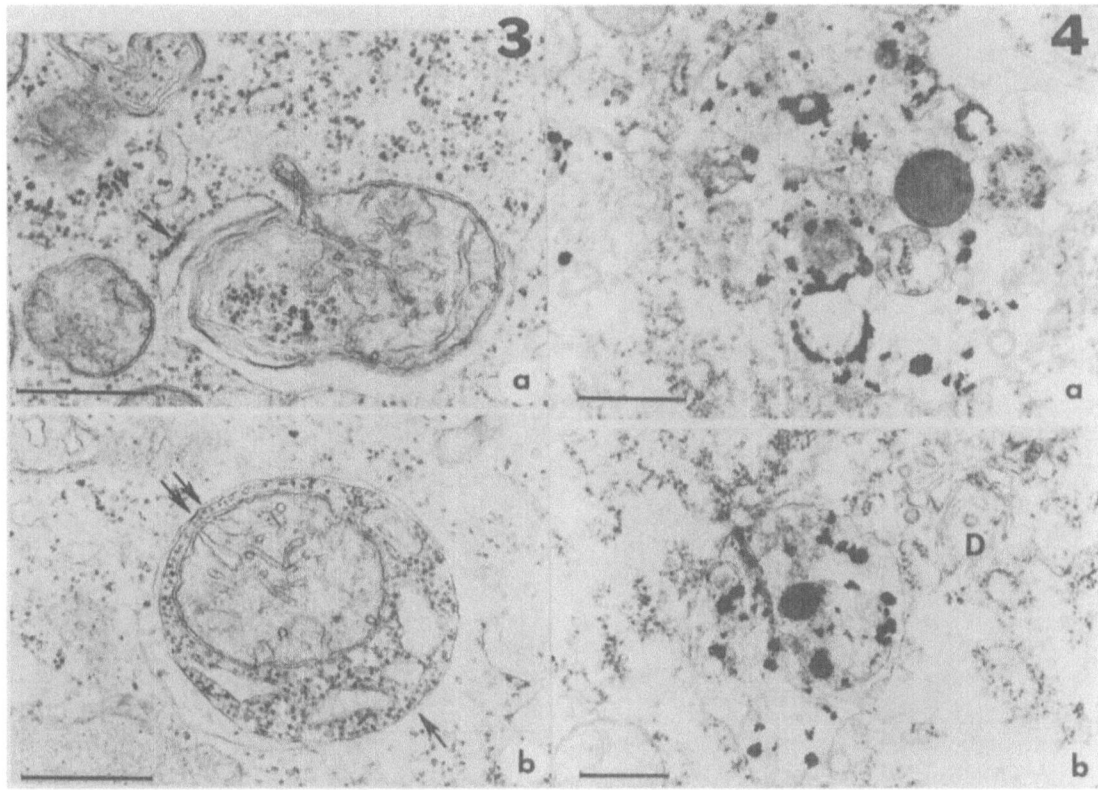

Fig. 3. A probable sequence in the entrapment of mitochondria in autophagosomes within the rudimentary gut of a newborn male. In a) an altered mitochondrion is partially invested by flattened cisternae. Cisterna at arrow bears ribosomes. In b) a relatively intact mitochondrion is completely surrounded by membranes along with vesicles and glycogen. Single membrane segment at arrow may have resulted from the fusion of 2 pre-existing membranes (double arrow).
Scalar indicates .5 micron.

Fig. 4. Acid phosphatase positive autolysosomes in the rudimentary gut of a newborn male. The large vacuole in a) contains a variety of inclusions. In b) the dictyosome (D) is nonreactive.
Scalar indicates .5 micron.

The interior of the dense bodies is almost entirely occluded by lipid deposits (Fig. 2). The RER is extremely reduced. A few intact mitochondria are retained.

Staining of glycogen by the PAS technique is most intense in the rudimentary gut of 24 hour old males, reflecting the highest concentration of glycogen particles per unit area. Further degradation of glycogen is presumed to be extra lysosomal, mediated by catabolic enzymes associated with the glycogen particle itself. The total disappearance of the rudimentary gut in extremely old males argues strongly in favor of the utilization of all glycogen reserves. However, the utilization of glycogen proceeds at a slower rate than that of membraneous or lipid constituents of the gut.

Summary

Major ultrastructural changes accompany the decrease in the size of the male rudimentary gut with age. Mitochondria, nuclei, endoplasmic reticulum and glycogen particles are destroyed through autophagy. There is a peak of Golgi activity in the embryo which is probably related to the heightened autolytic degradation which is observed in the newly-born male. Dense bodies containing lipid residues show a progressive increase in number as the animal ages. Glycogen particles, randomly scattered in the embryo, accumulate in the interior of the gut as a result of either accelerated degradation of other constituents in this zone, or as a result of the migration of organelles from the interior to the periphery.

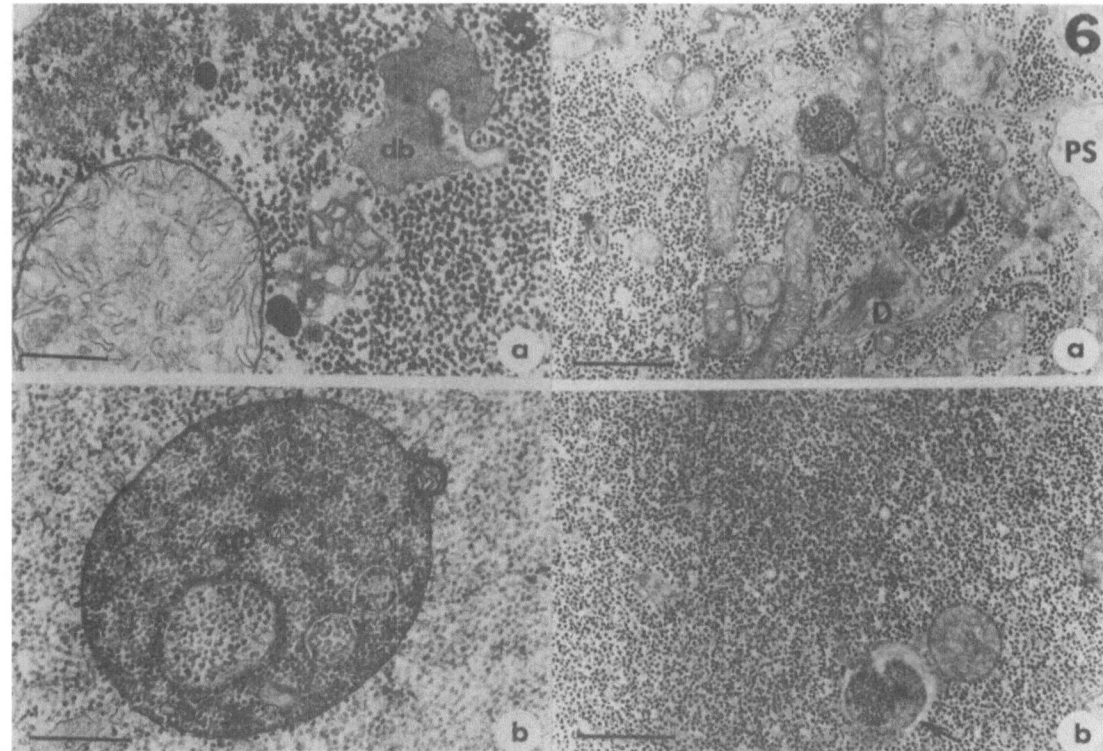

Fig. 5. Rudimentary gut of the male at 12 hours.
a) Normal mitochondrion, dense body (db) and numerous glycogen particles from the periphery of the gut.
b) A large, complex glycogen body (gb) from the interior.
Scalar indicates 5 micron.
Fig. 6. PATO staining of the rudimentary gut of a) newborn male and b) 12 hour old male. Both micrographs show the periphery of the gut. Glycogen particles seem to be more concentrated in b), whereas the number of mitochondria and ER vesicles is reduced. A glycogen body is shown at arrow in a) and b).
Scalar indicates .1 micron.

References

Buchner, H., Tiefenbacher, L., Kling, R. & Preissler, K. 1970. Über die physiologische Bedeutung des Magen-Darm-Rudimentes des Männchen von Asplanchna sieboldi (Rotatoria, Monogononta). Z. vergl. Physiol. 67: 453-454.

Gilbert, J. J. 1968. Dietary control of sexuality in the rotifer Asplanchna brightwelli Gosse. Physiol. Zool. 41: 14-43.

Gilbert, J. J. & Birky, C. W. Jr. 1971. Sensitivity and specificity of the Asplanchna response to dietary alpha tocopherol. J. Nutrition 101: 113-126.

Hudson, C. T. 1883. On Asplanchna ebbesbornii nov. sp. J. r. microsc. Soc.: 621-628.

Hudson, C. T. & Gosse, P. H. 1886. The Rotifera or Wheel-Animalcules, both British and Foreign. London Vol. I.: I-IV + 1-128; Vol. II.: 1-144.

Humason, G. L. 1962. Animal Tissue Techniques. W. H. Freeman and Co., San Francisco: 298-301.

Lange, A. 1911. Zur Kenntnis von Asplanchna sieboldii Leydig. Zool. Anz. 38: 433-441.

Luft, J. H. 1961. Improvements in epoxy embedding methods. J. Biophys. Biochem. Cytol. 9: 409-427.

Miller, F. & Palade, G. E. 1964. Lytic activities in renal protein absorption droplets. An electron microscopical cytochemical study. J. Cell Biol. 23: 519-552.

Powers, J. H. 1912. A case of polymorphism in Asplanchna simulating mutation. Am. Nat. 46: 526-552.

Schiaffino, S. & Hanzlikova, V. 1972. Autophagic degradation of glycogen in skeletal muscles of the newborn rat. J. Cell Biol. 52: 41-51.

Spurr, A. R. 1969. A low viscosity epoxy resin embedding medium for electron microscopy. J. Ultrastruct. Res. 26: 31-43.

Vye, M. V. & Fischman, D. A. 1971. A comparative study of three methods for the ultrastructural demonstration of glycogen in thin sections. J. Cell Sci. 9: 747-749.

Waniczek, H. 1930. Untersuchungen über einige Arten der Gattung Asplanchna Gosse (A. girodi de Guerne, A. brightwellii Gosse, A. priodonta Gosse). Ann. Mus. Zool. Pol. 8: 109-322.

AN ULTRASTRUCTURAL APPROACH TO FEEDING BEHAVIOUR IN PHILODINA ROSEOLA AND BRACHIONUS CALYCIFLORUS (ROTIFERS)

I. THE BUCCAL VELUM

P. CLÉMENT, J. AMSELLEM, A.-M. CORNILLAC, A. LUCIANI[1] & C. RICCI[2]

[1] Laboratoire d'Histologie et Biologie Tissulaire (LA 244 CNRS) et CMEABG, Université Claude Bernard (Lyon l); Villeurbanne, France, [2] Instituto di Zoologia, Universita degli studi, via Celosia, 10, Milano, Italy

Keywords: Rotifers, ultrastructures, feeding behaviour, digestive tract, myelin-like structure

Abstract

The structure situated in the lumen of the buccal tube, called here 'buccal velum', separates the buccal space from the pharyngeal space. The velum is a supple membrane-pile. Its structure resembles that of a fish-trap. The food, once ingested, cannot go back into the buccal lumen.

Introduction

De Beauchamp (1909) and Remane (1929-1932) noticed, in some rotifers, the presence of a buccal tube connecting the buccal field of the rotatory organ to the mastax.

The structure and function of this buccal tube are not well known, in spite of some ultrastructure studies in Monogonont rotifers (*Notommata copeus*: Clément & Pavans de Ceccatty, 1969; *Trichocerca rattus*: Clément, 1977a) and in a Bdelloid *Habrotrocha rosa*: Schramm, 1978). Our preliminary ultrastructural observations in *Brachionus calyciflorus* and more detailed ones in *Philodina roseola* indicate that the buccal tube is a mixed cell-structure: buccal and pharyngeal. Between the buccal and pharyngeal lumens there is a very original structure: a pharyngeal membrane-pile which we have called 'the buccal velum'. Schramm (1978) saw this membrane-pile but she did not reconstitute the structure as a whole, so she could not imagine its function like a fish-trap.

Material and methods

The animals tested are a clone of *B. calyciflorus,* from a strain given by R. Pourriot, kept at about 20°C and cultured on *Phacus pyrum,* and a clone of *Philodina roseola* provided by C. Ricci, kept at 24°C but fed on commercial fish food.

The animals, relaxed by 1% novocaine are fixed for two hours either by 2% OsO_4 or by 2% glutaraldehyde followed by 2% OsO_4. The fixative is buffered by sodium cacodylate giving 200 mosM osmotic pressure.

Fixed animals were embedded in epon, cut with a Reichert ultramicrotome and observed with an electron microscope Philips EM 300 or Hitachi HU 12 A at the 'Centre de Microscopie Electronique Appliquée à la Biologie et à la Géologie' (Lyon I, Université Claude Bernard).

Results

Philodina roseola

The buccal tube is formed by a simple epithelium. As in the buccal tube of *N. copeus* and *T. rattus* (Clément & Pavans de Ceccatty, 1969; Clément, 1977a, b) all the cells are mushroom-shaped: their caps are imbricated to make the tube-wall, and their stalks represent the cellbodies which contain the nuclei and an important part of the cytoplasm. The cell-bodies are interconnected by many gap-junctions.

Two easily recognized cell categories exist in the buccal tube. The first cell type bears buccal cilia characterized by the presence of an electron-dense substance at its apical extremity, while the other type has pharyngeal cilia limited by a double cytoplasmic membrane (Fig. 9 and 10). The apical side of pharyngeal cells also shows this double cytomembrane. The buccal velum consists of pile cytomembranes (Fig. 10) which are continuous with the apical membranes of the pharyngeal cells (Fig. 11). Transverse serial sections of the tube (Fig. 1 to 6, Fig. 8) show the distribution of these two kinds of cells. The pharyngeal cells go up into the buccal tube as far as the buccal velum.

The velum is a symmetrical structure. It begins at about

Hydrobiologia 73, 127-131 (1980).   *0018-8158/80/0732-0127$01.00.*
© *Dr. W. Junk b.v. Publishers, The Hague.*

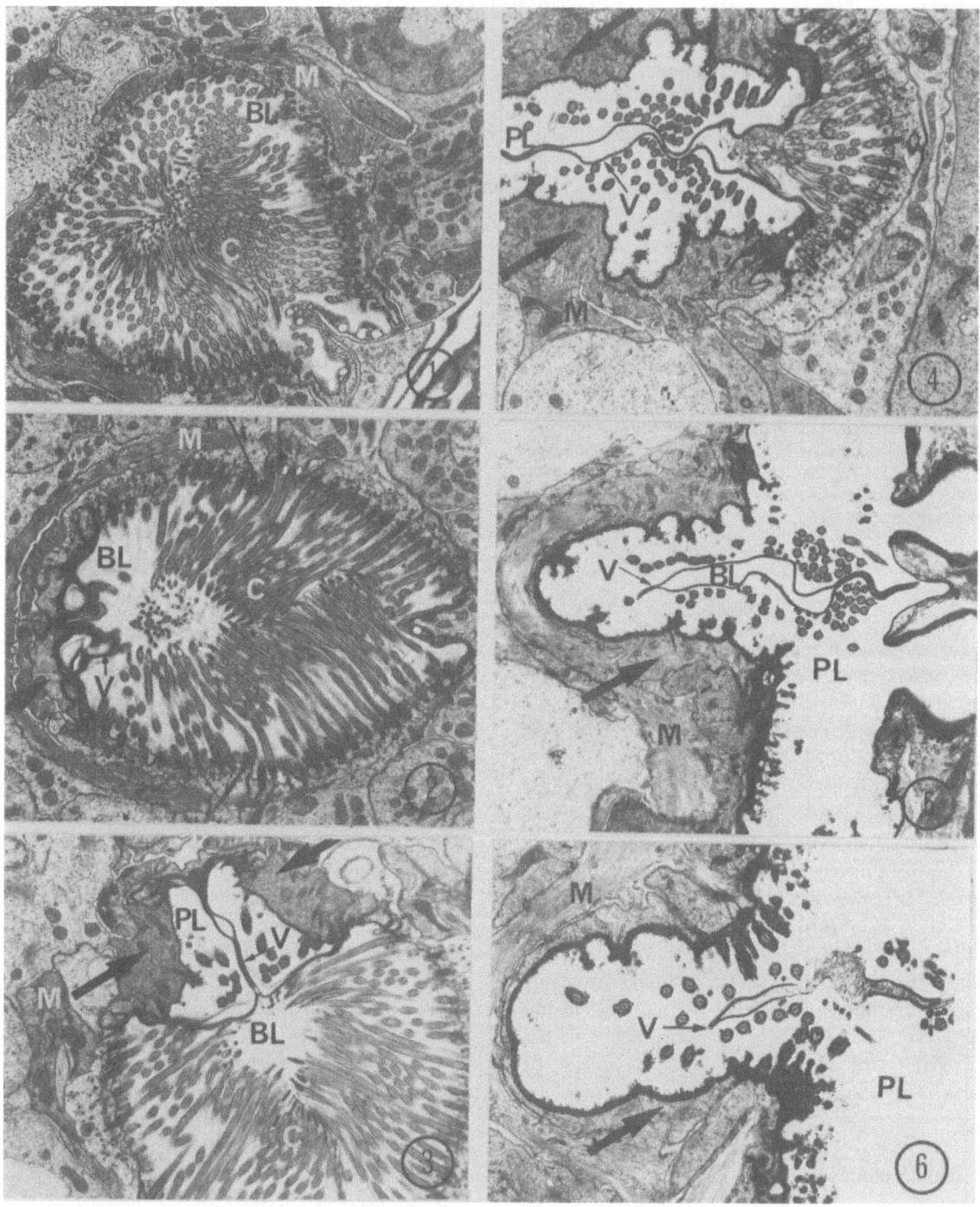

Fig. 1 to 6. Transverse sections of the buccal tube (see the localization of these sections in Fig. 7).
The epithelial wall is surrounded by muscles (M). A part of the tube wall is made of pharyngeal cells (arrows): they are caracterized by their myelin-like cuticle which also forms the buccal velum. In the buccal lumen, the buccal cilia (C) are characterized by an electrondense substance at their apical extremity.
Fig. 1. (x 5200). There is no pharyngeal cell; Fig. 2 (x 5300), Fig. 3. (x 6000), Fig. 4. (x 5700). There are pharyngeal and buccal cells; Fig. 5. (x 5200), Fig. 6. (x 8500). There are only pharyngeal cells.
BL: buccal lumen; C: buccal cilia; PL: pharyngeal lumen; V: buccal velum; arrows: pharyngeal cells.

the anterior third of the tube and ends near the trophi of the mastax. The buccal velum separates the buccal from the pharyngeal lumen. The latter gives two cone shaped extensions up into the buccal tube.

The buccal lumen narrows from its first third down to its base. At this point this lumen is just formed by the space between the two symmetrical parts of the velum (Fig. 5, 8i).

Brachionus calyciflorus

We have noted that in this species, there is a velum which

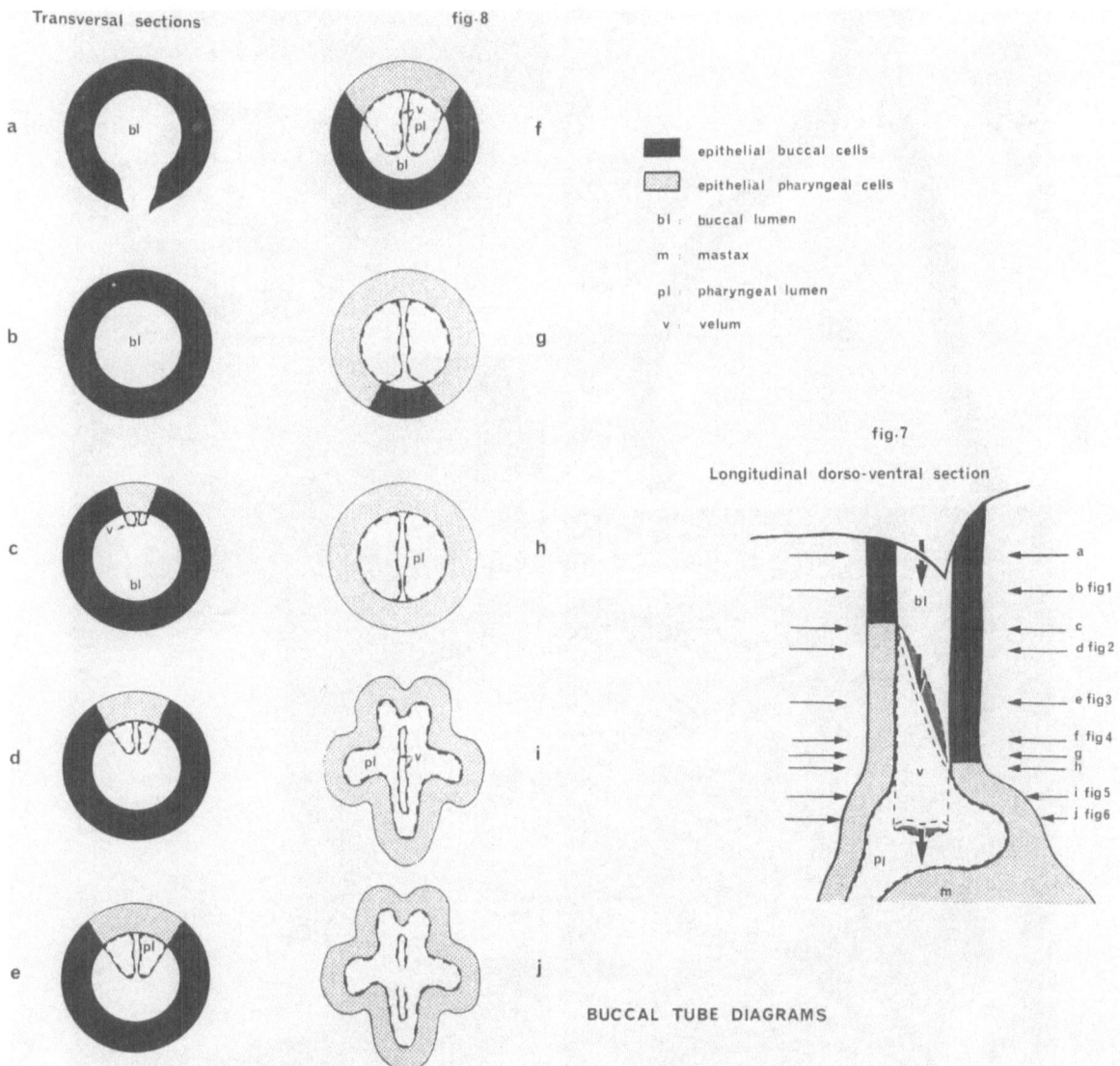

Fig. 7. Diagram of a longitudinal section of a buccal tube. The dorsal part is at left, the ventral one at right. The arrows show the transvere sections of Fig. 8. (a to j) and the transvere sections of the Fig. 1 to 6.
bl: buccal lumen; pl; pharyngeal lumen; m: mastax; v: buccal velum.
The large arrows indicate the entrance of food particles into the pharyngeal lumen.
Fig. 8. Diagram of transverse sections of the buccal tube. The level of each section is indicated in the Figure 7.
The legends are the same as in Figure 7.
In black/: epithelial buccal cells; In.....: epithelial pharyngeal cells.
The interrupted line is the cuticle of the pharyngeal cells; this cuticle forms the buccal velum.

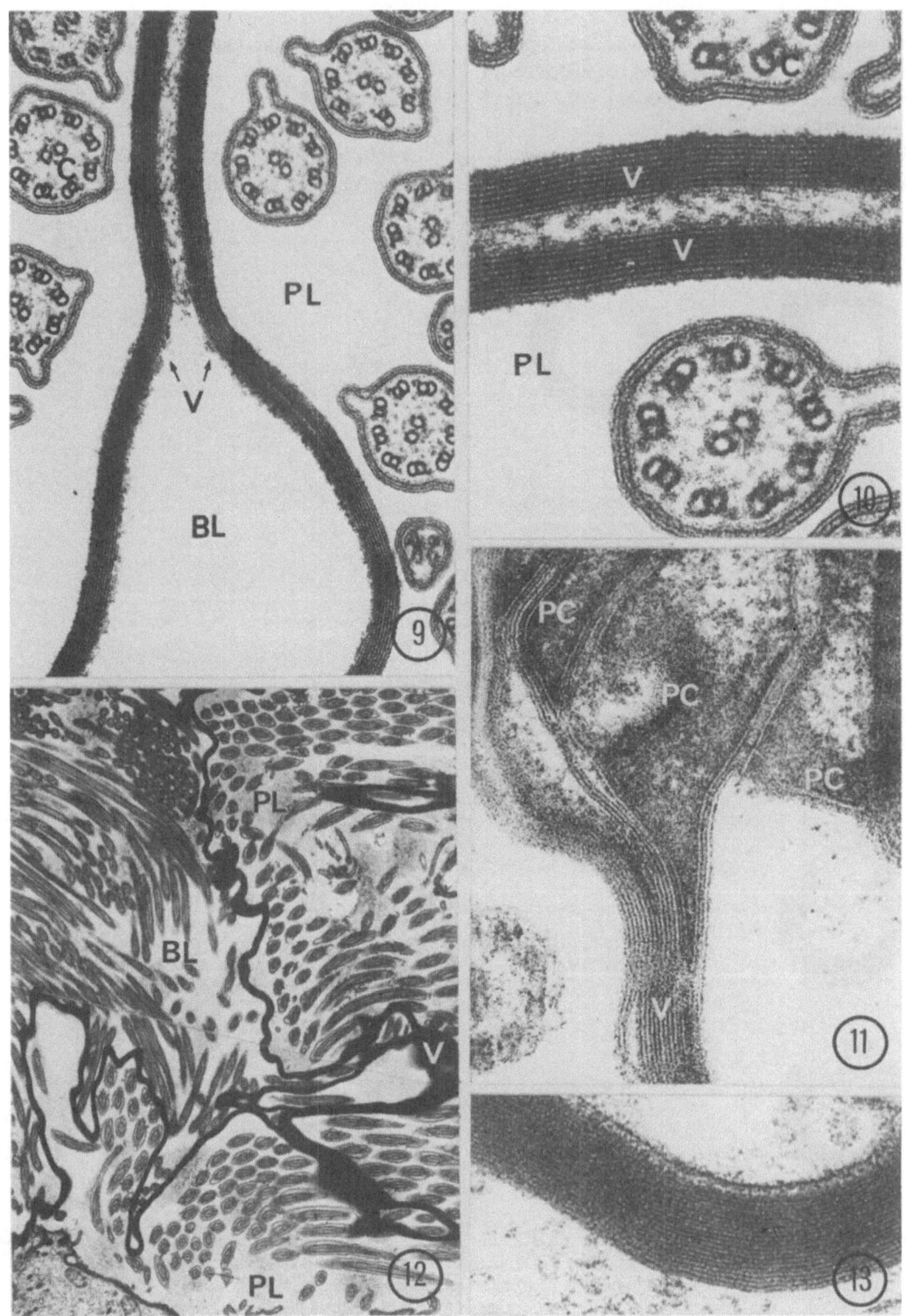

Figures 9 to 13. The structure of the buccal velum.

Fig. 9. (x 78 000) and Fig. 10. (x 150 000). *Philodina roseola*. Transverse sections of the buccal tube at the level of the
Figure 5. Note the myelin-like structure of the velum (V) which is formed by a membrane pile. In the pharyngeal lumen
(PL) the cilia (C) have a double cytoplasmic membrane. (BL) buccal lumen. Fig. 11. (x 90 000): *Philodina roseola* –
Formation of the velum. The apical membranes of several pharyngeal cells (PC) meet up to form a membrane pile: the
velum (V). Fig. 12. (x 4500): *Brachionus calyciflorus*. Transverse section of the buccal tube. The velum (V) is more
infolded than in *P. roseola*. Fig. 13. (x 134 000): *Brachionus calyciflorus*. The buccal velum is a membrane pile (a
myelin-like structure).

separates the buccal and the pharyngeal space. It shows the same structure (Fig. 13) and the same insertion on the pharyngeal cells as in *P. roseola* (Fig. 12). Nevertheless our observations are not sufficient to reconstitute the three-dimensionnal morphology of the velum. Our preliminary results show many velum folds (Fig. 12) indicating a greater complexity than in *P. roseola*.

Discussion

In *P. roseola* and probably in *B. calyciflorus* as well, the velum acts as a sort of fish-trap, preventing the alimentary particles from going back to the buccal lumen, once they have reached the mastax. Entrance into the pharyngeal lumen is thus made irreversible.

On the other hand, the buccal velum is a limp structure which cannot alone control ingestion of alimentary particles. If it prevents the return of the food particles from the pharyngeal lumen back to the buccal lumen, the admission of the food particles into the buccal tube and then into the pharyngeal lumen is already controlled by different receptors and effectors which we describe elsewhere (cf. Clément *et al.,* 1980, b and c).

References

De Beauchamp, P. 1909. Recherches sur les Rotifères: les formations tégumentaires et l'appareil digestif. Arch. Zool. exp. gén., 10: 1-410.

Clément, P. & Pavans de Ceccatty, M. 1969. Ultrastructures d'une surface sécrétante salivaire: le tube buccal d'un rotifere Notommata copeus. Morphologie, desmosomes septés, sécrétions. Bull. Soc. zool. Fr., 94: 599-612.

Clément, P. 1977a. Introduction à la photobiologie des rotiferes dont le cycle reproducteur est contrôlé par la photopériode. Approches ultrastructurale et expérimentale. Thèse n°7716. Université Claude Bernard (Lyon I), Fr.

Clément, P. 1977b. Ultrastructural research on rotifers. Arch. Hydrobiol. Bei.., 8: 270-297.

Clément, P., Amsellem, J., Cornillac, AM., Luciani, A. & Ricci, C. 1980b. An ultrastructural approach to feeding behaviour of P. roseola B. calyciflorus. (Rotifers). II- The oesophagus, In this volume, pp. 133-136.

Clément, P., Amsellem, J., Cornillac, AM., Luciani, A. & Ricci, C. 1980c. An utrastructural approach to feeding behaviour in Philodina roseola and Brachionus calyciflorus (Rotifers). III. Cilia and muscles. Conclusions. In this volume, pp. 137-141.

Remane, A. 1929-1932. Rotatoria. In: Dr. HG. Bronn's Klassen und Ordnungen des Tier-Reichs', IV (Vermes), 2 (Asschelminthes), 1 (Rotatorien, Gastrotrichen und Kinorhynchen), 3: 1-448.

Schramm, U. 1978. Studies on the ultrastructure of the rotifer Habrotrocha rosa Donner (Aschelminthes)–The alimentary system. Cell. Tiss. Res., 189: 525-535.

AN ULTRASTRUCTURAL APPROACH TO FEEDING BEHAVIOUR IN PHILODINA ROSEOLA AND BRACHIONUS CALYCIFLORUS (ROTIFERS)

II. THE OESOPHAGUS

P. CLÉMENT, J. AMSELLEM, A.-M. CORNILLAC, A. LUCIANI[1] & C. RICCI[2]

[1] Laboratoire d'Histologie et Biologie Tissulaire (LA 244 CNRS) et CMEABG, Université Claude Bernard (Lyon I), Villeurbanne, France; [2] Instituto di Zoologia, Universita degli studi, via Celosia, 10, Milano, Italy

Keywords: Rotifer, ultrastructures, feeding behaviour, digestive tract, cuticle

Abstract

The cuticular oesophagus is a simple expansion of the dorsal pharyngeal wall of the mastax. The ciliary oesophagus is the cellular anterior wall of the stomach lumen, but seems to have the same embryological origin as the pharynx.

In *Brachionus calyciflorus*, its cilia are surrounded by cuticular velums which have the same myelin-like structure and the same function as the buccal velum of *Philodina roseola*. In all cases, the oesophagus prevents the return of food particles from the stomach to the mastax lumen.

Introduction

Criteria to distinguish the end of buccal structures from the beginning of pharyngeal structures are: the characteristic buccal and pharyngeal cilia and the buccal velum, which originate from apical membranes of the anterior pharyngeal cells.

In this paper, we describe the end of the pharynx. Classical studies note the existence of an oesophagus between the mastax and the stomach (De Beauchamp, 1909, 1965; Remane, 1929-1932; Hyman, 1951). Its ultrastructure was briefly described in *Trichocerca rattus* by Clément (1977a, b) but Schramm (1978) did not find an oesophagus in *Habrotrocha rosa*. Thereby she incorrectly called a part of the buccal tube an oesophagus.

After meticulous observations which we think are not as widely known as they should be, De Beauchamp (1909) distinguished the cuticular oesophagus from the ciliary oesophagus. The former has the same cuticle as the pharynx. The latter is the upper part of the stomach between the cuticular oesophagus and the gastric glands. Each of these two sectors can be either short or long, depending on species. Unfortunately, classical drawings of Rotifers (Remane, 1929-1932; Hyman, 1951; Ruttner-Kolisko, 1974; Koste, 1978) show a uniform oesophagus,

and ignore the distinction made by De Beauchamp (1909).

Our ultrastructure observations show that the cuticular oesophagus is a pharyngeal expansion, and confirm the existence of a ciliary oesophagus whose features are very different from the stomach's.

Results (Material and methods: see Clément *et al.*, 1980a)

Philodina roseola: (Fig. 1 to 4)
The cells of the dorsal mastax roof extend backwards to constitute the thin epithelial wall of the cuticular oesophagus (Fig. 2). This oesophagus does not have its own epithelial cells. Its fine epithelial wall is doubled by a muff of muscle cells (Fig. 3). These muscle cells have only a few myofilaments, which are grouped together near the epithelial tube.

In *P. roseola,* the stomach is syncitial and its lumen is bordered by a special terminal web, described by Mattern & Daniel (1965), and by Schramm (1978) in *H. rosa.* Between the stomach and the cuticular oesophagus, there exists a short ciliary oesophagus, made of ciliated cells (Fig. 1 and 2). These cilia are very long. They form the vibratile flame beating in the stomach lumen (Fig. 2). The big gastric glands of *P. roseola* arrive in the ciliary oesophagus (Fig. 3 and 4).

Brachionus calyciflorus: (Fig. 5 to 8)
The cuticular oesophagus is shorter. It is also a simple pharyngeal epithelium expansion (Fig. 6). The cuticle is thicker than in *P. roseola:* it is made of a membrane-pile (Fig. 5 and 8). This membrane-pile separates from the pharyngeal epithelial cells, to form a cuticular velum which has the same structure as the buccal velum; it also has a fish-trap form (Fig. 6). Just under this 'cuticular velum', there are cilia borne by the cells of the very short ciliary oesophagus (Fig. 5 and 6). Meanwhile, the epi-

133

Hydrobiologia 73, 133-136 (1980). *0018-8158/80/0732-0133$00.80.*
© *Dr. W. Junk b.v. Publishers, The Hague.*

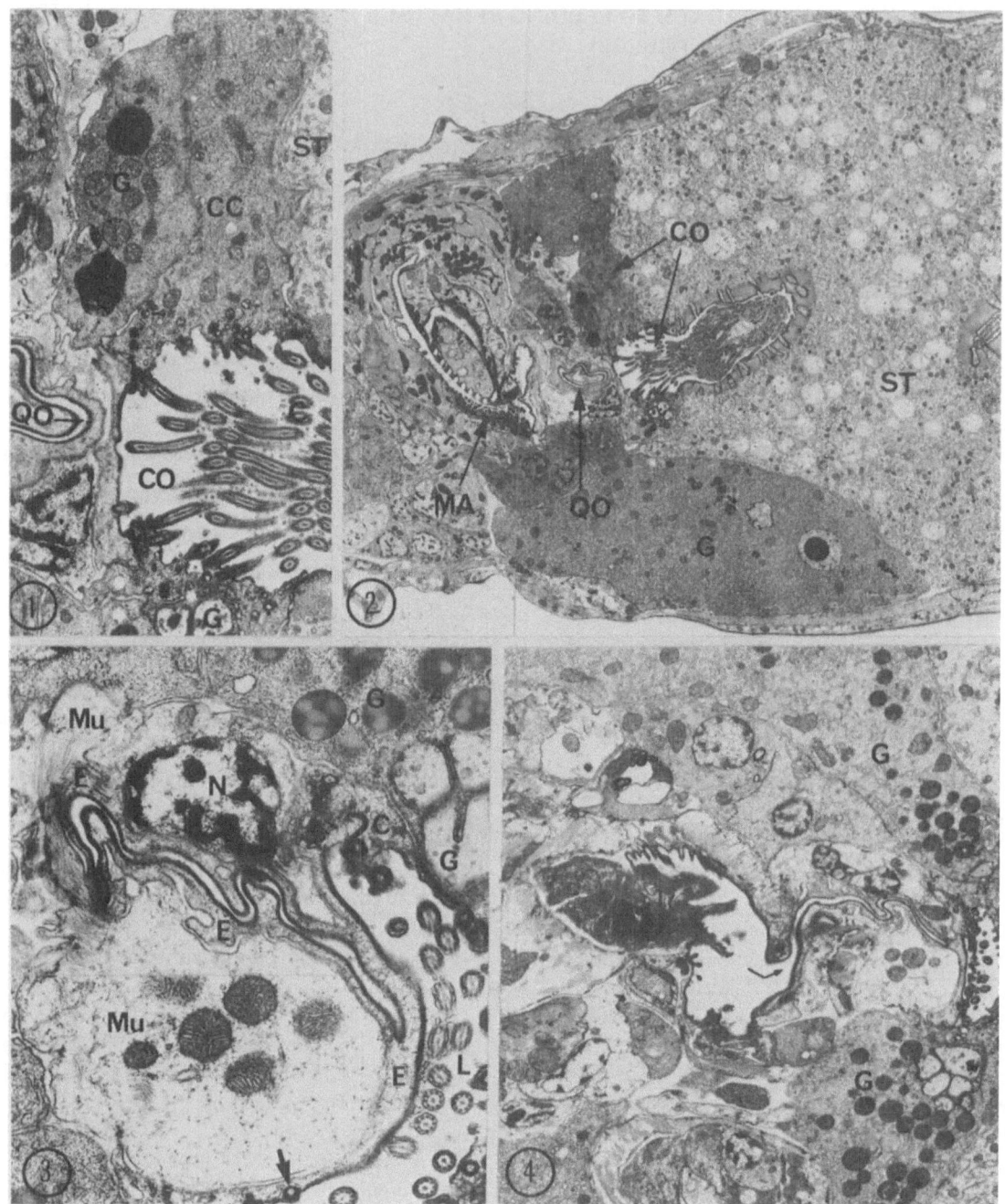

Figs. 1-4. *Philodina roseola.*

Fig. 1. x 10000. Cuticular oesophagus (QO) and ciliary oesophagus (CO): the cilia (C) are borne by ciliary cells (CC). The gastric gland (G) opens at the level of these cells. (ST) stomach syncitium; Fig. 2. x 3000. The lumen of the stomach syncitium (ST) contains the cilia which come from the ciliary oesophagus (CO). The cuticular oesophagus (QO) is in continuity with the dorsal pharyngeal epithelium of the mastax, which bears the striated pharyngeal cilia (MA). Syncytial gastric gland (G); Fig. 3. x 16400. A muscle cell (Mu), with its nucleus (N) is located around the epithelial wall (E) of the cuticular oesophagus. This cell (E) bears only a few cilia (arrow) when it is in contact with the upper part of the stomach lumen (L). See also Fig. 1. Most cilia are borne by epithelial cells (C) of the ciliary oesophagus. (G) = gastric gland; Fig. 4. x 6000. The arrow shows the communication between the mastax lumen (at left) and the cuticular oesophagus lumen: the epithelial cytoplasm and the cuticle are strictly the same. At right, a part of ciliary oesophagus and gastric glands (G).

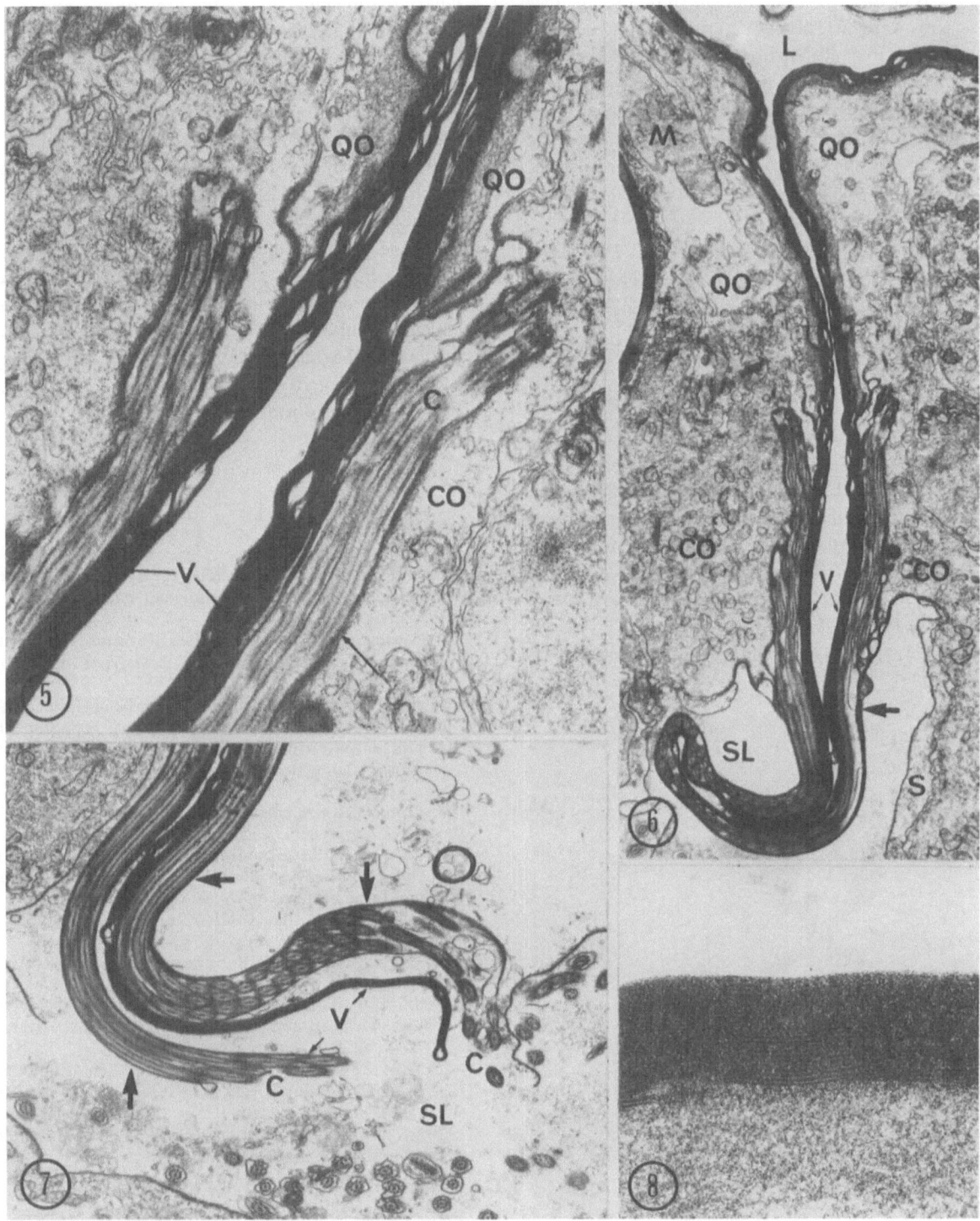

Figs. 5-8. *Brachionus calyciflorus.*
Fig. 5. x 20000. Detail of an axial section of the cuticular (QO) and ciliary (CO) oesophagus. At this level, the cilia (C) are located between the median velum (V) and the fine cuticle (arrow) of the ciliary cell; Fig. 6. x 6500. Axial section of the cuticular oesophagus (QO) and ciliary oesophagus (CO). The cuticle which lines the mastax lumen (L) extends in the cuticular oesophagus, and then forms a velum (V). Lining this velum are the cilia of the ciliary oesophagus; these cilia are separated from the stomach lumen (SL) by a new cuticilar velum (arrow) which comes from the ciliary oesophagus (CO). (S) = stomach cell. M = a muscular cell; Fig. 7. x 9500. Opening of the two velums in the stomach lumen (SL). The principal median velum (V) lines the end of the oesophagus lumen. The secondary cuticular velum (arrows) separates the oesophagus cilia from the stomach lumen, except at the extremity of the cilia (C); Fig. 8. x 127000. High power magnification of the cuticle of the cuticular oesophagus which shows a myelin-like aspect.

135

thelial cells of this ciliary oesophagus show also an apical cuticle (Fig. 5 and 6). This cuticle gives a second cuticular velum which surrounds the first. All this forms a vibratile flame with the cilia located between the two velums (Fig. 6). The internal velum opens into the stomachal lumen at the vibratile-flame extremity (Fig. 7): the food particles pounded by the mastax, go by this way. The cilia also end in the stomach lumen at the vibratile-flame extremity (Fig. 7).

The two parts of the oesophagus form a short epithelial structure not surrounded by a muscular layer: there is only a circular ring where the mastax lumen narrows to form the cuticular oesophagus lumen (Fig. 6).

Conclusions

Functions of the oesophagus:

In the two observed species, the food particles once in the stomach cannot go back into the pharynx. The cuticular oesophagus divides the digestive tube into compartments: it has the same function as the buccal velum in the more anterior region. Two different systems are used for this: muscles in *P. roseola,* and cuticular velums with a fish-trap form in *B. calyciflorus.*

On the other hand, the vibratile flame borne by the ciliary oesophagus, allows the food to progress into the stomachal lumen.

Nature of the oesophagus: (Fig. 9)

The cuticular oesophagus is only a pharyngeal expansion. The ciliary oesophagus is made of epithelial cells. These distinct cells do not look like the cells or the syncitium which form the stomach wall: in particular stomachal cilia are shorter and have no ciliary rootlets.

On the other hand, in *B. calyciflorus,* these epithelial cells look like the pharyngeal epithelial cells: they have the same multimembranous cuticle and they can form a cuticular velum. So it is possible that the ciliary oesophagus has the same embryological origin as the pharynx (stomodeal invagination), while the stomach seems to come from the dorsal ectoderm (Nachtwey, 1925; De Beauchamp, 1965).

References

De Beauchamp, P. 1909. Recherches sur les Rotifères: les formations tégumentaires et l'appareil digestif. Arch. Zool. exp. gén., 10: 1-410.

De Beauchamp, P. 1965. Classe des Rotiferes. In: 'Traité de zoologie, anatomie, systématique, biologie. P. Grassé', IV, 3: 1225-1379.

Clément, P. 1977a. Introduction à la photobiologie des rotifères dont le cycle reproducteur est contrôlé par la photopériode. Approches ultrastructurale et expérimentale. Thèse n° 7716, Université Claude Bernard (Lyon I): 262 pp.

Clément, P. 1977b. Ultrastructural research on Rotifers. Arch. Hydrobiol. Beih., 8: 270-293.

Clément, P., Amsellem, J., Cornillac, AM., Luciani, A. & Ricci, C. 1980a. An ultrastructural approach to the feeding behaviour of Philodina roseola and Brachionus calyciflorus. I. The buccal velum. In this volume, pp. 127-131.

Hyman, L. H. 1951. Class Rotifers. In: 'The invertebrates', MC Graw Hill, New York, vol. 3, pp. 59-151.

Koste, W. 1978. Rotatoria. Borntraeger, Berlin, 2 vols., 673 pp., 234 plates.

Mattern, C. F. T. & Daniel, W. A. 1965. The stomach cell of Rotifer. Electron microscope observations of the terminal. Web. J. Cell. Biol., 29: 547-551.

Nachtwey, R. 1925. Untersuchungen über die Keimbahn Organogenese und Anatomie von Asplanchna priodonta Gosse. Z. Wiss. Zool., 126: 239-492.

Remane, A. 1929-1932. Rotatoria. In: 'Dr. HG. Bronn's Klassen und Ordnungen des Tier-Reichs', IV (Vermes), 2 (Aschelminthes), 1 (Rotatorien, Gastrotrichen und Kinorhynchen), 3: 1-448.

Ruttner-Kolisko, A. 1972. Rotatoria. Binnengewässer, 26: 99-234.

Schramm, U. 1978. Studies on the ultrastructure of the Rotifer Habrotrocha rosa Donner (Aschelminthes). The alimentary system. Cell. Tiss. Res., 189: 525-535.

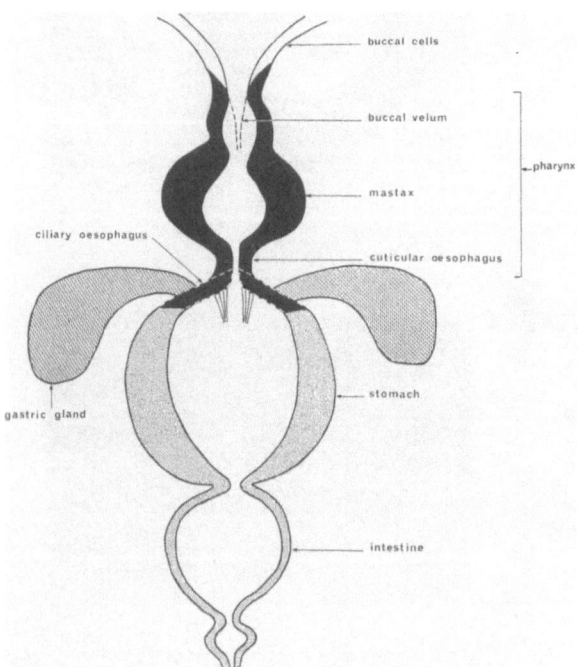

Fig. 9. Diagram of the digestive tract of a Rotifer.

buccal cells

buccal velum

pharynx

mastax

ciliary oesophagus

cuticular oesophagus

stomach

gastric gland

intestine

AN ULTRASTRUCTURAL APPROACH TO FEEDING BEHAVIOUR IN PHILODINA ROSEOLA AND BRACHIONUS CALYCIFLORUS (ROTIFERS)

III. CILIA AND MUSCLES. CONCLUSIONS

P. CLÉMENT, J. AMSELLEM, A.-M. CORNILLAC, A. LUCIANI[1] & C. RICCI[2]

[1] Laboratoire d'Histologie et Biologie Tissulaire (LA 244 CNRS) et CMEABG, Université Claude Bernard (Lyon l), Villeurbanne, France; [2] Instituto di Zoologia, Universita degli studi, via Celosia, 10, Milano, Italy

Keywords: Rotifers, ultrastructure, feeding behaviour, cilia

Abstract

The orientation of the cilia pseudotrochus is controlled by muscles inserted on their rootlets. The peculiar structures of buccal and pharyngeal cilia are described: their beating leads the food towards the mastax. The circular and longitudinal muscles of the buccal funnel allow peristaltic movements and probably the rejection of food items not accepted by the mastax receptors.

Introduction

In *Brachionus calyciflorus* and *Philodina roseola,* the acceptance or the rejection of food items seems to be made at least by the cilia of the pseudotrochus, the buccal funnel and the jaws of the mastax.

Optical microscope observations (Gilbert & Starkweather, 1977 on *B. calyciflorus*) are insufficient to precisely observe the structures involved in the feeding behaviour. In the two preceding papers (Clément *et al.,* 1980a and b) we described the buccal velum of *Philodina roseola* and *Brachionus calyciflorus,* and the oesophagial velum of *B. calyciflorus:* it is impossible to understand the structure and fish-trap function of these velums with an optical microscope.

Here we shortly describe the different effectors involved in the feeding behaviour: cilia and muscles. In conclusion, we try to make a synthetic scheme of the feeding behaviour, involving both preceding papers, and also information not yet published about the nature and the location of sensory receptors.

The material and methods are the same as in the preceding papers (see also Amsellem & Clément in this book). Our observations are more complete on *Philodina roseola* than on *Brachionus calyciflorus.*

Pseudotrochus

Cilia show the usual kinocil structure: 9 peripheral doublets, 2 central tubules, arms and intermediate structures; so they move actively and probably spontaneously.

They have, in their apical part, a cylindric dense structure which has the diameter of a microtubule, beside two central tubules (Fig. 1). The most interesting feature concerns the relation between the muscles and the ciliary rootlets. Each cilium is connected 1) to the neighbour cilium by an horizontal network of striated ciliary roots and 2) to desmosomes by a striated ciliary root which goes down into the epithelial cell (Fig. 1). The myofilaments of the longitudinal muscles are inserted on the other side of these desmosomes (Fig. 1). Thus the muscles can control cilia orientation.

In *Brachionus calyciflorus,* the cilia are grouped: each group forms a cirrus, visible in optical microscopy.

Buccal funnel

The cilia here are also active kinocils. At their apical extremity they have electron-dense material (Fig. 2). A similar differentiation has been observed in *Trichocerca rattus* and *Notommata copeus* (Clément, 1977a and b). Their striated rootlets go down to the perinuclear cytoplasm (Fig. 2) but have no relation with muscles. The buccal funnel wall is formed by three layers:

1. the closely joined apical parts of the epithelial ciliary cells; they bear the cilia. Only this layer is schematised in Fig. 7.

2. a double muscle layer formed by an inner longitudinal layer and an outer circular one.

3. and finally the thickest layer which is made up of the ciliated epithelial cell bodies. These cell bodies are joined together by gap-junctions (Fig. 2).

137

Hydrobiologia 73, 137-141 (1980). 0018-8158/80/0732-0137$01.00.
© *Dr. W. Junk b.v. Publishers, The Hague.*

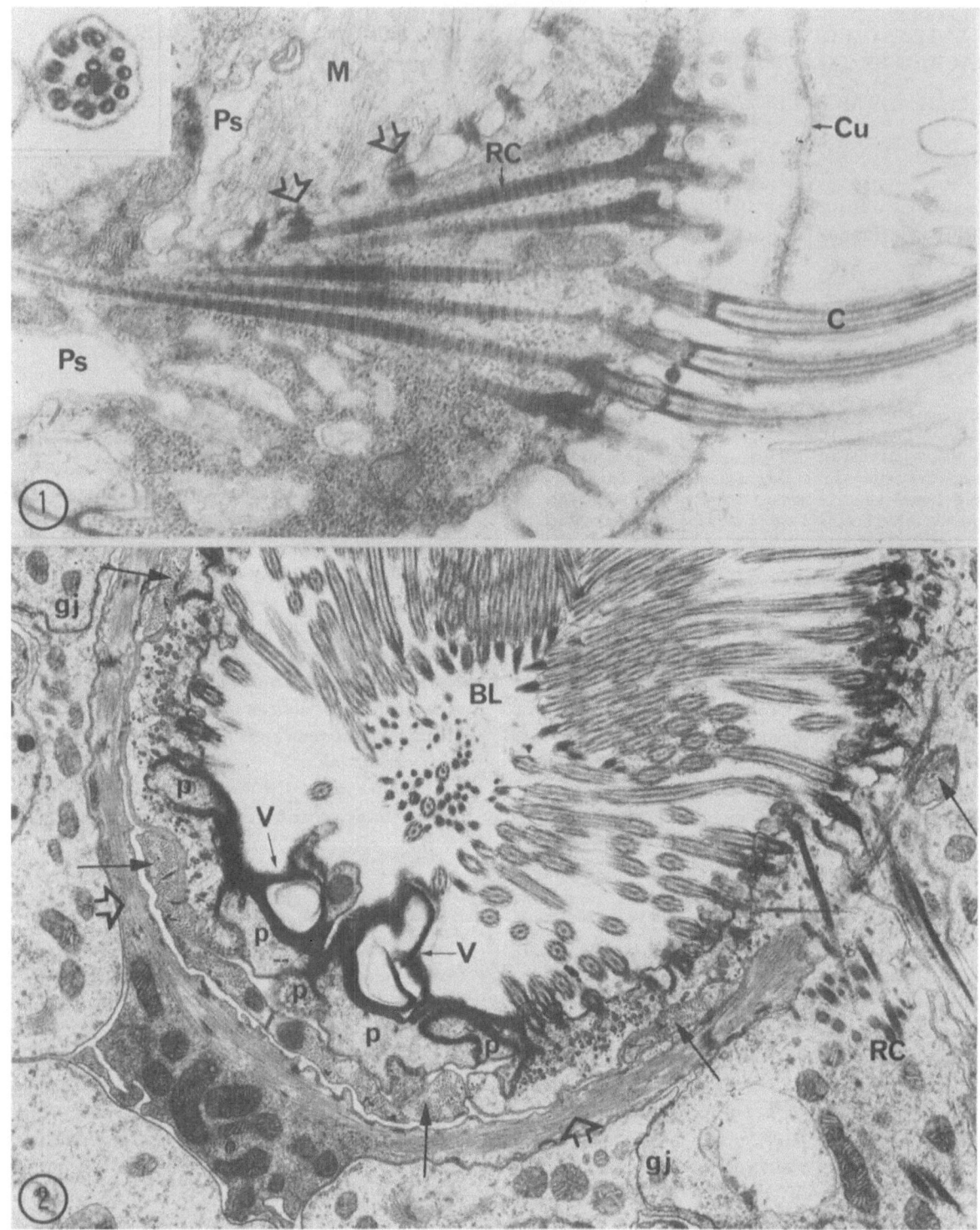

Fig. 1. 28000. Pseudotrochus cilia of *Brachionus calyciflorus*. Axial section. The cilia (C) go through a thin cuticle (Cu). The striated ciliary rootlets (RC) insert on epitheliomuscular desmosomes (arrows). A longitudinal muscle (M) inserts on the other side of the same desmosomes. Pseudocoel (Ps). Encart. x 100000. Transverse section of the extremity of a pseudotrochus cilium *(Brachionus calyciflorus)*. A cylindric dense structure is juxtaposed to the two central tubules;

Fig. 2. 13000. Transverse section of *Philodina roseola* buccal tube. The circular (large arrows) and longitudinal (fine arrows) muscles are regularly and symmetrically disposed all around the tube. In the middle of the buccal lumen (BL), the end of the buccal cilia contains dense-electron material. The rootlets (RC) of the buccal cilia are not attached to epitheliomuscular desmosomes: they originate inside the epithelial cell. The buccal velum (V) is in continuity with the pharyngeal cells of the tube (p). All the epithelial cells bodies are linked by gap junctions (gj).

138

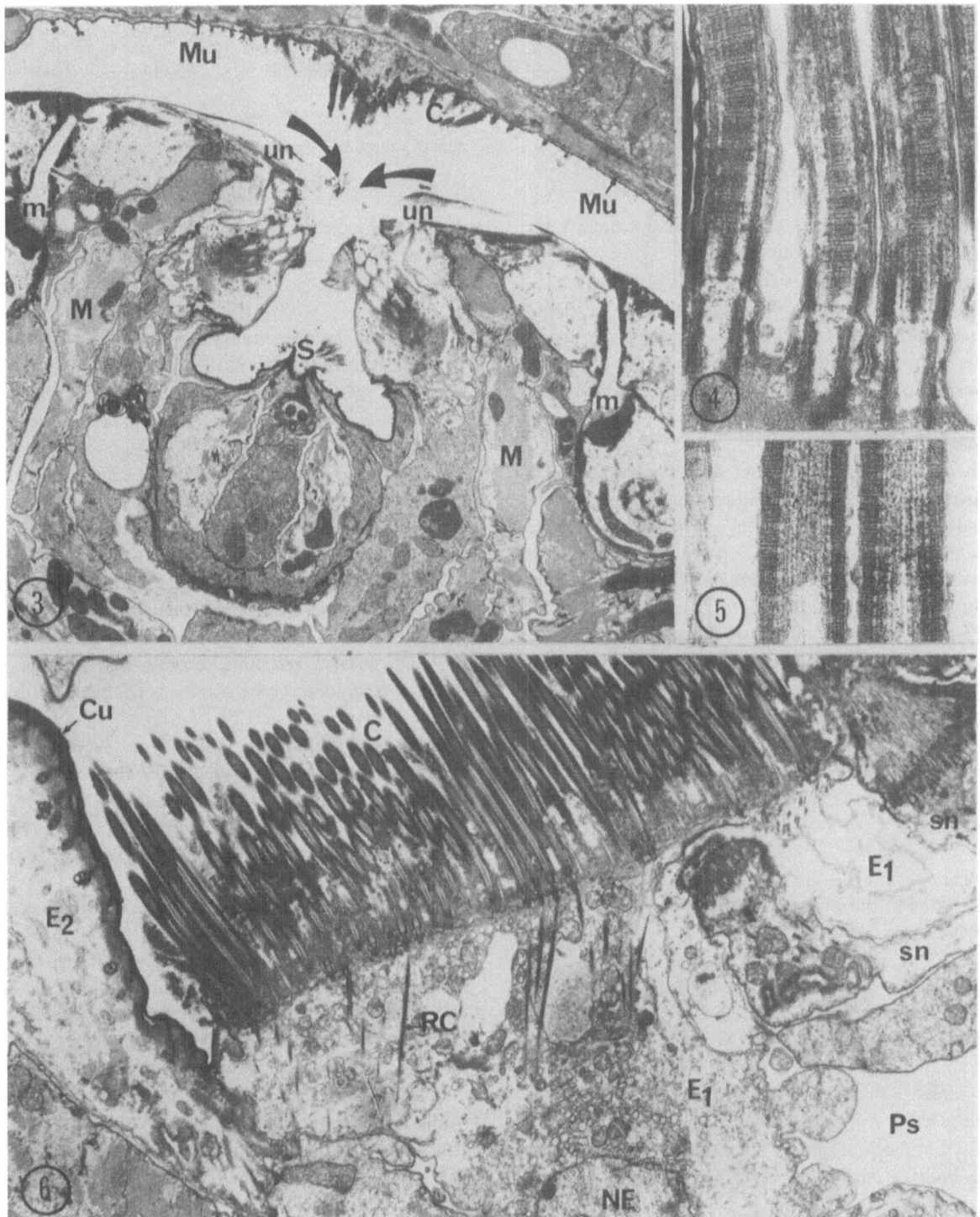

Fig. 3. x 6600. Transverse section of *Philodina roseola* mastax. The large curved arrows indicate the way taken by the food particles. The dorsal muscle (Mu) and the dorsal striated cilia (C) force them into contact with the uncus (un) and with the mastax sensory receptor (S); m: manubrium, M: mastax muscles; Fig. 4. x 60000. Axial section of three striated cilia of *Philodina roseola* (C in Fig. 3). Note their double cytomembrane and the localization of the dense transversal striated material, in the middle of the axonema; Fig. 5. x 60000. Axial section of two striated cilia of the mastax of *Brachionus calyciflorus* (C in Fig. 6). Note that the dense transversal striated material is situated between the axonema and the cytomembrane; Fig. 6. x 6600. Mastax of *Brachionus calyciflorus*. Striated cilia (C) have rootlets (RC) which go down into the epithelial cell (E1) whose nucleus (NE) is partly visible. Two sensory neurites (sn) each have a tuft of sensory cilia situated among striated cilia. On the left, another epithelial cell (E2) bears a myelin-like cuticle (Cu) which continues backwards to the cuticular oesophagus and forwards to the buccal velum.

The regular disposition of the muscles (Fig. 2) indicates possible peristaltic movements in the buccal tube.

Pharynx

The pharynx is the part of the digestive tract located between the buccal velum and the ciliary oesophagus (cf. the two preceding communications).

The pharyngeal cilia are found either just beneath the buccal velum or in a dorso-lateral position on the roof of the mastax (Fig. 7). They are short and have a larger diameter than those of the buccal funnel. They are active kinocilia (cf. Fig. 10 in the communication I on the buccal velum). In *Philodina roseola*, they are surrounded by two cytomembranes.

The cilia located on the roof of the mastax show a specialization previously unknown anywhere else in the animal kingdom: they contain a periodically striated dense material (Fig. 4 and 5; Cornillac *et al.*, 1979). This striated material is axial in *Philodina roseola* and peripheral in *Brachionus calyciflorus*.

The most important muscles are located in the mastax: they insert on the trophi (Fig. 3 and 7) and are innervated by the ganglion of the mastax (Fig. 7). There is also a muscular layer around the mastax, inserted on the mastax and partly on the anterior cells of the rotatory apparatus. This muscular layer is present around the roof of the mastax (Fig. 3 and 7); its innervation comes from the brain (Fig. 7).

Localization of the sensory receptors involved in feeding behaviour

In *Brachionus calyciflorus*, Clément (1977a and b) described chemoreceptors localized in the supple integument between the lorica and the anterior cilia of the rotatory apparatus. Clément (1977a and b) also described mechanoreceptor cilia which form tactile cirri in males and females of *Brachionus calyciflorus*.

Different receptors are also present in the anterior part of the body of *Philodina roseola*. In this species, no sensory receptor can be observed in the buccal tube. The only sensory structures between the pseudotrochus and the ciliary oesophagus are localized in the mastax (Fig. 7): two ciliary tufts among the periodically striated pharyngeal cilia on the roof of the mastax, and a more complex receptor on the floor of the mastax, between the trophi

(Fig. 3 and 7). These sensory cilia seem to be chemoreceptors, and perhaps also mechanoreceptors (Clément *et al.*, 1979).

In *Brachionus calyciflorus*, we find also two tufts of sensory cilia among the striated pharyngeal cilia (Fig. 6).

Conclusions on the phases and the mechanisms of the feeding behaviour in Philodina roseola and Brachionus calyciflorus (Fig. 7)

1. Food particles meet the anterior receptors of the animal by the beating of the cilia of the pseudotrochus.

2. The sensory structures are probably mechanoreceptive as well as chemoreceptive. Their neurons are cerebral.

3. The brain innervates muscles which are inserted on the pseudotrochus ciliary rootlets: a first selection can be made because the cilia may allow or not allow the entrance of food into the buccal funnel.

4. The food which goes through the buccal funnel then goes to the mastax owing to the action of buccal cilia, the buccal velum and the anterior pharyngeal cilia.

5. The mastax receptors (chemo- and perhaps also mechanoreceptors) sense the arriving food and release the trophi movements.

6. The food is pounded, filtered by special mastax cilia, then goes to the oesophagus. During the pounding and the filtration, mastax receptors can be informed by chemoreception if the food is acceptable or not and must thus be rejected or not. So, two types of behaviour are possible:

7. a) The food is considered suitable: the whole process runs on.

7. b) The food is considered unacceptable: the mastax receptors inform the mastax ganglion and the brain innervating the muscles of the mastax and those surrounding the buccal funnel: the mastax is projected forward to stop the food coming into the pharyngeal lumen, and the buccal funnel muscles reject this food towards the mouth owing to peristaltic contractions.

It seems that the food particles which first released the process in *P. roseola* are not rejected by this mechanism, because of the buccal velum position.

In rotifers, every muscle is innervated (Clément, 1977a and b); but we never observed an innervation of the ciliary epithelial cells.

So, in these conclusions, we suppose that ciliary beats are autonomous and not controlled by the nervous

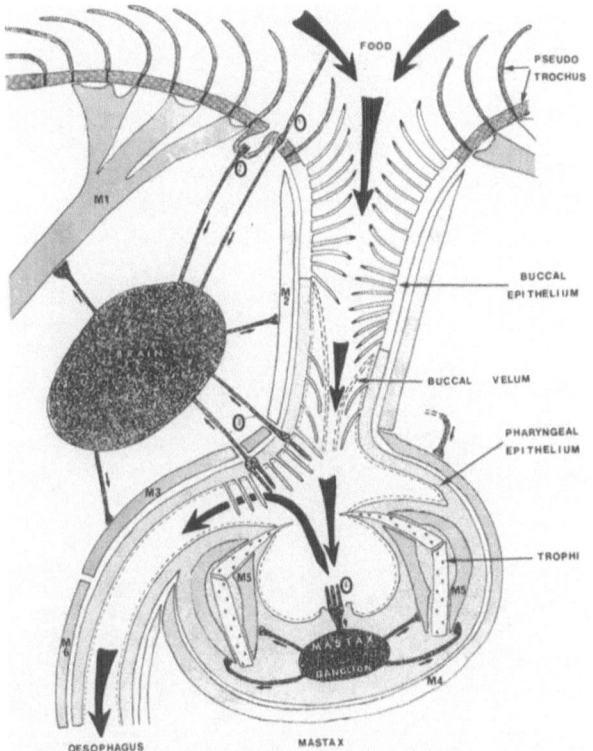

Fig. 7. Diagram of phases and mechanisms of the feeding behaviour in *Philodina roseola* (and perhaps in *Brachionus calyciflorus*).

See the text for the descriptions of the different phases.

There are three kinds of epithelial cilia:
- cilia of the pseudotrochus: in the pseudotrochus
- buccal cilia with their dense apical part: in the upper part of the buccal tube
- very special and periodically striated cilia of the pharynx: on the roof of the mastax.

The longitudinal retractor muscles (M1) are inserted on the rootlets of the cilia of the pseudotrochus.

Longitudinal and circular muscles (M2) surround the buccal and pharyngeal cells in the buccal tube. All these muscles, and those located around the dorsal (M3) and ventral (M4) part of the mastax, and around the cuticular oesophagus (M6) are innervated by the brain. The muscles (M5) inserted on the trophi of the mastax are innervated by the mastax ganglion.

The different sensory receptors which seem involved in this behaviour are:
1 Anterior mechanoreceptors
2 Anterior chemoreceptors
3 Sensory receptors of the dorsal wall of the mastax
4 Sensory receptors of the ventral part of the mastax.

system, except in the pseudotrochus: at this level, the nervous system controls the cilia by way of the longitudinal rootlets inserted on the striated rootlets of the cilia.

These conclusions (Fig. 7) are a series of hypotheses based on ultrastructural observations; it is now necessary to test them by direct behavioural observations.

References

Amsellem, J. & Clément, P. 1980a. A simplified method for the preparation of rotifers for transmission and scanning electron microscopy. In this volume, pp. 119-122.
Rotifer Symp., Hydrobiologia.

Clément, P. 1977a. Introduction à la photobiologie des rotifères dont le cycle reproducteur est contrôlé par le photopériode. Approches ultrastructurale et expérimentale. Thèse n° 7716. Université Claude Bernard Lyon I, France.

Clément, P. 1977b. Ultrastructural research on rotifers. Arch. Hydrobiol. Beih. 8: 270-297.

Clément, P., Cornillac, A. M. & Luciani, A. 1979. Cils chémorécepteurs dans le mastax des Rotifères. Biol. cell., 35, p. 25a.

Clément, P., Cornillac, AM., Amsellem, J., Luciani, A. & Ricci, C. 1980a. An ultrastructural approach to the feeding behaviour of Philodina roseola and Brachionus calyciflorus (Rotifers) I. The buccal velum. In: H. J. Dumont & J. Green (eds.) Proc. II Int. Rotifer Symp., Hydrobiologia.

Clément, P., Cornillac, AM., Amsellem, J., Luciani, A. & Ricci, C. 1980b. An ultrastructural approach to the feeding behaviour of Philodina roseola and Brachionus calyciflorus (Rotifers) II. The oesophagus. In: H. J. Dumont & J. Green (eds.) Proc. II Int. Rotifer Symp., Hydrobiologia.

Cornillac, AM., Luciani, A. & Clément, P. 1979. Des cils au contenu dense strié périodiquement. Biol. Cell., 35:

Gilbert, J. J. & Starkweather, P. L. 1977. Feeding in the rotifer Brachionus calyciflorus. I. Regulatory mechanisms. Oecologia, 28: 125-131.

WORKSHOP ON THE DETERMINATION OF POPULATION PARAMETERS

W. HOFMANN

Max-Planck-Institut für Limnologie, Abt. Allgemeine Limnologie, Plön, BRD

Keywords: population parameters, population dynamics, population ecology

Introduction

The moderator started by focussing upon the dynamics of the individual population parameters or structure elements. Any parameter, such as abundance, dispersion, fertility, mortality, age structure, mictic rate, habit, morbidity, and constitution, may change with time. In order to characterize population dynamics, all the parameters should be taken into account. However, it is difficult to get information on, for instance, the age distribution, morbidity, and constitution, of natural populations. Furthermore, it is impossible to estimate the fertility of non-egg-carrying species. For the calculation of mortality and production on the basis of the generation time determined in the laboratory at different temperatures, a constant fertility is assumed. However, the data from egg-carrying rotifers show that this assumption is not correct.

A rather complex population structure was exemplified by a *Filinia* population from Plußsee: Many parameters did not only change with time but also with depth. These changes in the population parameters in time and space reflect responses of the population to changing ecological conditions. Hence, the study of such responses throws light upon the relationship between a population and its environment.

Discussion

The discussion focussed on the difficulties of quantifying the results of field studies: the calculation of fertility on the basis of the egg ratio is difficult owing to the very loose relationship between duration time of embryonic development and temperature (in the case of low temperatures) and because of diurnal changes in egg production (A. Ruttner-Kolisko, J. J. Gilbert).

In general, it was considered difficult to study natural populations because of the environmental complexity (C. King).

The discussion showed different views between workers dealing with laboratory cultures under controlled conditions and those who work in the field. In the laboratory, things can be observed very precisely and the results can be quantified, whereas in the field the patterns of distributions of population parameters in time and space give only a rough idea of what may happen within a population. Likewise, the relationships between populations and their environment can be characterized on a qualitative or semi-quantitative basis, only (W. Hofmann). This view was supported by C. King who stressed the complementary functions of laboratory and field studies.

Hydrobiologia 73, 143 (1980). 0018-8158/80/0732-0143$00.20.
© Dr. W. Junk b.v. Publishers, The Hague.

METABOLIC UNIFORMITY OVER THE ENVIRONMENTAL TEMPERATURE RANGE IN BRACHIONUS PLICATILIS (ROTIFERA)

Robert W. EPP & William M. LEWIS, Jr.

Department of Environmental, Population and Organismic Biology, University of Colorado, Boulder, Colorado 80309, U.S.A.

Keywords: Rotifers, Metabolism, Metabolic Rate, Respiration, Zooplankton

Abstract

Organisms inhabiting small water bodies frequently encounter radical, short-term fluctuations in temperature. The population of *Brachionus plicatilis* that was used in this study encounters temperature changes as great as 12°C in 24 hours. In this paper we describe the metabolic response to temperature change in this eurythermal rotifer population. Metabolism was determined over the environmental range (15°C to 32°C) in intervals of 2°C. Utilization of such narrow temperature intervals allowed us to approximate the instantaneous Q_{10}, which we define as the Q_{10} over an infinitely small temperature interval.

Our results show the existence of a plateau in the curve of respiration against temperature ($Q_{10} = 1$) over the range 20-28°C. The plateau is bounded on either side by temperature ranges over which metabolism is temperature sensitive (Q_{10} values from 3.4 to 4.8). It is significant that the plateau occurs over the environmental temperature range for the major portion of the growing season. This population is thus programmed to hold a constant (preferred?) metabolic rate in spite of diel temperature fluctuations when the environment is otherwise favorable.

Introduction

Although the effects of temperature on metabolism have been investigated intensively in many animal groups, the ecological implications of the temperature-metabolism relationship are often difficult to assess. This is especially true of aquatic invertebrates. Part of the problem stems from the fact that metabolic studies are frequently performed with little regard for the environmental temperature range. Studies of Q_{10} in particular typically lack sufficient resolution within the environmental range to provide significant insight into metabolism under field conditions.

In some recent studies of copepod metabolism (Epp & Lewis, 1979a), we have examined Q_{10} over narrow temperature intervals within the environmental range in an attempt to avoid this shortcoming. In the present study, we have taken the same approach to metabolism in a eurythermal population of *Brachionus plicatilis*.

Our work on the tropical copepod *Mesocyclops brasilianus* documented the existence of a respiratory plateau within the environmental temperature range. Over this plateau, temperature changes have essentially no effect on metabolic rate. This plateau covers the thermal range encountered by the copepods during diel vertical migration. We have argued that the plateau is adaptively significant in maintaining a preferred level of metabolism in spite of the diel fluctuations in temperatures that result from migration. For our present work, we hypothesize that rotifer populations may also show a thermally-independant metabolic rate *within the environmental temperature range* and that this may function as a buffer to maintain a preferred metabolic rate. We have chosen for our study a population that is exposed to rather extreme temperature fluctuations ($> 12°$ per day). If rotifers benefit by maintaining a constant metabolic rate in spite of fluctuating temperatures, then rotifers that are subjected to pronounced short-term thermal variation should show the most obvious respiratory plateau.

Materials and methods

All experimental animals were derived from a single resting egg. The egg was collected from the sediment of a small meromictic pond near Gaynor Lake in Boulder

Hydrobiologia 73, 145-147 (1980).   0018-8158/80/0732-0145$00.60.
© *Dr. W. Junk b.v. Publishers, The Hague.*

County, Colorado (Pennak, 1949). The pond is typically frozen solid until early spring and usually dries by early summer. Throughout the ice-free period, a stable stratification exists. Water of high salinity is typical of the bottom layer, and water of a lower salinity is typical of the surface layer. During the study period, the specific conductivity of the bottom ranged from 25,000 to 47,500 μmho/cm (25°C) while that near the surface ranged from 11,000 to 45,000 μmho/cm (25°). Solar heat is trapped in the bottom layer where temperatures exceed 20°C early in the ice-free period. Bottom temperatures are at times 12°C higher than those at the surface. Because the pond is shallow (max. depth = 20 cm), its thermal buffering capacity is minimal. Consequently, there are sizable short-term temperature fluctuations which parallel atmospheric thermal changes. This is especially true of the water near the surface, where a change of 12°C was recorded over a five-day interval during the study period. Bottom temperature remains slightly in excess of 30°C for several weeks in the early summer.

The pond has supported large populations of *B. plicatilis* annually for at least the last five years. These populations are subjected to radical short-term temperature fluctuation as the result of the rapid heat loss and gain at the surface. Additionally, we have indirect evidence for vertical migration. Migration of only 12 cm can result in a temperature change of 10° or more.

All experimental animals were cultured in full strength sea water (35‰) and were fed *Dunaliella salina*. Subsamples from the main culture were maintained at the desired experimental temperature for a minimum of one generation prior to the respiratory measurements. Metabolic rates of individual rotifers were determined with a Cartesian diver micro-respirometer following the precautions described in Epp & Lewis (1979b).

Results

Sixty-one animals were tested at nine temperatures (15°C through 32°C). The effect of temperature on metabolism was tested for significance by analysis of variance (Snede-

cor & Cochran, 1976). The null hypothesis is that metabolism is uniform across temperature. This hypothesis was rejected ($P < 0.001$) indicating that the temperatures used in this experiment have a significant effect on metabolism. In the next step of the analysis, the method of a posteriori contrasts (Student-Newman-Keuls: Snedecor & Cochran, 1976) was used to determine which metabolic rates were distinct from each other and which are not. The results of the analysis are shown in Table 1. Metabolic rates that are not significantly different are joined by a dotted line. A noticeable respiratory plateau exists over the temperature range 20° to 28°C. Metabolic rates change significantly with temperature below 20°C and above 28°C.

The Q_{10} for any temperature interval can be calculated according to the equation of Prosser (1973). For purposes of comparison, we have computed Q_{10} values for intervals varying in breadth from 2°C to 17°C. Some of these are shown in Figure 1 along with the deduced Q_{10} vs temperature relation for instantaneous Q_{10} based on an infinitely small interval. Q_{10} is assumed equal to 1 when the difference between metabolic rate at different temperatures is not significant (Table 1). Q_{10} is equal to 1 over any temperature interval that is totally within the range 20° to 28°C. Q_{10} values for restricted temperature intervals which are totally or partially outside the range 20° to 28°C are very high (i.e., 3.4 to 4.8). Q_{10} values for broad temperature intervals which also span the range 20° to 28°C are intermediate (i.e., 1.9 to 2.4) and within the range that is considered typical for many metabolic reactions (Prosser, 1973).

Discussion

Our populations of *B. plicatilis* encounter temperatures within the thermal range 20° to 28°C over the major portion of the growing season. Over this thermal range, metabolism is not affected by temperature change. Metabolism is a reflection of the energy-requiring processes within an animal and the respiratory plateau indicates the constancy of these processes over the environmental thermal range. The data thus confirm our general hypothesis

Table 1. Results of the Student-Newman-Keuls multiple range test for the effect of temperature on metabolism. Means that are statistically indistinguishable at P = 0.05 are joined by a dotted line.

Temperature, °C	15	18	20	22	24	26	28	30	32
Mean Oxygen Consumption nl/ind/h	1.124	1.732	2.373	2.826	2.706	2.544	2.705	3.765	4.905

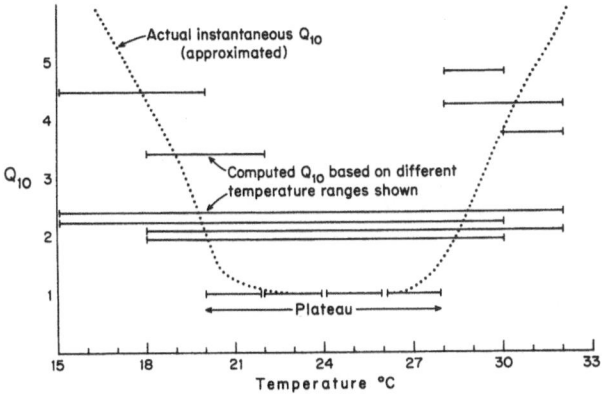

Fig. 1. Q_{10} values of *Brachionus plicatilis* over selected temperature intervals. The constancy of metabolism over the range 20° to 28°C is illustrated by the Q_{10} value of 1 over this thermal range.

that freshwater invertebrates tend to maintain a constant metabolic rate over temperature ranges that are typical of the environment.

Metabolism is affected significantly by temperature change below 20°C and above 28°C. Temperatures below 20°C are encountered only early in the growing season. High Q_{10} over the range 15° to 20°C (Fig. 1) allows metabolism and associated energy-requiring processes to be accelerated up to the preferred level when conditions permit. Stable metabolic rates between 15° and 20°C are either not possible or not desirable. Temperature above 32°C is indicative of the desiccation of the pond. The rapid increase in metabolism with temperature in this thermal range may signal this event and thereby play a role in resting egg production. Alternatively, this may simply be outside the zone of feasible metabolic control.

Q_{10} is frequently determined over relatively large temperature ranges (i.e., > 5°C) with little regard for the environmental range of the animal in question. The present study illustrates how such determinations can lead to erroneous conclusions concerning the effect of temperature on metabolism. In the present case, the Q_{10} is actually 1 over an extended thermal range which coincides with the thermal range encountered most frequently by the animals, but this would not necessarily be evident without documentation of metabolic rates at numerous temperatures within the normal environmental range.

Acknowledgements

This work was supported by National Science Foundation grant # DEB 78 05324.

References

Epp, R. W. & Lewis, W. M. Jr. 1979a. Metabolic responses to temperature change in a tropical freshwater copepod (Mesocyclops brasilianus) and their adaptive significance. Oecologia (in press).

Epp, R. W. & Lewis, W. M. Jr. 1979b. Sexual dimporhism in Brachionus plicatilis (Rotifera): Evolutionary and adaptive significance. Evolution (in press).

Pennak, R. W. 1949. Annual limnological cycles in some Colorado reservoir lakes. Ecol. Monogr. 19: 235-267.

Prosser, C. L. 1973. Comparative Animal Physiology. Phildelphia: W. B. Saunders.

Snedecor, G. W. & Cochran, W. G. 1976. Statistical Methods. Iowa State University Press, Ames, Iowa.

DENSITY-DEPENDENT SEXUAL REPRODUCTION IN NATURAL POPULATIONS OF THE ROTIFER ASPLANCHNA GIRODI

Charles E. KING[1] & Terry W. SNELL[2]

[1]Department of Zoology, Oregon State University, Corvallis, Oregon 97331, U.S.A.
[2]Division of Science, University of Tampa, Tampa, Florida 33606, U.S.A.

Received December 29, 1979

Keywords: Rotifers, *Asplanchna,* sexual reproduction

Abstract

The monogonont rotifer *Asplanchna girodi* was continuously present in daily and bi-daily plankton samples of Golf Course Pond during the spring of 1977. Two cycles of sexual reproduction occurred during this period. By isolation and culture of females it was possible to determine the reproductive type of collected individuals. Data thus obtained suggest that environmental cues associated with population density are responsible for the production of sexual females.

Introduction

Two distinct patterns of reproduction occur in monogonont rotifers. Amictic females reproduce by diploid, ameiotic parthenogenesis. Mictic females form haploid eggs and reproduce by sexual reproduction. If unfertilized, these haploid eggs develop into haploid males; if fertilized, they develop into thick-shelled resting eggs that hatch to produce amictic females (Birky & Gilbert, 1971; Ruttner-Kolisko, 1972; Robotti, 1975; Jones & Gilbert, 1976; King & Snell, 1977). Whether a given female will be mictic or amictic is generally determined during egg formation or early embryonic development (Gilbert, 1968), and is a response to environmental cues. Although it is frequently possible to identify male rotifers in the plankton (Carlin, 1943), the elucidation of these environmental cues is made difficult in nature because of the morphological homogeneity of mictic and amictic females, because of the short generation time of many species (2-4 days), and because of the difficulty in isolating effects of individual environmental variables. We present in this paper the first clear evidence from a field population that production of mictic females in *Asplanchna girodi* is associated with population density.

Laboratory studies have indicated that sexual reproduction is associated with photoperiod, or photoperiod modified by density in *Notommata copeus* (Clement &

Pourriot, 1975; Pourriot, 1963), population density in *Brachionus calyciflorus* (Gilbert, 1963), and dietary tocopherol (vitamin E) in *Asplanchna sieboldi, A. intermedia,* and *A. brightwelli* (Gilbert, 1977). In contrast, studies of *A. girodi* have concluded that mictic female production is either spontaneous (Beauchamp, 1935; Birky, 1964), or related to environmental deterioration from a variety of sources (Lechner, 1966). In our own laboratory studies of this species, mictic females have been consistently produced on a diet of tocopherol-free *Paramecium tetraurelia* under conditions of high population density. Our attempts to produce mictic females in short-term experiments at low rotifer density by using tocopherol-containing algal foods *(Chlamydomonas reinhardti, Chlorella vulgaris,* and *Euglena gracilis)* mixed with *P. tetraurelia* or with solutions of tocopherol have been unsuccessful. These results confirm the investigation of Gilbert & Litton (1978). Clearly, the initiation of sexual reproduction differs in *A. girodi* from that of the other three species of *Asplanchna* mentioned above.

The field analysis of mictic female production presented in this paper has been made possible by (1) sampling at intervals of one week or less, and (2) the cloning of 72-144 individuals from each collection to determine female types. All plankton collections were made in triplicate with a quantitative (Miller) sampler from Golf Course Pond near Tampa, Florida.

Results and discussion

Table 1 presents a summary of field data collected during the first 10 months of 1977. Density of *A. girodi* in Golf Course Pond varied from 0 to 94.2 females per liter in the 112 samples taken during this period. All samples with a density of 1.25 individuals per liter or less contained only amictic females. With a single exception, all samples having

Hydrobiologia 73, 149-152 (1980).   0018-8158/80/0732-0149$00.80.
© *Dr. W. Junk b.v. Publishers, The Hague.*

Table 1. Minimum densities at which sexual reproduction was observed and maximum densities at which no sexual reproduction was observed (± SE) by month (1977) for *Asplanchna girodi* in Golf Course Pond. Densities are expressed as females per liter.

Month	Sexual	Peak density	Minimum Sexual density	Maximum Non-sexual density
Jan.	no	0.9 ± 0.06	–	0.9 ± 0.06
Feb.	yes	2.4 ± 0.23	2.4 ± 0.23	1.2 ± 0.20
Mar.	–	0	–	–
Apr.	yes	94.2 ± 13.53	1.5 ± 0.57	0.7 ± 0.23
May	yes	19.2 ± 1.50	2.4 ± 0.62	1.8 ± 0.50
June	yes	15.4 ± 2.50	2.3 ± 0.47	1.0 ± 0.24
July	yes	5.6 ± 1.13	1.6 ± 0.28	0.8 ± 0.20
Aug.	no	0.1 ± 0.07	–	0.1 ± 0.07
Sep.	no	0.9 ± 0.14	–	0.9 ± 0.14
Oct.	yes	3.4 ± 0.31	3.4 ± 0.31	0.8 ± 0.20
		MEANS:	2.3 ± 0.28	0.9 ± 0.15

densities of 1.5 or greater contained mixtures of mictic and amictic females. Sexual reproduction thus occurs over a wide range of temperatures (11.5-31.9 C), pH (7.0-8.8), conductivity (250-380 μmhos/cm), Secchi disc readings (0.25-2.25 m), and both phytoplankton and zooplankton composition. These data suggest, but do not prove, that the occurrence of sexual reproduction in *A. girodi* is related either directly or indirectly to population density.

In April and May of 1977 plankton samples were taken at intervals of one to two days. Two cycles of mictic female production occurred during this period (Fig. 1). In each case mictic females were first observed when total female density (amictic + mictic) exceeded 1.5 individuals per liter. As density increased, the proportion of unfertilized mictic females and then fertilized mictic females increased. Population density was much higher in the April cycle (94 individuals per liter on April 15) than in the May cycle (19 individuals per liter on May 18). This difference may explain the observation that at one point 48% of the April individuals were mictic, whereas 26% was the highest frequency of mictic females in May. Subsequently, as total female density decreased, the proportion of mictic females also decreased. These observations further suggest that mictic female production in *A. girodi* is dependent upon environmental cues related to population density.

The frequencies of different female types from 38 collections are graphed in Fig. 2 as a function of population density. The linear regression of proportion amictic females has a slope of -1.314 and a coefficient of determination of 61.4%; that is, the regression accounts for 61.4% of the sums of squares attributable to population density. Linear regressions of proportions of unfertilized and fertilized mictic females have positive slopes of 0.432 and 0.882 and coefficients of determination of 34.0% and 52.1%. The differences between the slopes and intercepts of the two regressions (2.5% for the unfertilized mictic curve, and -1.0% for the fertilized mictic curve) are a consequence of the time lag occurring between first production of males and first production of resting eggs. Initially, all mictic females are unfertilized, but as their male offspring increase in number the proportion of fertilized mictic females increases.

Mictic female production in clones of *A. girodi* collected from Golf Course Pond and No Name Lake (also near Tampa, Florida) has recently been investigated in the laboratory by Gilbert & Litton (1978). Evidence from these studies confirms the non-reactivity of *A. girodi* to tocopherol. Further, data obtained by Gilbert and Litton demonstrate that mixis rates are more uniform and much higher at experimental densities of 1 individual/15 ml (67 individuals/l) than at 15 individuals/15 ml (1000/l). The lower of these two densities is within the observed range of *A. girodi* abundance in nature (see Table 1), whereas it is unlikely that *A. girodi* reaches natural densities as high as 1000/l. Unfortunately, it is not feasible to perform experiments of this type at densities lower than one individual per liter which, as suggested by the data in Table 1, would be necessary in order to avoid the production of mictic females.

These data are of considerable interest as an illustration of the diversification that can take place in stimuli inducing sexual reproduction within a single genus. *As-*

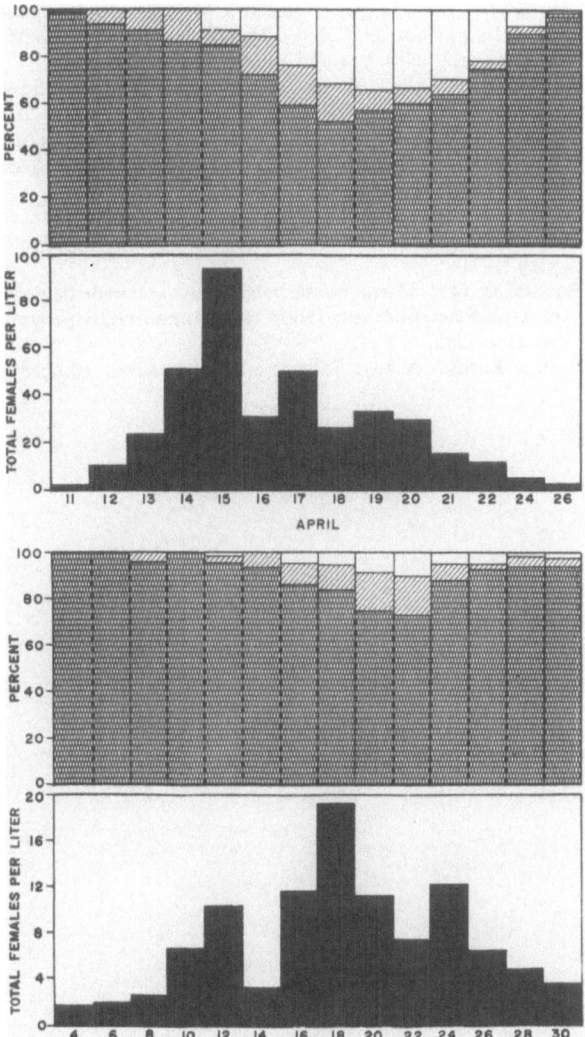

Fig. 1. Population composition and density (total females per liter) of *A. girodi* during two periods of sexual reproduction in April and May, 1977. Amictic females are indicated by cross-hatching, unfertilized mictic females by diagonal stripes, and fertilized mictic females by unmarked columns. Note scale differences in the April and May density panels.

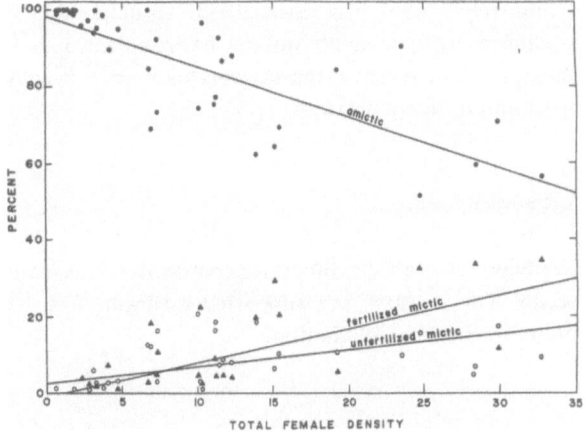

Fig. 2. Linear regression of population composition against total female density. Amictic females are designated by shaded circles, unfertilized mictic females by unshaded circles, and fertilized mictic females by shaded triangles.

planchna girodi is the first reported species in this genus whose mictic female production is known to be associated primarily with population density rather than tocopherol. In contrast, although sensitivity to dietary tocopherol may vary with population density (Birky, 1969), mictic females in both *A. sieboldi* and *A. brightwelli* appear to be produced only in the presence of tocopherol. Does the divergence in factors affecting mictic female production in *A. girodi* on one hand, and *A. sieboldi* and *A. brightwelli* on

the other reflect a dietary difference in nature? The answer to this question is probably 'no' because although all members of this genus are predatory, their diets include tocopherol-containing algae as well as animal prey (Hurlbert *et al.*, 1972; Guiset, 1977). It therefore seems unlikely that feeding differentiation is responsible for the divergent evolution in the two groups.

Why should sexual reproduction be associated with population density? There are at least three reasonable explanations. First, before fertilization can occur, male and female must meet. Since male *Asplanchna* are not strong swimmers, they cannot search for females over large areas. Rather, male-female encounter appears to be a chance event and the likelihood of encounter will increase as density increases. This mechanism was suggested by Gilbert (1967).

A second explanation is based on the fact that rotifer populations are able to grow rapidly when conditions are favorable. Conditions which are favorable for survival and reproduction of amictic females should also be propitious for mictic females. Thus by initiating mictic female production as density increases, it is likely that sexual reproduction can be completed before the environment deteriorates. In this case increasing densities are indicators of satisfactory environments for successful sexual reproduction.

Finally, production of mictic females early in the growth cycle enhances the probability that sexual reproduction will occur before the strong selection associated with parthenogenesis in this group truncates genetic variability

(King, 1977, 1980). It is reasonable to speculate that for organisms living in highly variable environments individual fitness is increased by reproducing before loss of genetic variation in the population.

Acknowledgments

We thank Dr. John J. Gilbert for comments on the manuscript. This research was supported by grants from the National Science Foundation.

References

Beauchamp, P. de. 1935. Sur la proportion numérique des deux sortes de femelles chez Asplanchna girodi. C. r. Soc. Biol. 120: 128-130.

Birky, C. W., Jr. 1964. Studies on the physiology and genetics of the rotifer, Asplanchna I. Methods and physiology. J. exp. Zool. 155: 273-292.

Birky, C. W., Jr. 1969. The developmental genetics of polymorphism in the rotifer Asplanchna. III. Quantitative modification of developmental responses to vitamin E, by the genome, physiological state, and population density of responding females. J. exp. Zool. 170: 437-448.

Birky, C. W., Jr. & Gilbert, J. J. 1971. Parthenogenesis in rotifers: The control of sexual and asexual reproduction. Am. Zool. 11: 245-266.

Carlin, B. 1943. Die Planktonrotatorien des Motalastrom. Zur Taxonomie und Ökologie der Planktonrotatorien. Medd. Lunds Univ. Limnol. Inst. 5: 1-260.

Clement, P. & Pourriot, R. Influences du groupement et de la densité de population sur le cycle de reproduction de Notommata copeus (rotifere). I. Mise en évidence et essai d'interpretation. Arch. Zool. exp. gen. 116: 375-422.

Gilbert, J. J. 1963. Mictic-female production in the rotifer Brachionus calyciflorus. J. exp. Zool. 153: 113-124.

Gilbert, J. J. 1968. Dietary control of sexuality in the rotifer Asplanchna brightwelli Gasse. Physiol. Zool. 41: 14-43.

Gilbert, J. J. 1977. Mictic-female production in monogonont rotifers. Arch. Hydrobiol. Beih. 8: 142-155.

Gilbert, J. J. & Litton, J. R., Jr. 1978. Sexual reproduction in the rotifer Asplanchna girodi: effects of tocopherol and population density. J. exp. Zool. 204: 113-122.

Guiset, A. 1977. Stomach contents in Asplanchna and Ploesoma. Arch. Hydrobiol. Beih. 8: 126-129.

Hurlbert, S. H., Mulla, M. S. & Wilson, H. R. 1972. Effects of an organophosphorus insecticide on the phytoplankton, zooplankton, and insect populations of fresh-water ponds. Ecol. Monogr. 42: 269-299.

Jones, P. A. & Gilbert, J. J. 1976. Male haploidy in rotifers: relative DNA content of nuclei from male and female Asplanchna. J. exp. Zool. 198: 281-285.

King, C. E. 1977. Effects of cyclical ameiotic parthenogenesis on gene frequencies and effective population size. Arch. Hydrobiol. Beih. 8: 207-211.

King, C. E. 1980. The genetic structure of zooplankton populations. In: Kerfoot, W. C. (ed.), The Structure of Zooplankton Communities. New England University Press, in press.

King, C. E. & Snell, T. W. 1977. Genetic basis of amphoteric reproduction in rotifers. Heredity 39: 357-360.

Lechner, M. 1966. Untersuchungen zur Embryonalentwicklung des Rädertieres Asplanchna girodi de Guerne. Roux. Arch. Entwickl. Mech. Org. 157: 117-173.

Pourriot, R. 1963. Influence du rythme nyethémeral sur le cycle sexual de quelques rotifères. C. r. Acad. Sci. Paris 256: 5216-5219.

Robotti, C. 1975. Chromosome complement and male haploidy of Asplanchna priodonta Gosse 1850 (Rotatoria). Experientia 31: 1270-1272.

Ruttner-Kolisko, A. 1972. Rotatoria. Binnengewässer 26: 99-233.

THE ROTIFER FAUNA OF THE RIVER LOIRE (FRANCE), AT THE LEVEL OF THE NUCLEAR POWER PLANTS

N. LAIR

Université de Clermont II, 63170 Aubière, France

Keywords: Rotifers, thermal pollution, running water

Abstract

In the heated affluents of nuclear power plants of the river Loire, rotifers are abundant. Cosmopolitan species are numerically dominant, but a tropical fauna is also present, among which the genus *Brachionus,* representing 20% of the total species, is best represented. From a comparison between water upstream and downstream of the power plants, it further appears that in downstream warmed-up waters, some species show an important development, but not in colder upstream waters.

Introduction

Since 1976, I have studied the zooplankton of the river Loire near nuclear power plants in project, in construction or in place (Fig. 1). Rotatoria and Cladocera are abundant, but Copepoda are rare.

During the three years of this study – in 1976 series of 12 samples were taken at 10 stations during summer; in 1977 series of 20 samples were taken, and in 1978, 25 samples, representing in all 508 analyzed samples, – were studied.

The site of Belleville-sur-Loire is situated in a sector actually without power plants (alt 136 m, width of river 300 m). The site of Dampierre-sur-Burly was canalized in the 19th century; the town of Gien lies 10 km upstreams. The river is now partially altered by the building of the central plant (alt. 123 m, width 300 m). The nuclear plant of St-Laurent-des-Eaux at 30 km from Orleans is situated on a sector, which was canalized during the building of the central (alt 85 m, width 200-325 m). The nuclear plant of Chinon is situated after the confluence of Loire and Indre. This site is now damaged by extraction of sediments (alt. 33 m, width 60-375 m).

Sampling

The problem of sampling rivers is a difficult one. However, the plankton of the Loire, before crossing the cooling system of a central, where it is exposed to thermal and mechanical stress, spends time in the calm water near the water-intake, and is allowed to reproduce in calm water again at the level of the effluents. During 1976 (10 stations studied each 2 days at St-Laurent-des Eaux during summer) results have shown that rotifera are very rare in the running water in between the two calm waters.

Flow variations renew the water of the calm biotopes slowly. A quantitative study was carried out in order to find the total number of species, both upstream, downstream, and in the lotic zones of the river as well. Quantitative samples were taken in lentic facies only by filtering 100 liters of water over a mesh of 45 μm.

Looking at the list of species, and without claiming to define indicator-species, it appears that of the great number of species found, the majority lives in eutrophic areas; few prefer oligotrophic or hypereutrophic waters, showing that the Loire is a moderately eutrophic river. The abundance of species living on substrates, like macrophytes, periphyton, psammon etc..., shows the importance of hydraulic phenomena accompanied by the carrying down of particulate material.

Species list

Most species found are planktonic, living also in littoral areas.

Asplanchnidae
- *Asplanchna brightwelli* (Gosse, 1850)
  Asplanchna herricki (De Guerne, 1888)
  Asplanchna priodonta (Gosse, 1850)
- *Asplanchnopus sp.* (De Guerne, 1888)

Hydrobiologia 73, 153-160 (1980).   *0018-8158/80/0732-0153$01.60.*
© *Dr. W. Junk b.v. Publishers, The Hague.*

Fig. 1. Situation of the nuclear power plants of the river Loire.

Brachionidae
– Anuraeopsis fissa (Gosse, 1851)
– Brachionus angularis (Gosse, 1851)
  Brachionus bennini (Leissling, 1924)
  Brachionus bidentata (Anderson, 1889)
  Brachionus calyciflorus (Pallas, 1766)
  Brachionus diversicornis (Daday, 1883)
  Brachionus dolabratrus (Harring, 1915)
  Brachionus havanaensis (Rousselet, 1911)
  Brachionus leydigi (Cohn, 1862)
  Brachionus nilsoni (Ahlstrom, 1940)
  Brachionus quadridentatus (Hermann, 1783)
  Brachionus rubens (Ehrenberg, 1838)
  Brachionus urceolaris (O. F. Muller, 1773)
– Epiphanes macrourus (Daday, 1894)
– Euchlanis dilatata (Ehrenberg, 1832)
– Kellicottia longispina (Kellicott, 1879)
– Keratella cochlearis (Gosse, 1886)
  Keratella quadrata (O. F. Muller, 1786)
  Keratella tecta (Gosse, 1886)
  Keratella valga (Ehrenberg, 1834)
– Lepadella ovalis (O. F. Muller, 1826)
  Lepadella patella (O. F. Muller, 1786)
– Mytilina crassipes (Lucks, 1912)
  Mytilina mucronata (O. F. Muller, 1773)
– Notholca acuminata (Ehrenberg, 1832)
  Notholca caudata (Carlin, 1943)

Notholca labis (Gosse, 1887)
Notholca squamula (O. F. Muller, 1786)
– Platyas quadricornis (Ehrenberg, 1832)
– Rhinoglena sp. (Ehrenberg, 1853)

Conochilidae
– Conochilus sp. (Ehrenberg, 1834)

Lecanidae
– Lecane bulla (Gosse, 1886)
  Lecane closterocerca (Schmarda, 1895)
  Lecane flexilis (Gosse, 1889)
  Lecane luna (O. F. Muller, 1776)
  Lecane lunaris (Ehrenberg, 1832)
  Lecane papuana (Murray, 1913)
  Lecane sp. (Nitzsch, 1827)

Notommatidae
– Cephalodella gibba (Ehrenberg, 1832)

Synchaetidae
– Polyarthra dolichoptera (Idelson, 1925)
  Polyarthra vulgaris (Carlin, 1943)
– Synchaeta sp. (Ehrenberg, 1832)

Testudinellidae
– Filinia brachiata (Rousselet, 1901)
  Filinia cornuta (Weisse, 1847)
  Filinia longiseta (Ehrenberg, 1834)
  Filinia terminalis (Platte, 1834)
– Horaella sp. (Donner, 1949)
– Pompholyx sulcata (Hudson, 1885)
– Testudinella discoidea (Ahlstom, 1938)
  Testudinella mucronata (Gosse, 1886)
  Testudinella patina (Hermann, 1783)

Trichocercidae
– Trichocerca brachyura (Gosse, 1851)
  Trichocerca elongata (Gosse, 1886)
  Trichocerca longiseta (Schrank, 1802)
  Trichocerca pusilla (Jennings, 1903)
  Trichocerca similis (Wierzejski, 1893)
  Trichocerca stylata (Gosse, 1851)
  Trichocerca taurocephala (Hauer, 1931)
  Trichocerca tenuior (Gosse, 1886)
  Trichocerca sp. (Lamarck, 1801)
Some species living in littoral areas, or on the bottom, have also been collected:

Brachionidae
- *Colurella adriatica* (Ehrenberg, 1831)
  Colurella colurus (Ehrenberg, 1830)
  Colurella sp. (Bory de St-Vincent, 1824)
- *Trichotria pocillum* (O. F. Muller, 1776)
  Trichotria tetractis (Ehrenberg, 1831)
  Trichotria sp. (Bory de St-Vincent, 1827)

Dicranophoridae
- *Encentrum lutetiae* (Harring u. Myers, 1928)
  Encentrum sp. (Ehrenberg, 1838)

Flosculariidae
- *Sinantherina socialis* (Linne, 1758)

Notommatidae
- *Notommata sp.* (Ehrenberg, 1830)

Qualitative data

If we consider only the upstream stations, the number of species (monthly means) is low to zero in the beginning of 1978; it increases from Belleville to Dampierre from June onwards, while at St-Laurent it shows a clear tendency to become low during summer; this phenomenon is accentuated at Chinon. There is a maximum in June, then a little depletion in July, and another maximum at the end of July (St-Laurent, Chinon) or in early August (Belleville, Dampierre). Probably favored by an extra long low water level, a certain number of species live until December, contrasting with their early disappearance the winter before. Each site studied has a relative originality in the phenology of its species.

Among the species found, some were collected in a small number of samples only. In the interpretation of results, concerning both seasonal variation and variation between upstream and downstream stations, only those species sufficiently common in my samples have been

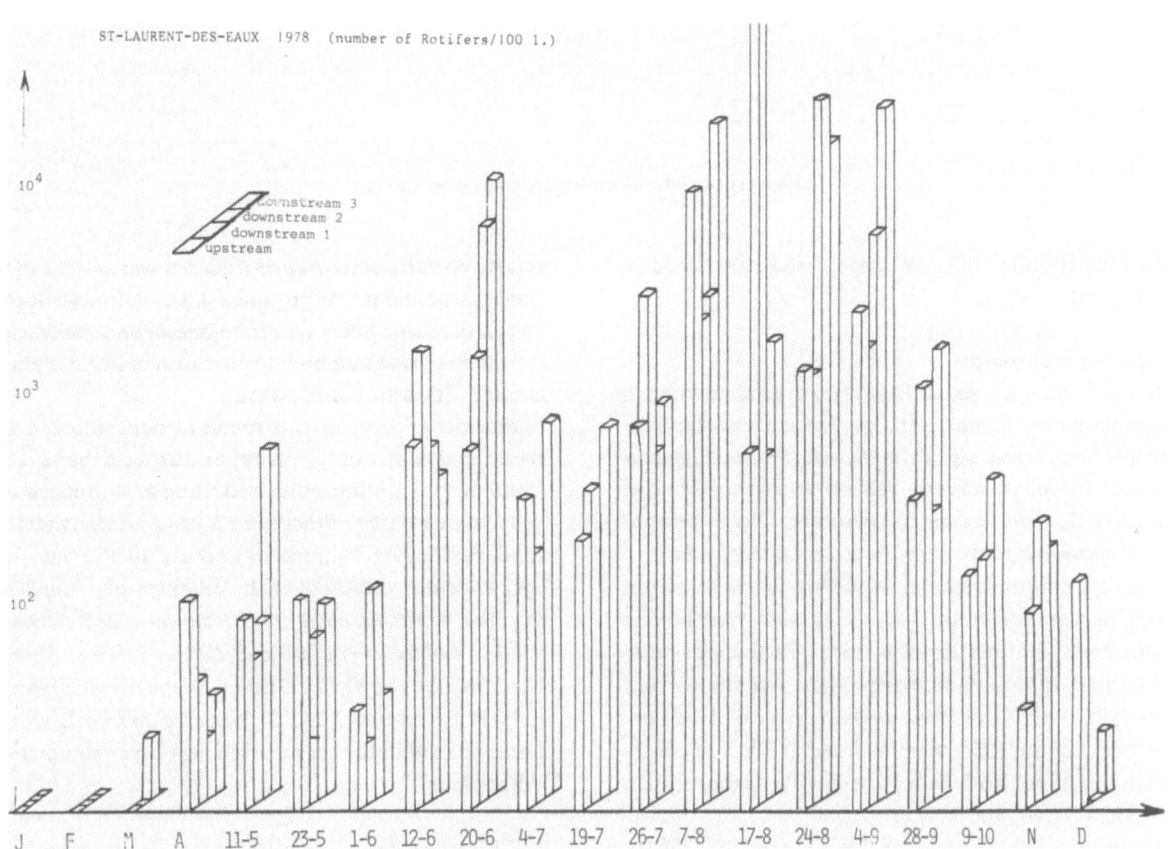

Fig. 2. Population density of Rotifers at Saint-Laurent-de-Eaux.

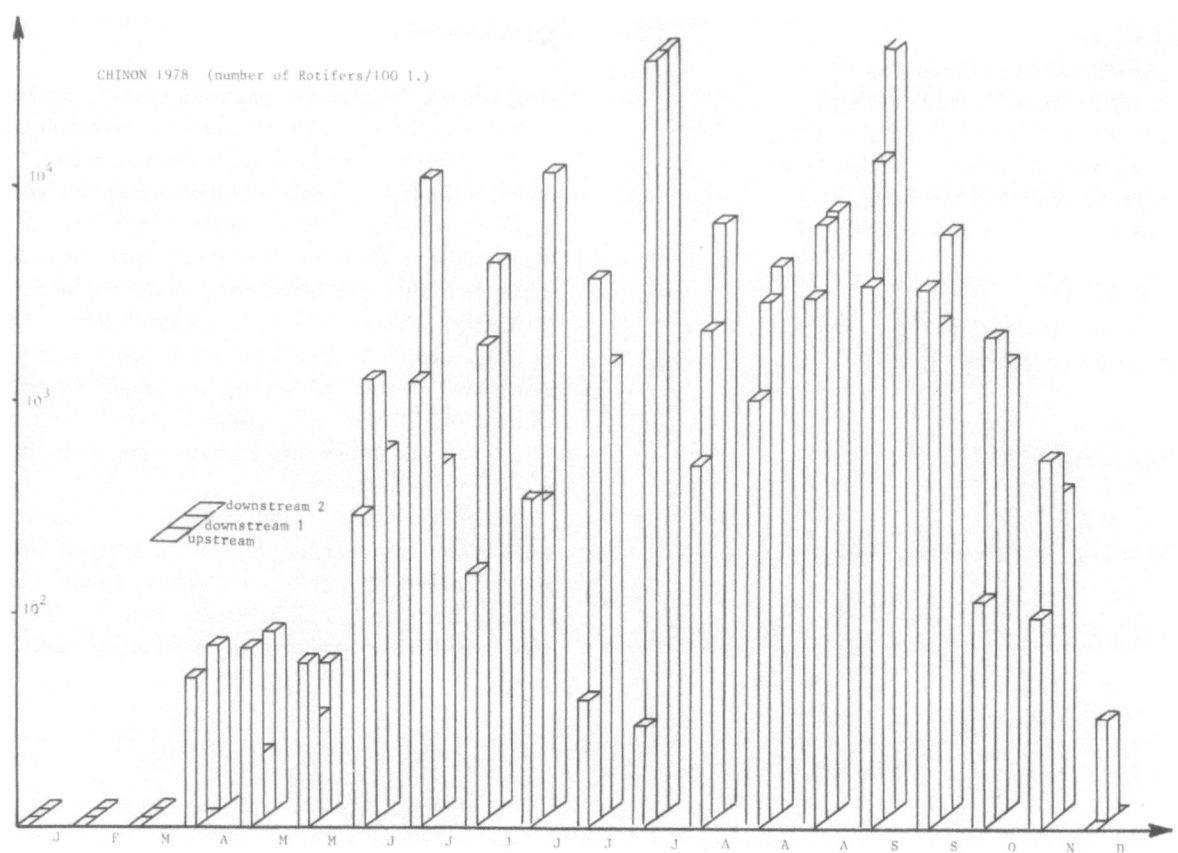

Fig. 3. Population density of Rotifers at Chinon.

taken into account. Most of these species also occur at each central.

a. Seasonal occurrence

Although sometimes sparse in winter, a certain number of common species is practically present around the year. They are: *Brachionus angularis, B. calyciflorus, B. quadridentatus, Euchlanis dilatata, Filinia longiseta, Keratella cochlearis, K. tecta, Lecane closterocerca, Polyarthra vulgaris, Pompholyx sulcata, Trichocerca taurocephala.*

These are eurytopic species, found in eutrophic environments, except *Polyarthra vulgaris* which lives in oligtrophic areas (Ruttner-Kolisko, 1974). *Trichocerca taurocephala* lives in psammon, periphyton, and among vegetation (Koste, 1978); these biotopes are typical of the Loire river. Most of the rotifers enumerated above, are cosmopolitan but have also been reported in tropical or equatorial areas (Green, 1967).

In summer, only 2 species are found at the four nuclear power plants simultaneously: *Anuraeopsis fissa* and *Sinan-*

therina socialis. Anuraeopsis fissa is a warm stenothermal species, abundant in tropical lakes (Ruttner-Kolisko, 1974); the same holds true for *Sinantherina socialis* which is found in great numbers downstream of nuclear plants in summer (Pourriot *et al.,* 1972).

Brachionus urceolaris, is found at Belleville and Dampierre in summer only, but appears around the year at St-Laurent and Chinon, thus reflecting an influence of the warming of waters. *Brachionus leydigi* is totally restricted to Belleville. At Dampierre appear, all the year, some species scarce or absent at the other statons: *Cephalodella gibba, Keratella quadrata, Lecane luna,* and *Trichocerca similis. Colurella colurus* and *Lecane lunaris,* abundant here, are rare elsewhere. Further, *Filinia terminalis* appears only at St-Laurent, and *Trichocerca pusilla* lives at St-Laurent around the year and is seen only in summer at the other places.

b. Differences between upstream and downstream

During the exceptional low-water of the summer of 1976

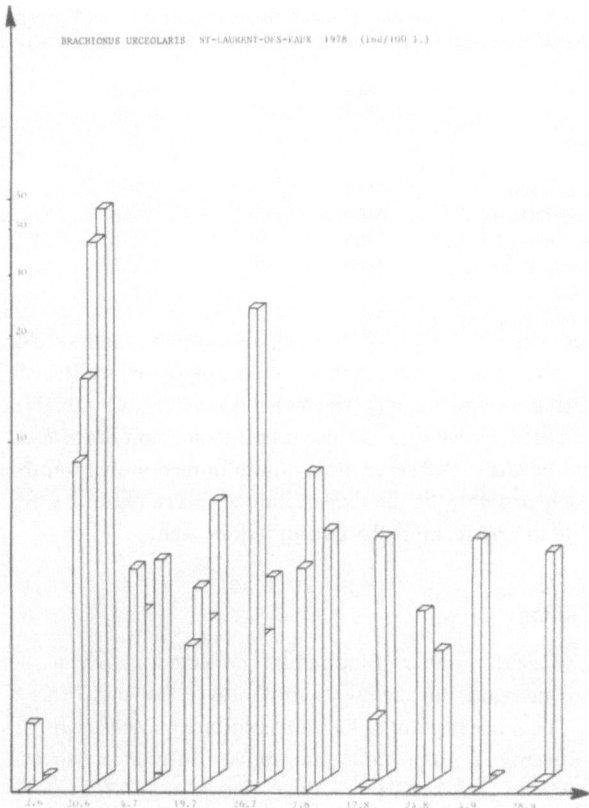

BRACHIONUS URCEOLARIS ST-LAURENT-DES-EAUX 1978 (ind/100 l.)

Fig. 4. Temporal distribution of *Brachionus urceolaris* (St-Laurent-des-Eaux).

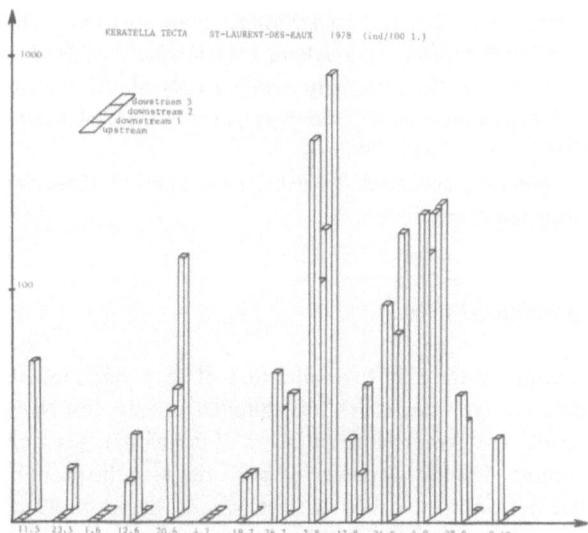

KERATELLA TECTA ST-LAURENT-DES-EAUX 1978 (ind/100 l.)

Fig. 5. Temporal distribution of *Keratella tecta* (St-Laurent-des-Eaux).

the Loire plankton at St-Laurent contained *Brachionus dolabratus* and *Brachionus bidentata,* which are warm water species (Lair *et al.,* 1978). We also observed here that *Filinia terminalis* was abundant at the upstream station, while *Brachionus bennini, Lecane papuana, Lecane lunaris, Conochilus* sp. and *Sinastherina socialis* were abundant downstream. In 1977 and 1978, we have neither collected *Brachionus dolabratus* (American species!) nor *Lecane papuana* (pantropical species) and we have rarely taken *Brachionus bidentata. Lecane bulla* has not been collected during these two years; it lives in a temperature range of 10°C to 32°C (Koste, 1978), and is also abundant in equatorial lakes where it may represent up to 40% of all rotifers (Green, 1967). It is frequent in Europe, but its bloom in the Loire was probably caused by the hot summer of 1976.

Qualitative data show that some species first appeared downstream, and (or) where waters are warmed up disappeared upstreams first. This phenomenom is very clear in *Brachionus.* At St-Laurent, *Brachionus urceolaris* first appeared downstream in 1977; in 1978, it disapperared up-

stream in mid-August and persisted downstream till October. At Chinon, a station situated more downstream, *Brachionus angularis* was present downstream in February and March 1978, to appear upstream in April only. In the same way, *Brachionus calyciflorus* appeared downstream in February and disappeared first in waters which had not been warmed up. *Brachionus quadridentatus* similarly lived longer downstream. *Brachionus urceolaris* has been taken downstream from September to November 1978, while it disappeared upstream at the end of August.

At St-Laurent, *Filinia terminalis* disappeared upstream mid-August and persisted downstream till the end of September. Similar observations were made in *Keratella cochlearis* (St-Laurent), *Lecane closterocerca* (St-Laurent), *Euchlanis dilatata* (Chinon), *Filinia longiseta* (Chinon) and *Trichocerca taurocephala* (Chinon).

With regard to *Filinia terminalis,* observations made in summer 1976 do not agree with those of 1977 (when its distribution was the same up and downstream) and those of 1978. However, the hydraulic regime was very different during these three years: exceptional low water during 1976, but high waters from 1977 till July 1978. As a consequence, it was not possible to compare the thermal optimum. This species was inhibited by the elevated temperatures reached downstream during 1976, while this was no longer true in the two following years.

Among typical summer species, either at St-Laurent or at Chinon, *Anuraeopsis fissa* (scarce in 1977) appeared downstream first and lasted longer than upstream. The same observation was made in *Sinastherina socialis*. *Trichocerca pusilla,* present in summer only at Chinon (as well as at Belleville and Dampierre) occurred at St-Laurent from April to October.

Generally speaking, a much greater number of species occurred downstream.

Quantitative data

Because of the excessive water flow of 1977, quantitative data are expressed as relative abundances only. For 1978, results are expressed as numbers of individuals per unit volume. The low water period coincides with the productive period, so that two distinct periods, based on water flow can be distinguished for that year.

a. Population density
Species that had been developing since June showed low numbers during high waters periods. Average counts at upstream stations are shown in Table 1.

Table 1. Average counts at upstream stations (explanation in text).

	High water level	Low water level	
		− S	+S
Belleville	343	7970	8935
Dampièrre	175	11480	11980
St-Laurent	471	10450	25900
Chinon	233	8450	11660

Number. m^{-3} with (+S), and without (−S) *Sinantherina socialis*.

Densities are small, if compared with data of Ruttner-Kolisko (1974) who gives 200 to 500 ind/l. During low water, density is of the order of 10^4 ind/m^3 (Figs. 2 and 3), and it is even higher if we take into account *Sinantherina socialis*. Such a densities are of the same order as those found in Lake Erie by Davis (1969).

b. Variation between upstream and downstream stations

St-Laurent des eaux

The mean number of individuals collected is presented in Table 2. There is a difference between upstream and the point immediately downstream the nuclear power plant (1): here planktonic Rotifers are only half as numerous as upstream, reflecting a probable inhibition in the sector of

Table 2. Mean number of individuals in upstream and downstream stations at St. Laurent des Eaux (explanation in text).

	mean ind/m^3 − S	%	mean ind/m^3 + S	%
upstream	10450	21	25900	17
downstream 1	6160	12	47500	32
downstream 2	9900	19	27970	19
downstream 3	14800	48	47280	32

the effluent. After this decrease, the number of individuals climbs back to the number found upstream, while even further downstream (3) the mean number of individuals is doubled. However, if we also take into account *Sinantherina socialis,* we observe a maximum immediately after the nuclear plant, the number of individuals remaining superior to upstream at the two others as well.

Chinon

The mean number of individuals collected upstream and downstream, during low water period is shown in Table 3.

The nuclear plant of Chinon is situated downstream St-Laurent; no important quantitative difference between upstream and downstream data appears here if *Sinantherina socialis* is disregarded. However, there is an important quantitative difference if we include *S. socialis*, the maximum being situated immediately after the nuclear power plant.

Specific variations
At St-Laurent, *Brachionus urceolaris* (Fig. 4), *Keratella tecta* (Fig. 5), and *Euchlanis dilatata* appear and disappear in a remarkable succession. At Chinon, this is the case for *Brachionus calyciflorus, B. urceolaris* (Fig. 6), *Keratella cochlearis* and *K. tecta*. The distribution of *Sinantherina socialis* is remarkable also: at the two nuclear power plants, it is rare upstream compared to downstream (Figs. 7 and 8).

Table 3. Mean number of individuals in upstream and downstream stations at Chinon (explanation in text).

	mean ind/m^3 − S	%	mean ind/m^3 + S	%
upstream	8450	32	11660	8
downstream 1	8550	33	79540	55.5
downstream 2	9120	35	52310	36.5

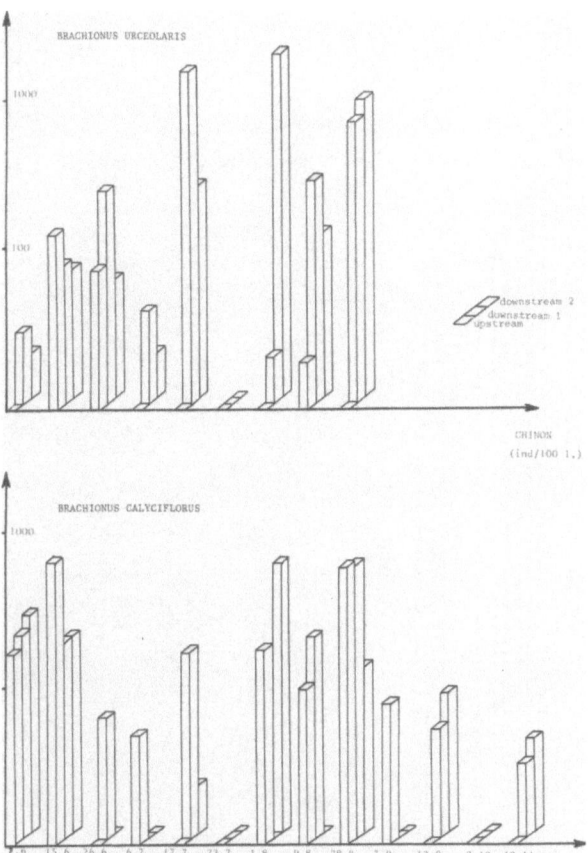

Fig. 6. Temporal distribution of *Brachionus urceolaris* and *Bra-chionus calyciflorus* (Chinon).

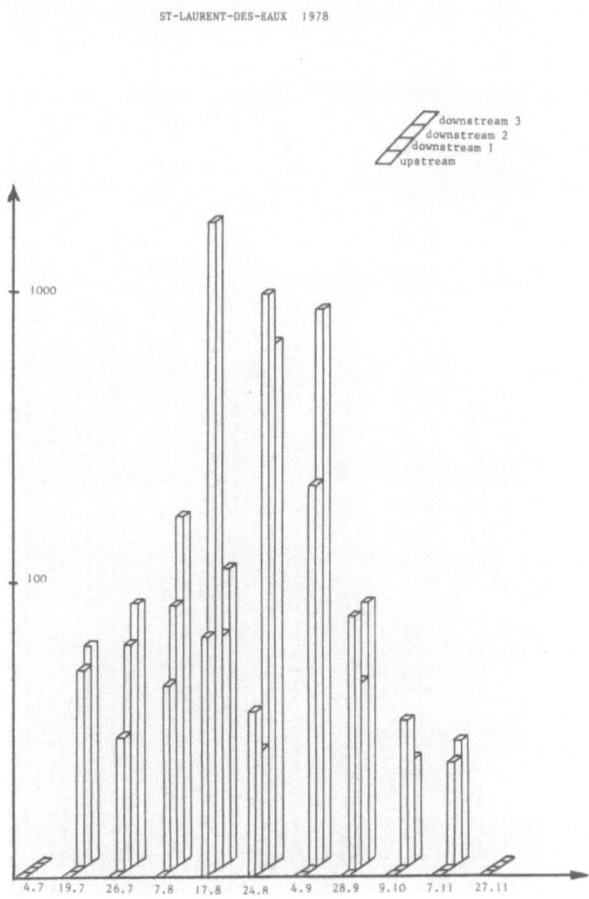

Fig. 7. Temporal distribution of *Sinantherina socialis* (St-Laurent-des-Eaux).

Conclusion

The river Loire, near the nuclear power plants contains planktonic Rotifera, with a majority of cosmopolitan species; along the course of the river, there is a specific distribution between stations. Some of the species encountered are abundant in warm areas only; e.g. *Lecane papuana,* limited to the tropics and subtropics (Edmondson & Hutchinson, 1934). Green (1972), in a study on latitudinal zonation of planktonic rotifers, indicates that *Brachionus* is more important in the tropics than in temperate and northern areas. In the Loire, *Brachionus* represents 20% of the species total. This is a comparable to the river Sokoto (Nigeria) (Green, 1960). In the Loire, this genus was very sensitive to increases in temperature; this was also true for *Keratella* and, perhaps, for other species as well. *Sinanthe-rina socialis* and *Anuraeopsis fissa* are always found during summer and at each station.

From a comparison between upstream and downstream,

it appears that in warmed-up waters, some species have a more important development, the phenology of the populations increases and their diversity as well. We even found three species which had never been reported from Europa: *Lecane papuana, Brachionus dolabratus,* and *Brachionus havanaensis.* They constitute a 'man-made' fauna. Resting eggs transported passively may explain their presence in this river (the same observations have been made in macrophytes). The river Loire locally provides a preadapted 'tropical' environment for them but further observations are necessary, in different meterological and hydraulic conditions, in order to confirm these observations. This also explains the important development of *Sinantherina socialis* with major outbursts downstreams.

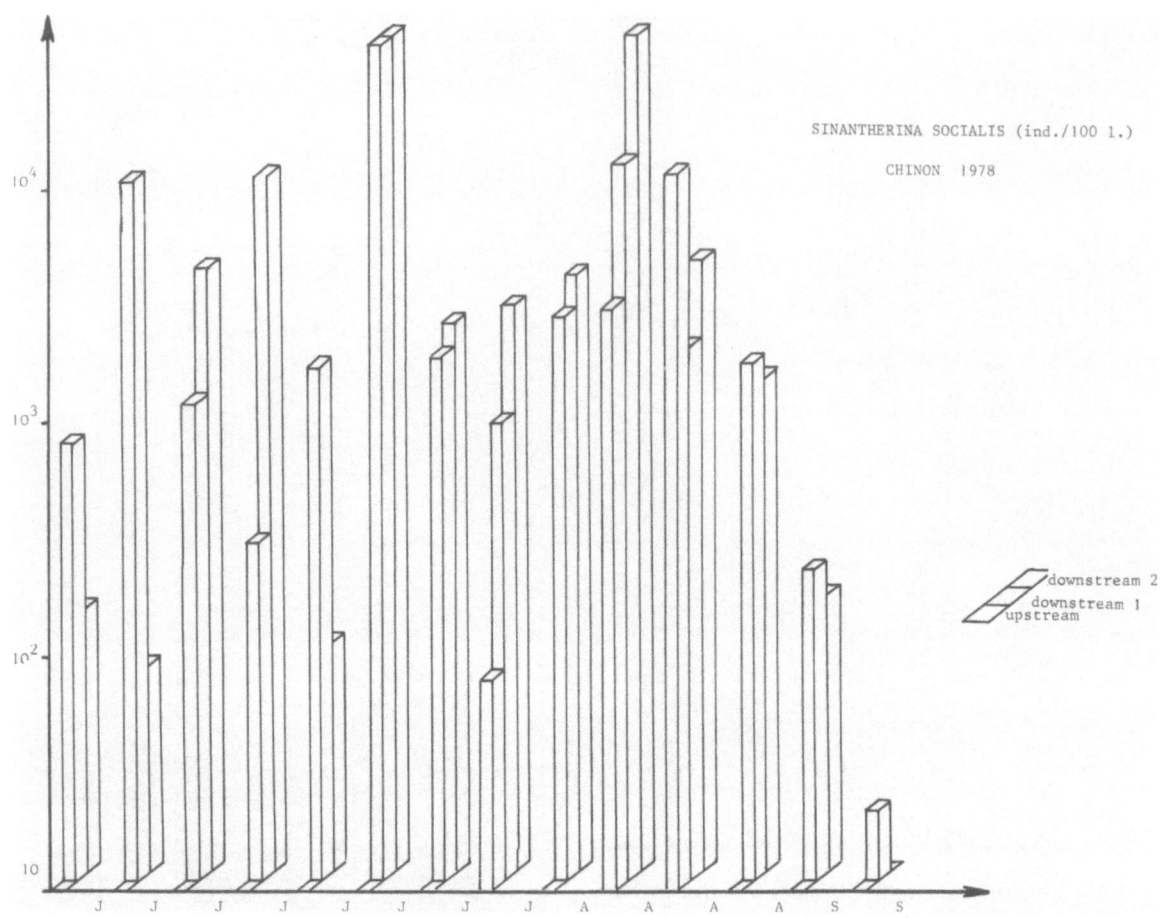

Fig. 8. Temporal distribution of *Sinantherina socialis* (Chinon).

Acknowledgements

This study was requested by E.D.F. and financed by 'Region d'equipement de Tours'. I thank Orleans's S.R.A.E. for taking samples, M. Monpeyssin for technical collaboration and R. Pourriot for help in identification of species.

References

Davis, C. C. 1969. Seasonal distribution, constitution and abundance of zooplankton in Lake Erie. J. Fish. Res., 26: 2460-2476.

Edmondson, W. T. & Hutchinson, G. E. 1934. Report on Rotatoria. Yale North India Expedition. Mem. Conn. Acad. Sci., 10: 153-186.

Edmondson, W. T. 1944. Ecological studies of sessile Rotatoria. Part I. Factors affecting distribution. Ecol. Monogr., 11: 21-66.

Green, J. 1960. Zooplankton of the river Sokoto. The rotifera. Proc. zool. Soc. Lond., 135: 491-523.

Green, J. 1967. Association of rotifera in zooplankton of the lake sources of the white Nile. J. Zool., 151: 343-378.

Green, J. 1972. Ecological studies on crater lakes in west Cameroon. Zooplankton of Barombi Mbo, Mboandong, Lake Kotto and Lake Soden. J. Zool., 166: 283-301.

Lair, N., Millerioux, G. & Restituito, F. 1978. Examen physicochimique et répartition du plancton de la Loire, en période d'étiage (été 1976) au niveau de la centrale nucléaire de St-Laurent-des-Eaux. Cah. Hydrobiol. Montereau, 6: 53-80.

Pourriot, R., Rouyer, G. & Peltier, M. 1972. Prolifération des Rotifères épiphytes et pollution thermique dans la Loire. Bull. fr. Piscic., 244: 111-118.

Koste, W. 1978. Rotatoria. Borntraeger. Berlin., 2 vols: 673 pp., 234 plates.

Ruttner-Kolisko, A. 1972. Rotatoria. Binnengewässer, 26: 99-234.

TEN YEARS QUANTITATIVE DATA ON A POPULATION OF RHINOGLENA FERTÖENSIS (BRACHIONIDAE, MONOGONONTA)

A. HERZIG

Institut für Limnologie, Österreichische Akademie der Wissenschaften, A-1090 Wien, Berggasse 18/19, Austria

Keywords: Rhinoglena, winter plankton, shallow lake, abundance, resting eggs, distribution

Abstract

Some information, based upon a ten years study, is given on the ecology of *Rhinoglena fertöensis,* a cold stenothermic rotifer. The largest numbers were always found during winter months under ice. A remarkable increase in population density is recorded throughout the investigation and is explained by the good quality of the food supply and a parallel increase in its abundance. The dynamics of this species appear to be closely connected with water temperature gradients. The distribution of *Rhinoglena fertöensis* is discussed in connection with water chemistry.

Introduction

The cold stenothermal genus *Rhinoglena* inhabits the pelagial of small and/or shallow lakes (Ruttner-Kolisko, 1974). It appears from September/October to May and, up to now, only in athalassic, saline waters of low concentration. The first description is found in Varga (1929) from Neusiedlersee. In Germany, it was first found by Remane between 1934-1936 in Süsser See, Central Germany according to Althaus (1957), who confirmed this record during her studies on the rotifer fauna of nine athalassic, saline water bodies in the surroundings of Halle/Saale. Löffler (1959) and Peschek (1961) observed *R. fertöensis* in small, shallow and saline lakes east of Neusiedlersee (Seewinkel, Burgenland, Austria) and Zakovsek (1961) noticed it again in Neusiedlersee during her studies from 1950-1952. According to Bartoš (1959) it is probable that *R. fertöensis* could be found in water bodies of Southern Slovakia.

All authors agree that *R. fertöensis* can be defined as a cold stenothermal species with a main occurrence in the winter months. In Neusiedlersee it is a typical planktonic animal found in the open lake, rarely occurring near the *Phragmites* belt and occasionally appearing in small ponds within the reed belt, as long as these are connected by channel to the open lake.

A short survey of the species composition and abundance of rotifers in Neusiedlersee (1968-1975) already exists (Herzig, 1979). The aim of this paper is to report on the changes in the population dynamics of *R. fertöensis* over ten years together with possible reasons for these.

Material and methods

The population studied has formed the major component of the winter plankton of Neusiedlersee since 1973. A detailed description of the limnology of this lake can be found in Löffler (1979). Some chemical date can be seen in table 2.

The 1968/69 winter samples were taken at three different depths using a Ruttner sampler but in 1970/72 a hand pump was used and since 1975 vertical net hauls (30 µm mesh size) through 1.55 m maximal depth. The latter had the advantage of an integrated sample without clogging of the net since most of the particles are smaller than 30 µm. The sample interval varied between 2 weeks in autumn and spring and 4 weeks in winter. All samples were counted in petri dishes under a dissecting microscope. More details about sampling, subsampling and counting techniques can be found in Herzig (1974, 1979).

In order to obtain a reasonable estimate of the population in the lake 8-20 separate stations have been sampled

Hydrobiologia 73, 161-167 (1980). 0018-8158/80/0732-0161$01.40.
© Dr. W. Junk b.v. Publishers, The Hague.

and from these counted samples a mean for the whole lake was calculated. For the years 1968-1972, samples were pooled, but for most sampling occasions since 1972 95% confidence limits can be fitted to the calculated mean population estimates. All calculations were performed with log-transformed values.

Results and discussion

Abundance

Data on the abundance of *R. fertöensis,* its numbers of amictic and mictic eggs, the prevailing water temperature and the duration of ice cover are summarized in Fig. 1. The points for 1951/52 are values calculated from Zakovsek (1961). 1967-1979 data resemble the results of the entire study period. In 1951/52, 1967/68 and 1971/72 *R. fertöensis* can be found mainly from January to April with the maximum occurrence in February or April. The highest densities reached are between 300 and 3500

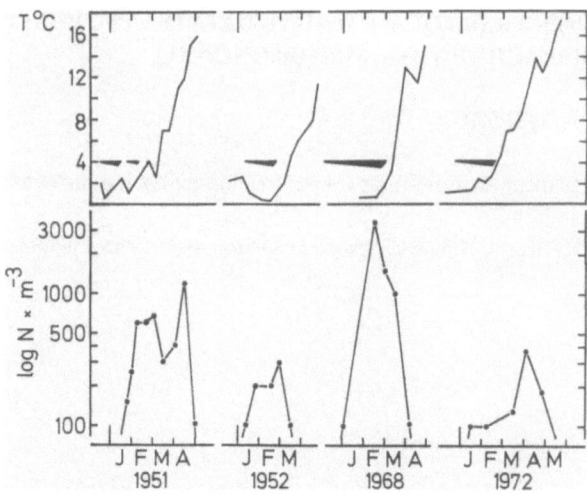

Fig. 1. Water temperature, duration of ice cover, numbers of amictic and mictic eggs, and population densities of *Rhinoglena fertöensis* (vertical lines indicate the 95% confidence interval, horizontal bars the duration of ice cover; shaded area = number of mictic eggs; solid line = total numbers; dotted line = number of amictic eggs).

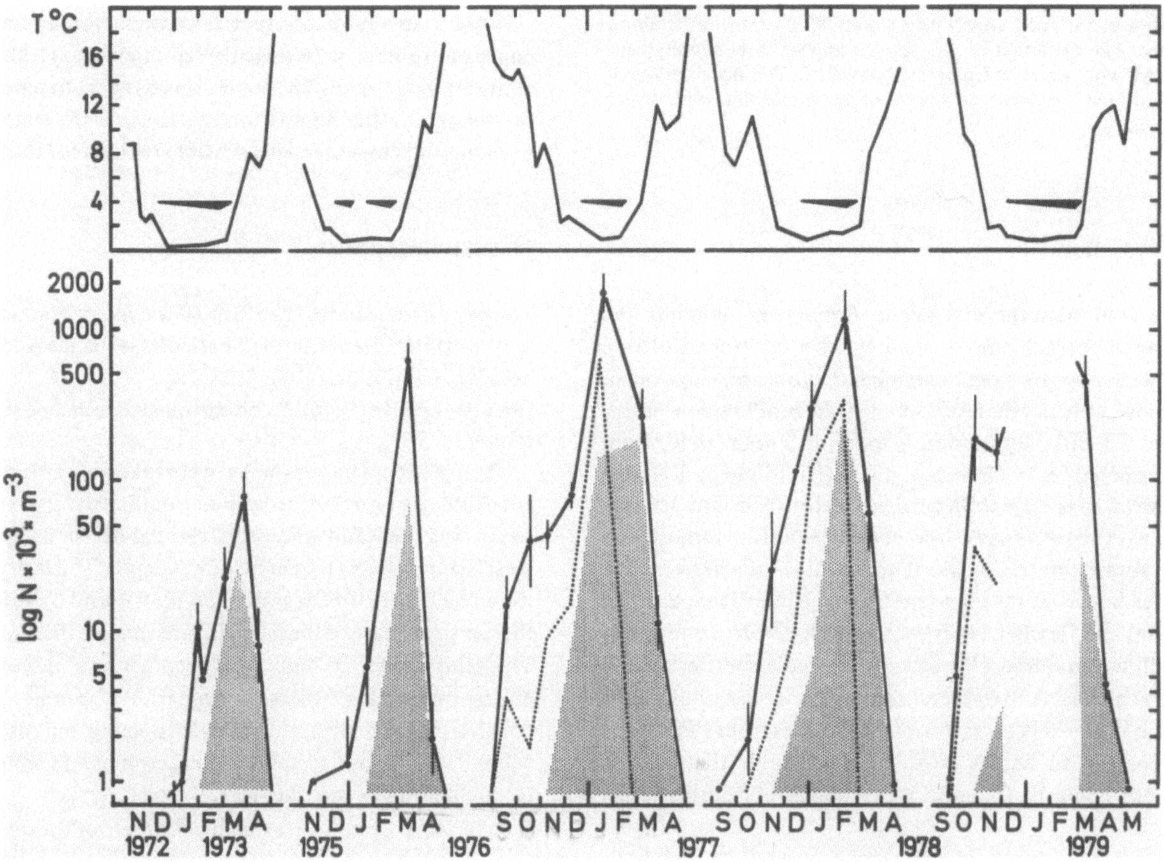

ind·m^{-3}, which represents between 0.8 and 3.3% of the total rotifer fauna. Since winter 1972/73 *R. fertöensis* has become more and more abundant and nowadays it dominates the winter plankton and is even of some importance in autumn and spring (see Table 1, Fig. 1). As is obvious from Fig. 1 the maximum densities vary between 600 and 1800 ind.l^{-1} during the last years and this represents a 600-6000 fold increase over earlier years. Peak numbers usually occur during the period January-March, but actually depends on the time of onset of the hatching of the resting eggs as well as the environmental conditions during the initial growth.

Fig. 2 demonstrates that the tremendous increase in population numbers of *R. fertöensis* over the last ten years parallels a similar increase in Chlorophyll a biomass. *Rhinoglena* is described as a filter feeder which prefers small, unicellular algae such as *Cryptomonas* spp. and detritus (Varga, 1929; Ruttner-Kolisko, 1974; Pourriot, 1977). During the winter months in Neusiedlersee the phytoplankton consists mainly of *Cyclotella* spp. plus small *Navicula*, small *Nitzschia* spp., *Oocystis lacustris*, *Chroococus minutus*, *Cryptomonas* spp. and *Rhodomonas lacustris*. These algae range in size between 5-21.8 μm in length and 2.8-13 μm in width (Dokulil, 1979) and seem to form a good supply for *R. fertöensis*, especially at the higher concentrations occurring since winter 1975/76. Moreover, an increasing amount of bacteria (acc. to Gunatilaka pers. comm., *Rhodopseudomonas gelatinosa*: 0.5-1.2 μm) is evident since 1977. That means the good food quality, the improving food quantity, and the absence of any predator, especially since 1975 (acc. to Herzig, 1979, great decrease of cyclopoids) has formed favourable conditions for the growth of *R. fertöensis*.

In parallel to the higher population abundances, an increasing number of resting eggs is produced, forming a powerful and vital input for the following autumnal amictic generation.

Population dynamics

The development of the *Rhinoglena* population can be split into 3 phases:
  (a) hatching of the resting eggs,
  (b) exponential growth phase until reaching the maximum density and switching to sexual reproduction, and
  (c) rapid decrease and complete disappearance.
Using the instantaneous coefficient of increase or decrease of a population (r) derived by applying the continuous model

$$N_T = N_0 e^{rt} \qquad\qquad r = \frac{\ln N_T - \ln N_0}{t}$$

where N_T and N_0 are densities at time T and 0, t = time interval between time T and 0. Fig. 3 illustrates these three phases: high rates of increase in September, October or as

Table 1. Relative abundance of *Rhinoglena fertöensis* (percentage of total rotifer numbers).

	autumn	winter	spring
1950/51/52	–	1.0	0.2
1968/69	–	0.8	–
1970/71	–	0.3	0.08
1971/72	–	3.3	3.0
1972/73	0.4	20.0	9.8
1975/76	1.8	71.2	16.7
1976/77	30.4	99.6	14.2
1977/78	27.5	97.5	7.6
1978/79	22.2		11.0

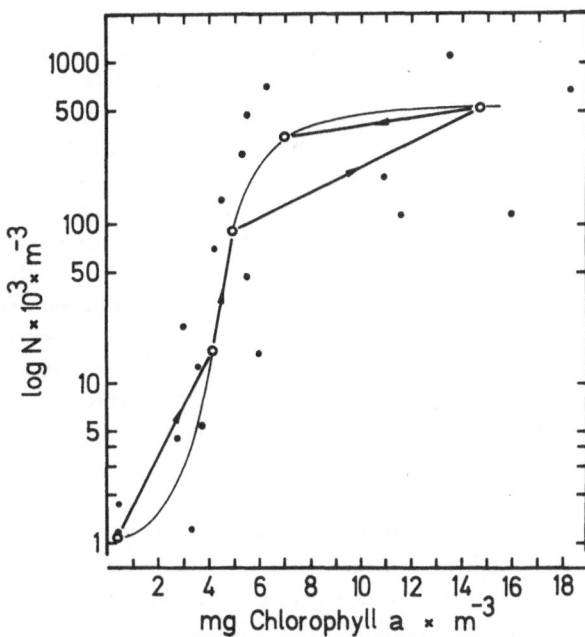

Fig. 2. Relationship between Chlorophyll *a* content and total numbers of *Rhinoglena fertöensis* (1967/68, 1972/73, 1975/76, 1976/77, 1977/78) (o mean values for the period December-March, ● values observed at the different sampling dates; arrows indicate the trend from 1967/68-1977/78).

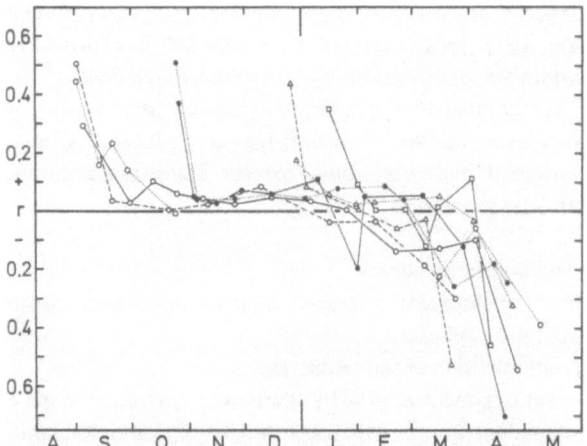

Fig. 3. *Rhinoglena fertöensis:* intrinsic rate of population increase and decrease (o———o 1977/78, o – – – o 1976/77, o...o 1978/79, ●———● 1972/73, ●...● 1975/76, △– – –△ 1967/68, △...△ 1971/72, □———□ 1951.

late as December correspond to the hatching of the resting eggs; this is followed by a rather long period of positive r-values resulting from parthenogentic reproduction; in spring time a rapid decrease with high negative r-values can be seen.

Data on the developmental times of the amictic eggs are so far not available, but are required to get an estimate of birth-and death rates. Applying literature data on the duration of the embryonic developmental time (pooled regression for rotifers in Bottrell *et al.,* 1976) for the calculation of these instantaneous coefficients, one obtains high mortality rates during the first period of autumnal growth and extremely low ones during the period of ice cover. Assuming that these derived values are in the right order of magnitude, the high death rates could be attributed to mechanical demage caused by the suspended material originating from wind-induced mixing. However, this must remain speculative until measured developmental data for the species are available.

Hatching of the resting eggs

Up to now the factors controlling the hatching of resting eggs of rotifers are largely unknown and critical examinations under controlled experimental conditions have not been performed (Gilbert, 1974). Nipkow (1961) was able to induce hatching by transferring eggs from cold to warmer media. The same treatment was applied to *R. fertöensis.* Resting eggs collected from the natural population were incubated at 3.8°C and, after a dormancy of 153-180

days, 32% to 63% hatching was induced by a change in temperature from 3.8°C to 4.5°C or from 3.8°C to 6.3°C respectively. According to Ruttner-Kolisko (1974), the duration of the resting period does not appear to be fixed; she has noted that some cultured resting eggs of approximately the same age will hatch after a minimum period of dormancy whereas the rest hatch gradually and irregularly or not at all. For *R. fertöensis* 5-6 months could be considered as such a minimum latent period. The observations are in good agreement with Nipkow (1961), who hatched the resting eggs of 30 species of planktonic rotifers from Zürichsee and found that the majority needed a dormancy of 4-6 months.

The development time of the *R. fertöensis* resting eggs was 12-16 days in 4.5°C and 6-8 days at 6.3°C. For comparison, Nipkow (1961) found in his experiments development times ranging from 1-8 days at a temperature of 12-14°C.

In 1972 and 1975 hatching of the resting eggs occurred in November, but from 1976-1978 it had already started at the beginning of September. It is evident from field observations, that hatching of the resting eggs is closely related to the rapid decrease of water temperature (below 15°C) in autumn (see Fig. 1). A slower increase in population density corresponds with a longer period (not less than 10 days) of water temperatures higher than 15°C (see Fig. 1). So it seems, that this change in temperature which normally takes place within one week (e.g. 1.9.1976 - 19.9.°C → 5.9.1976 - 13.3°C) represents an environmental signal inducing the hatching process. Evidence for such temperature induced hatching can be seen in Fig. 1 for other years.

The exponential growth phase

The first part of this phase is dominated by the parthenogenetic reproduction by amictic females. *R. fertöensis* is viviparous and usually one or two embryos can be found simultaneously inside a female. Only rarely were 3 embryos found, although Varga (1929) described 3-4 embryos, sometimes even 6, inside the mother animal. Althaus (1957) found mostly one embryo per female, but never several. Very often it was observed that the animal inside the mother itself carried an embryo; this observation is in agreement with the data of Varga (1929).

The length of the females varies between 350 and 520 μm, a range which is in close agreement with the measurements of Althaus (1957). No significant differences between autumnal and winter or spring animals are found. The length measurements given by Varga (1929) are somewhat higher (400-600 μm) but values such as 800-850 μm, which he measured on animals appearing just after ice break were never recorded in Neusiedlersee during the last 10 years.

The highest numbers of amictic eggs can be counted during the time of exponential growth and reached 350-625 eggs l^{-1}. The maxima in egg numbers and individual numbers are found during the time of ice cover or towards the end of this period.

At the end of this phase of exponential growth maximum sexual reproduction can be observed. During the investigation period it started in November at the earliest, with a maximum in January, February or March. Varga (1929) and Althaus (1957) found the sexual periods in spring time.

The first records of males date from the 8th of November, representing less than 2% of the total individual numbers. In late winter, beginning of spring the percentage of males increases up to 10-12%, which is in accordance with findings of Varga who never found more than 15%. The males exhibit only a little structural degeneration (Varga, 1929; Ruttner-Kolisko, 1974); their length vary between 263 and 357 μm (1975/78) (Varga, 1929: 300-500 μm). The mating behaviour is extensively described by Varga (1929) and he mentions that 5-6 hours after copulation the formation of the resting egg can be seen. Normally one resting egg is produced per mictic female, although once a 'twin' was found. The resting egg itself is covered all over with spines. The shape of the eggs shows some variation in length (L) and width (W) (including the spines) and in spine length (Lsp) (mean with ranges in parenthesis: L = 106 μm (91-132 μm), W = 67 μm (59-80 μm), n = 120; Lsp = 27 μm (16-36 μm), n = 146). They are spherical or ellipsoid in shape. This polymorphism of the resting eggs seems to be quite common among rotifers (Gilbert 1974), e.g. resting eggs of *Brachionus calyciflorus* could vary considerably in different localities, at different times of the season in the same water body, and even when produced by the same female.

Resting eggs of *R. fertöensis* are not deposited and are set free only by death and decomposition of the mictic female. The deposited resting eggs are difficult to find because they are covered with detrital material and mud. The colour varies from brown to black. As can be seen from Fig. 1 the production of resting eggs can occur over a period of 3-5 months, forming a reasonably secure foundation for the next amictic generation next autumn.

In Neusiedlersee the sexual period is associated with the population maxima, which possibly is a strategy of maximizing the efficiency of sexual reproduction and hence resting egg production. This timing of the sexuality is also shown, e.g. for *Brachionus plicatilis* (Ito, 1969; in cultures) or *Brachionus calyciflorus* (Halbach, 1970; in cultures). Halbach states that under limited conditions of food and space, peaks of mictic female production tended to coincide with population maxima. In the case of *Rhinoglena fertöensis*, food availabilty could also play a role; this can be seen from Fig. 2 which illustrates the close relationship between algae biomass and rotifer density. It could happen quite easily that the food supply could run short at the time of peak individual numbers.

After having reached peak densities and the highest degree of sexuality phase (c) is following: rapid decline and disappearance (see Fig. 1, 3). This happens at the time of ice break, which always is followed by a period of rapidly increasing water temperatures. In this 10 years survey it happens second half of March, April or at latest beginning of May, depending upon the actual temperature situation. During this population decline 80-100% are mictic females carrying resting eggs.

Assuming that a mictic female has a life expectancy of 10 days after the formation of the resting egg, *R. fertöensis* would has produced 1.9×10^6 resting eggs per m^2 in 1976/77 and $1.65 \times 10^6.m^{-2}$ in 1977/78. These values are very similar to those found for *Keratella quadrata* by Nipkow (1961).

Some notes on the distribution of Rhinoglena fertöensis

As mentioned earlier, this rotifer is an inhabitant of athalassic, saline and rather shallow water bodies. The pH of these lakes varies between 6.9 and 8.5 (sometimes even higher). A summary of the chemical characteristics of such lakes is given in Table 2. Hence it follows, that the occurrence of this species is given at alkalinity values ranging from 6 to 36 mequ.l^{-1}. Mass occurrence are recorded from water bodies with an average alkalinity of 10 meq.l^{-1} (Neusiedlersee), 20 meq.l^{-1} (Darscho, Peschek 1961) as well as at a range of 24-36 meq.l^{-1} (Süsser See, Althaus 1957). Herzig (unpubl. res.) found densities of 2-100 ind.l^{-1} in lakes with an alkalinity range of 12-35.2 meq.l^{-1} where the mean values for alkalinity are higher than 30 meq.l^{-1}, then it seems that *R. fertöensis* cannot develop successfully (Table 2 b). The water chemistry of the lakes inhabited by *Rhinoglena fertöensis* is found to be rather similar apart from the higher Ca^{++} and Cl^- values observed in Süsser See, but these did not have any negative effect on the population according to the results of Althaus (1957). Apart from higher alkalinity values (mean

Table 2. Chemical situation of lakes with (a) and without (b) *Rhinoglena fertöensis* (mean for the time November-March, range in parenthesis).

Lake	meq.l⁻¹ Alkalinity	mg.l⁻¹ Ca++	Mg++	Na+	K+	Cl⁻	SO₄--
(a)							
Seewinkel 1956/57[1] LÖFFLER 1957, 1959	17.0 (11.0-22.5)	12.5 (6.0-23.5)	50.8 (33.0-80.0)	404 (220-604)	11.4 (9.0-14.5)	74.5 (21-135)	166 (96-214)
Seewinkel 1970/72[2] HERZIG unpubl. res.	19.8 (12.0-35.2)						
Neusiedlersee 1968/73 NEUHUBER, pers. comm.	9.4 (6.2-11.4)	28 (20-40)	107.4 (73.2-130.5)	216 (147-304)	24.6 (19.6-31.3)	131.4 (85.2-174)	341 (235-418)
Neusiedlersee 1973/79	10.7 (8.7-16.2)						
Süßer See 1954/55 ALTHAUS 1957	24-36	282-327	82-103			560-896	704-856
(b)							
Seewinkel 1956/57[3] LÖFFLER 1957, 1959	67 (32.0-142.8)	1.5 (0-5.0)	2.7 (0-6.0)	2050 (1050-4500)	12 (8-20)	359 (165-865)	753 (528-1410)
Seewinkel 1970/73[3] HERZIG unpubl. res.	93 (14.1-320)						

[1]Hallabernlacke, Xixsee, Darscho, Untere Höllacke
[2]Darscho, Stundlacke
[3]Birnbaumlacke, Fuchslochlacke, Obere Halbjochlacke

} shallow lakes ($\bar{z} < 0.5$ m), east of Neusiedler-see (Seewinkel, Burgenland, Austria)

> 30 meq.l^{-1}) a high sodium content of the lake water seems to be a limiting factor. This idea can be confirmed by the findings of Löffler (1958) and Herzig (unpubl. res.) who did not find *R. fertöensis* in lakes with a high sodium content, even although these lakes were situated near to known *Rhinoglena* habitats. The papers of Nógradi (1956, 1957) who describes the sodium lakes of the Great Hungarian Plain in which *R. fertöensis* is not found, provide additional evidence.

In conclusion, the following should be stated; there seem to exist negative correlations between the occurrence of *R. fertöensis* on the one hand and higher alkalinity values and a high sodium content of the lake water on the other.

Summary

1. On the basis of a ten year study on Neusiedlersee (Burgenland, Austria) the changes in abundance and population dynamics of *Rhinoglena fertöensis* are described and discussed. This cold-stenothermic species occurs maximally during the winter months, when the lake is covered by ice. Its occurrence is restricted to water temperatures below 15°C. Higher temperatures are tolerated for a short time only (1-2 days), but result in low population growth. It is shown, that *R. fertöensis* is increasing in numbers throughout the 10-year period of observation, which happens to parallel an increase in algae biomass and the occurrence of high numbers of bacteria. At the moment this species is dominating the winter rotifer plankton and is of some importance in late autumn and spring too.

2. 3 phases can be determined in the population dynamics, the hatching of the resting eggs in autumn, a phase of exponential growth until reaching the peak density and at this time showing the highest degree of sexuality, and finally a rapid decrease in individual numbers and disappearance of the population after the ice break in spring.

3. Hatching of the resting eggs takes place at the time of the steepest temperature decline in autumn, the highest percentages of mictic females with resting eggs occur at the time of rapid temperature increase in spring.

4. The relation between water chemistry and the occurrence of *Rhinoglena fertöensis* is discussed. It is shown that the animal cannot develop successfully in waters with higher alkalinities (mean higher than 30 meq.l^{-1}) and is avoiding lakes which show a high sodium content.

Acknowledgements

The author is indebted to Dr. H. Metz, Biological Station Illmitz and colleagues of the Zentralanstalt für Meteorologie und Geodynamik, Vienna, for providing temperature readings. Dr. A. Duncan is sincerely thanked for improving the English and critical comments. This survey was done in the scope of an IBP (1967-1974) and MaB-study (1975-1979) on Neusiedlersee and hence was sponsored by these.

References

Althaus, B. 1957. Faunistisch–ökologische Studien an Rotatorien salzhaltiger Gewässer Mitteldeutschlands. Wiss. Z. Univ. Halle, Math.-Nat. 6: 117-158.

Bartoš, E. 1959. Vírnicí-Rotatoria. Fauna ČSR, Svazek 15, 969 pp.

Bottrell, H. H., Duncan, A., Gliwicz, Z. M., Grygierek, E., Herzig, A., Hillbricht-Ilkowska, A., Kurasawa, H., Larsson, P. & Weglenska, T. 1976. A review of some problems in zooplankton production studies. Norw. J. Zool. 24: 419-456

Dokulil, M. 1979. Seasonal pattern of phytoplankton. In: Neusiedlersee, the limnology of a shallow lake in Central Europe. (Ed. H. Löffler, Junk, The Hague, pp. 203-231.

Gilbert, J. J. 1974. Dormancy in rotifers. Trans. am. microsc. Soc., 93: 490-513.

Halbach, U. 1970. Einfluss der Temperatur auf die Populationsdynamik des planktischen Rädertieres Brachionus calyciflorus Pallas. Oecologia, 4: 176-207.

Herzig, A. 1974. Some population characteristics of planktonic crustaceans in Neusiedlersee. Oecologia, 15: 127-141.

Herzig, A. 1979. The zooplankton of the open lake. In: Neusiedlersee, the limnology of a shallow lake in Central Europe. (Ed. H. Löffler), Junk, The Hague, pp. 281-335.

Ito, T. 1960. On the culture of mixohaline rotifer Brachionus plicatilis O. F. Müller in the sea water. Rep. Fac. Fish. Prefect. Univ. Mie, 3: 708-740.

Löffler, H. 1959. Zur Limnologie, Entomostraken- und Rotatorienfauna des Seewinkelgebietes (Burgenland, Österreich). S.-B.öst. Akad. Wiss. math. nat. Kl. Abt. I, 168: 315-362.

Löffler, H. (Ed.) 1979. Neusiedlersee, the limnology of a shallow lake in Central Europe, Junk, The Hague, 543 pp.

Nipkow, F. 1961. Die Rädertiere im Plankton des Zürichsees und ihre Entwicklungsphasen. Schweiz. Z. Hydrol., 23: 398-461.

Nógradi, T. 1956. Limnologische Untersuchungen an Natrongewässern der ungarischen Tiefebene. Hidrólogiai Közlöny 36: 130-137.

Nógradi, T. 1957. Beiträge zur Limnologie und Rädertierfauna ungarischer Natrongewässer. Hydrobiologia 9: 348-360.

Peschek, E. 1961. Beiträge zur Biologie der Salzlacken im Neusiedler-See-Gebiet. Verh. int. Ver. Limnol., 14: 1124-1131.

Pourriot, R. 1977. Food and feeding habits of Rotifera. Arch. Hydrobiol. Beih. 8: 243-260.

Ruttner-Kolisko, A. 1974. Plankton rotifers. Die Binnengewässer, 26, 1. Suppl., 146 pp.

Varga, L. 1929. Rhinops fertöensis, ein neues Rädertier aus dem Fertö (Neusiedlersee). Zool. Anz., 80: 236-253.

Zakovsek, G. 1961. Jahreszyklische Untersuchungen am Zooplankton des Neusiedlersee. Wiss. Arb. Burgld., 27: 1-85.

THE ABUNDANCE AND DISTRIBUTION OF FILINIA TERMINALIS IN VARIOUS TYPES OF LAKES AS RELATED TO TEMPERATURE, OXYGEN, AND FOOD

A. RUTTNER-KOLISKO

Biol. Station d. Osterr. Akad. d. Wissensch., A-3293 Lunz am See, Austria

Keywords: Rotatoria, population dynamics, vertical distribution, autecology, lake types

Abstract

Although morphologically variable, *Filinia terminalis* constitutes a well defined ecological unit living below 12-15°C, and forming large maxima up to several thousand ind./l at an oxygen content $<$ 2 mg/l. Field investigations and laboratory experiments indicate facultative anaerobiosis and food sources of chemosynthesizing or plankton decomposing bacteria. On the grounds of these physiological requirements a hypothetical schema is presented attempting to predict the occurrence of the two species, *F. terminalis* and *F. longiseta* in oligotrophic, eutrophic, meromictic, and polymictic, shallow lakes.

Introduction

Two distinct ecological units exist within the genus *Filinia:* one living in the epilimnion of summer warm lakes, of ponds and river backwaters up to a temperature of 24° and even 28°C; the other living under apparently various conditions but never in the epilimnion at high temperatures. The morphological features used to distinguish the two units vary independently of each other from population to population to such a degree that they may overlap, and even within the same population these features vary with time. Considerable confusion exists therefore as to taxonomical boundaries and denominations. It is not possible to discuss here the validity of the individual distinguishing features; they will be dealt with elsewere (Ruttner-Kolisko, in prep.). In accordance with most other authors I use in the present text the name *F. longiseta* for the epilimnetic thermophile form and *F. terminalis* for the unit whose ecological requirements I am discussing.

Results

In the lakes of the English Lake District the *Filinia* populations do not differ morphologically from each other, and correspond exactly to the *F. terminalis* photographed and described by Pourriot (1965b). Their ecological behaviour, however, is different in the two eutrophic lakes Esthwaite Water (Est.) and Blelham Tarn (Bl.) on the one hand and the two mesotrophic Windermere Basins (W.N., W.S.) on the other hand (Fig. 1). The pattern of abundance is remarkably similar within both pairs of lakes: in Est. and Bl. the spring maximum is followed by a complete gap during summer and autumn, whereas in W.S. and W.N. the population is perennial and the maxima are lower by one, and nearly two orders of magnitude respectively.

This difference can be explained by the physiological requirements of *F. terminalis* and by the physico-chemical properties of the particular lakes.

F. terminalis is known to occur at low temperatures only – in my experience not over 15°C; a few records of higher temperature exist in the literature but passive transport due to unusual turbulence may be involved in these cases (Buikema & Löffelman, 1978; Maeseneer *et al.,* 1978). The population maximum which occurs mostly in a very narrow, well defined stratum, lies always below 10°C (Hofmann, 1972; Larsson, 1971; Pejler, 1961; Rutt.-Kol., 1972).[*] It is also well known from the same sources that *F. terminalis* can not only stand a very low oxygen content of the water but virtually thrives under such conditions; it has even been found in complete anaerobic layers.

In order to assess how long *F. terminalis* can survive

---

[*] Attempts to culture *F. terminalis* at various temperatures in our laboratory – so far never successful over a long time – failed immediately at temperatures of 15° and 20°C but succeeded for a few days at 10° and 5°C.

Hydrobiologia 73, 169-175 (1980).   0018-8158/80/0732-0169$01.40.
© *Dr. W. Junk b.v. Publishers, The Hague.*

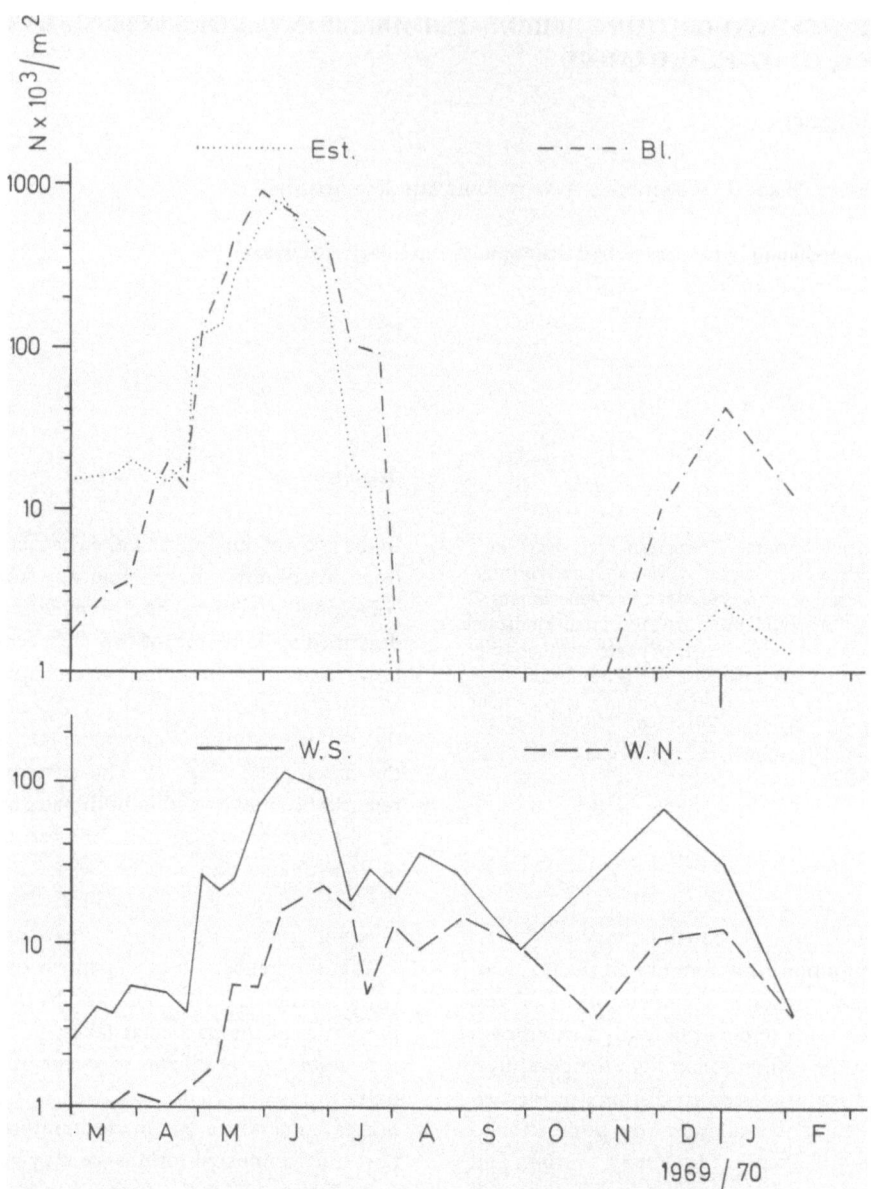

Fig. 1. Abundance of *Filinia terminalis* in 4 lakes of the English Lake District during the year 1969/70. Est. = Esthwaite Water; Bl. = Blelham Tarn; W.S. = Windermere South-Basin; W. N. = Windermere North-Basin.

under anaerobic conditions we carried out the following experiment using the population of the Lunzer Obersee (L.O.S.):

Water from 9 m depth ($O_2 = 0.09$ mg/l) was pumped through a 10 l bottle, so that the bottle was 10 times flushed while – by means of a 40 μ net inserted into the pumping system – the plankton was 10 times concentrated After careful sealing for transport the bottle was stored in the laboratory at 5°C in darkness. Every day 0.5 l water were drained from the bottle, and automatically replaced

with desoxygenated water. The O_2-content of the sample was determined, and the plankton examined. The experiment was repeated twice with similar results.

The plankton sampled at the start of the experiment (Tab. 1) consisted, apart from *F. terminalis,* of some *Cyclops* nauplii, of a dense population of the oligoaerobic ciliate *Coleps,* of the putriphile ciliate *Caenomorpha,* and mainly of the sulphur bacterium *Chromatium* causing a

Table 1. Population dynamics of *Filinia terminalis* in a laboratory experiment under anaerobic conditions.

	start	after 24h	48h	96h	7 days
Temperature (°C)	4.6	5.0	5.0	5.0	5.0
Oxygen (mg/l)	0.09	0.10	0.39	0.68	0.07
F.term. (N/l)	400	220	480	260	30
F.term.+egg(% of N)	~ 10	< 10	~ 50	–	–
Nauplii	++	++	++	+	–
Coleps sp.	+++	++	few	–	–
Caenomorpha sp.	+	+	+	+++	++
Chromatium (watercolour)	slightly pinkish	clearly pink	pink sediment	colourless	colourless

Table 2. Abundance of *Filinia terminalis* and productivity (indicated by chlorophyll a) in Blelham Tarn (Bl.) Esthwaite Water (Est.), Windermere South- and North-Basin (W.S., W.N.).

	Bl.	Est.	W.S.	W.N.
Maximal densities of F.term. (ind./l)	142	135	9	3
Chlorophyll-a values (mg/m³)*	10-30	10-30	10-13	4-7

* from JONES 1972.

pinkish colour of the water. With the initial increase of oligoaerobic bacteria, visible by the intensified Chromatium colour, *F. terminalis* also increased and reproduced in the practically oxygen-free water. Only at the onset of processes of putrification, indicated by the increase of *Caenomorpha* and the disappearance of *Coleps* and *Chromatium* the *Filinia* population also died off. Thus, the experiment shows that *F. terminalis* is capable of facultative anaerobiosis over a considerable period of time; similar results were reported recently for an increasing number of invertebrates (Saz, 1971; Hochachka & Mustafa, 1972).

In Fig. 2 the temperature- and oxygen-isopleths of two of the English lakes are shown together with the distribution in depth and time of *F. terminalis*.* In Esthwaite Water, only 15 m deep and highly eutrophic, the desoxygenated zone comes up high enough to meet the 10° and 12°C isotherme, thus leaving no space for *F. terminalis* to exist in summertime. In both the deep Windermere basins, however, where no anaerobic zone occurs, there is room enough for *F. terminalis* to live below 12°C all the year round. Similar distribution patterns have been recorded from many other eutrophic and oligotrophic lakes (Pejler, 1957, 1961; Hofmann, 1972, among others). A different pattern, however, has been described from meromictic lakes, where *F. terminalis* is also perennial but lives exclusively just above the oxygen-free zone under oligoaerobic conditions (Larsson, 1971; Rutt.-Kol., 1974; Schaber, unpubl. data).

The temperature and oxygen requirements of *F. terminalis* do not, however, explain the time, location, and intensity of the population maxima in various lakes. These are obviously related to food as indicated by the findings in the 4 English lakes investigated during the same time, and situated close enough to each other for climatic differences to be negligible. Here, the densities (mean values/l derived from 3 to 3 m, and 5 to 5 m sampling respectively) correspond clearly to the productivity of the lakes indicated by the chlorophyll-*a* values (Tab. 2).

Yet, the actual food source has still to be established experimentally. It is generally assumed from field experience and from cultures (*F. 'passa'*, Pourriot, 1965) that *Filinia* is a micro-filterfeeder (food particles < 10 μ) living on decaying phytoplankton and bacteria associated with it. A preliminary experiment of my own seems to corroborate this view: out of 3 different food items offered to *F. terminalis* specimens from Est. and W.S., namely fresh nannoplankton, pure bacteria, and decaying plankton, only the latter caused a few animals to produce an egg.* The assumption that decaying phytoplankton plus bacteria is the main food source for *F. terminalis* would also explain the time of its highest abundance in the 4 English lakes which was shortly after the spring phytoplankton peak. Moreover, evidence from plate counts of a metalimnetic peak of decomposing bacteria in both Windermere basins (Jones, 1977) corresponds in depth to the highest densities of *F. terminalis* (see Fig. 2).

On the other hand, the preference of *F. terminalis* for oligoaerobic water would suggest that chemosynthesizing specialists among the bacteria who live in narrow strata just above the O₂-free zone also provide a suitable food for this species. The pattern of occurrence in meromictic

---

* Only one example for each of the similar pairs is represented. Temperature and O₂-data from Windermere Laboratory routine records.

* For technical reasons the experiment lasted only for a few days: due to their jumping motions the animals were all caught sooner or later in the surface film of the culture vessels.

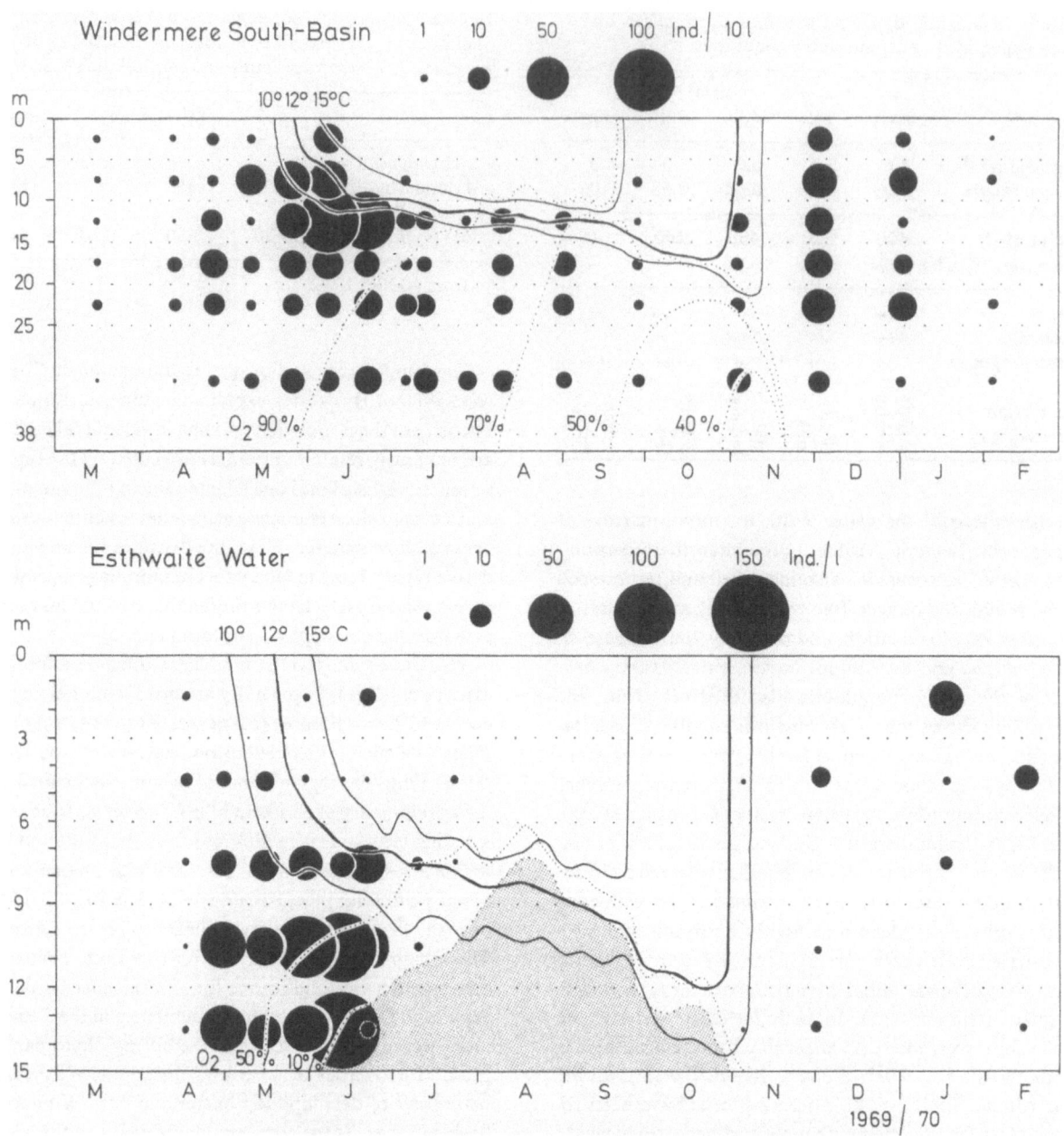

Fig. 2. Vertical distribution and density of *Filinia terminalis* in the mesotrophic Windermere South-Basin and in the eutrophic Esthwaite Water. Area of circles proportional to numbers in 10 l, and in 1 l derived from samples taken in 5 m, and 3 m steps respectively. Isothermes and O_2-% are indicated; shaded covers $O_2 < 10\%$ (~1 mg/l).

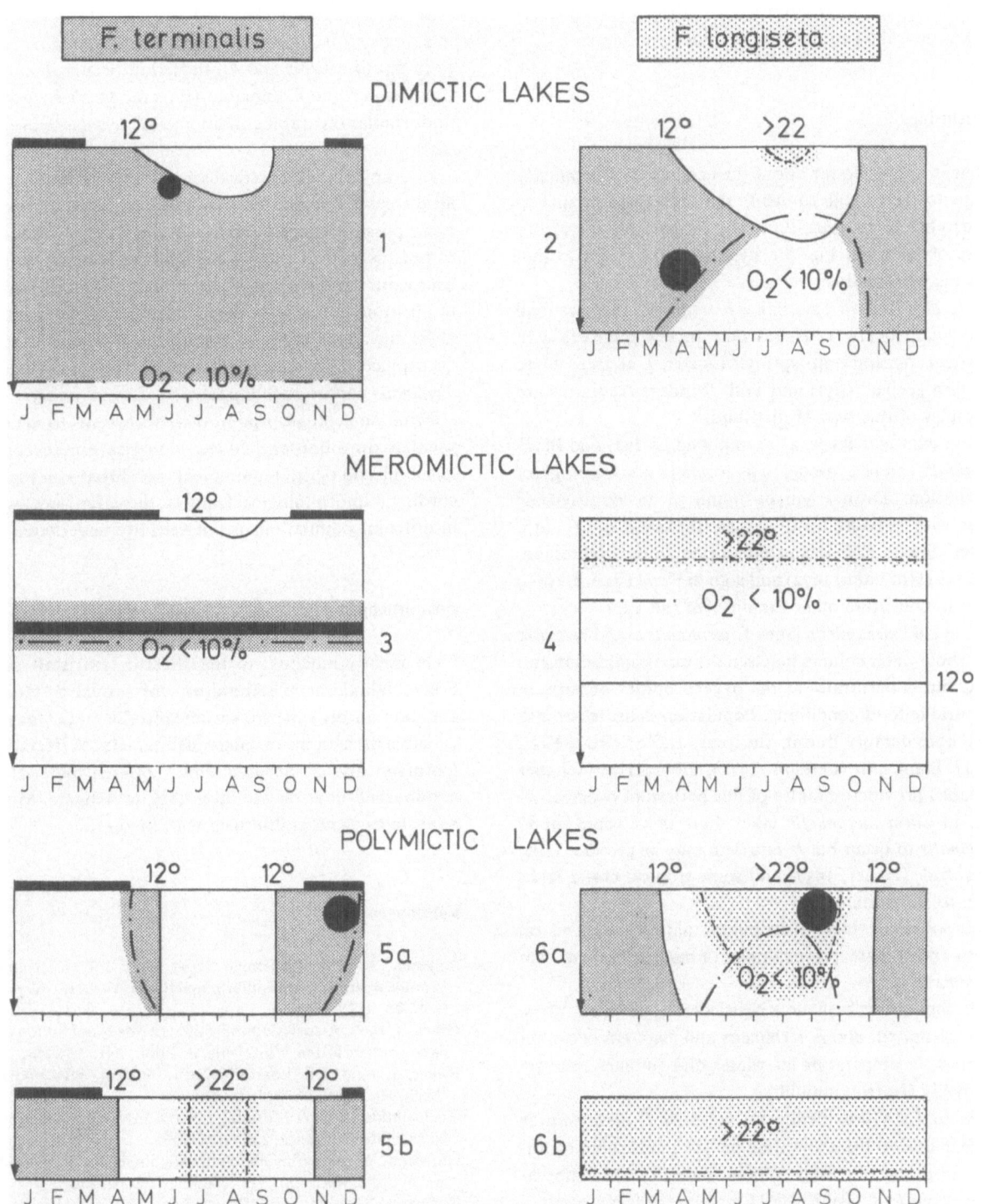

Fig. 3. Hypothetical distribution pattern of *Filinia terminalis* and *F. longiseta* in various types of lakes. *F. term.* shaded, *F. long.* stippled area; Black circles(bar)indicate time and location of highest population densities. —— = 12°C; - - - - = 22°C; -·-··-·- = 10% O_2. For further explanation see text.

lakes as well as our experiments with the population of the L.O.S. described earlier support this hypothesis.

Conclusions

What is known so far about the ecology of *F. terminalis* seems to me enough to justify the attempt to predict its occurrence or its substitution by *F. longiseta* in various types of lakes. In Fig. 3 a hypothetical schema to that effect is presented:

1. in *oligo(meso)trophic* lakes *F. terminalis* may occur all year round below 12°C but only in low numbers with maximal densities in spring. Lunzer Untersee, Övre Laksjøn (Pejler, 1957) and both Windermere basins are examples of this type of distribution.

2. in *eutrophic* lakes, as exemplified by Est. and Bl. *F. terminalis* can be expected to be absent in summer; highest population densities will be found in the oligoaerobic layer with maxima up to several thousand ind./l. In a warm climate *F. longiseta* may occur in the epilimnion. Plußsee (Hofmann, 1972) and lago di Nemi (Parise, 1960) seem to constitute other examples of this type.

3. in *cold meromictic* lakes *F. terminalis* could colonize the whole water column between the monimolimnion and 12°C but concentrates at the oxygen border because of favourable food conditions. Population densities do not vary considerably during the year. L.O.S. (Rutt.-Kol., 1974), Blankvatn (Larsson, 1971), Piburgersee (Schaber unpubl.) provide examples of this pattern of occurrence.

4. in *warm meromictic* lakes there is no room for *F. terminalis* to occur but *F. longiseta* may be present. Lake Singkarak (Hauer, 1937) and some tropical crater lakes seem to show this type.

In polimictic lakes distribution patterns depend on depth and climate; two examples of *moderate climate* are presented:

5b. the extremely shallow Neusiedlersee never stratifies, is O$_2$-saturated, and *F. terminalis* and *longiseta* occur according to temperature in winter and summer time respectively (Herzig, unpubl.).

5a. in Ösbysjøn (depth 3.5 m) only *F. terminalis* is present during the short time between anaerobic conditions in the whole water column under ice and temperatures above 12°C (Pejler, 1961); maximal densities occur in autumn when decaying phytoplankton is available.

6b. in a shallow unstratified *tropical* lake only *F. longiseta* can be expected, and is in fact found for instance in Lake George (Green, 1967).

6a. an ecological niche may be available for both species in a *subtropical* climate; if the lake is deep enough for short term stratification, and O$_2$-depletion occurs in summer time as in Lake Commabio (Ravera, 1977) *F. longiseta* finds similar favorable conditions as *F. terminalis* in cooler eutrophic lakes.

Like any other theoretical schema Fig. 3 does not cover all the cases known to occur in nature. Since there exists every possible transition between the types of lakes mentioned – as well as other types – intermediate or different behaviour of *Filinia* is also to be found. The graph is meant to illustrate the hypothesis that by knowing temperature, O$_2$-regime, and phytoplankton dynamics of a lake the occurrence, abundance, and vertical distribution of *F. terminalis* can be roghly predicted – and inversely.

From an ecological point of view it seems to me more useful to draw borders between biological units according to eco-physiological requirements and behaviour than according to morphological features, the variation of which in different populations is not yet fully understood.

Acknowledgements

I am greatly indebted to the Director and staff of the F.B.A. Windermere Laboratory for providing facilities and data for my work, in particular to Dr. J. G. Jones for valuable discussions on bacterial food. Drs. A. Herzig, W. Hofmann and P. Schaber kindly offered material and unpublished data on the lakes they investigate. Miss E. Kronsteiner gave skilful help with the graphs.

References

Buikema, A. L. & Löffelman, P. H. 1978. Effect of pumped storage operations on rotifer populations. Verh. Int. Ver. Limnol., 20: 1597-1603.

Green, J. 1967. Associations of rotifera in the zooplankton of the lake sources of the White Nile. J. Zool., 151: 343-378.

Hauer, J. 1937. Die Rotatorien von Sumatra, Java und Bali. Arch. Hydrobiol. Suppl. 15: 296-384.

Hochachtka, P. W. & Mustafa, T. 1972. Invertebrate facultative anaerobiosis. Science, 178: 1056-1060.

Hofmann, W. 1972. Zur Populationsökologie des Zooplanktons im Plußsee. Verh. int. Ver. Limnol., 18: 410-418.

Hofmann, W. 1977. The influence of environmental factors on population dynamics in planktonic rotifers. Arch. Hydrobiol. Beih., 8: 77-84.

Jones, J. G. 1972. Studies on freshwater microorganisms: Phosphatase activity in lakes of differing degree of eutrophication. J. Ecol., 60: 777-791.

Jones, J. G. 1977. The effect of environmental factors on estimated viable and total populations of planktonic bacteria in lakes and experimental enclosures. Freshw. Biol., 7: 67-91.

Larsson, P. 1971. Vertical distribution of planktonic rotifers in a meromictic lake: Blankvatn near Oslo, Norway. Norw. J. Zool. 19: 47-75.

Maeseneer, J. de, de Pauw, M. & Waegeman, D. 1978. Influence of the mud layer of the 'Watersportbaan' at Ghent on some aquatic life forms, especially chironomid larvae and Filinia spp. Hydrobiol., 60: 151-158.

Parise, A. 1960. I rotiferi del lago di Nemi. Arch. oceanogr. Limnol., 12: 1-99.

Pejler, B. 1957. Taxonomical and ecological studies on planktonic rotatoria from Northern Swedish Lapland. Kungl. Svensk. Vetensk., Handl., 6, 5: 1-68.

Pejler, B. 1961. The zooplankton of Osbysjøn, Djursholm. I. Seasonal and vertical distribution of the species. Oikos, 12: 225-248.

Pourriot, R. 1965a. Recherches sur l'ecologie des Rotifères. Vie Milieu, suppl., 21: 1-224.

Pourriot, R. 1965b. Notes taxonomiques sur quelques rotifères planktoniques. Hydrobiol., 26: 579-604.

Ruttner-Kolisko, A. 1974. The vertical distribution of plankton rotifers in a small alpine lake with a sharp oxygen depletion (L.O.S.). Verh. int. Ver. Limnol., 19: 1286-1294.

Ruttner-Kolisko, A. Der longiseta-terminalis Komplex der Gattung Filinia unter Berücksichtigung autökologischer Gebenheiten. (in prep.).

Saz, H. J. 1971. Facultative anaerobiosis in invertebrates: Pathways and control systems. Am. Zool., 11: 125-135.

ON THE ECOLOGY OF NOTHOLCA SQUAMULA MÜLLER IN LOCH LEVEN, KINROSS, SCOTLAND[1]

Linda MAY

Institute of Terrestrial Ecology, 78 Craighall Road, Edinburgh, Scotland

[1] Part of a dissertation for the Degree of Doctor of Philosophy to the Council for National Academic Awards at Paisley College, Scotland, in conjunction with the Institute of Terrestrial Ecology, Edinburgh, Scotland

Keywords: Rotifers, temperature, diatoms, grazing

Abstract

Notholca squamula was rarely found in Loch Leven when the water temperature rose above 10°C. Under favourable temperature conditions its abundance appeared to be closely related to that of *Asterionella formosa*. In the laboratory the animal was seen to feed on this diatom by breaking open the frustule and ingesting the cell contents.

Introduction

Notholca spp. are generally considered to be cold stenotherms. They commonly occur in arctic and temperate localities but are seldom found in tropical and subtropical zones (Pejler, 1977). In the plankton of Loch Leven the genus is represented solely by *Notholca squamula* Müller (Fig. 1a) which appears in early December and persists until the late spring. This paper considers the influence of food and temperature on the abundance of the species, and presents observations on the behaviour of animals feeding on large species of diatoms.

The Site

Loch Leven, near Kinross, Scotland, is a shallow eutrophic loch with a mean depth of 3.9 m. Descriptions of its physical and chemical characteristics are given by Smith (1974) and Holden & Caines (1974) respectively. The loch is situated in an area with temperate climate and is generally isothermal.

Materials and methods

The plankton was sampled at weekly intervals, from January 1977 to August 1979. Samples were taken with a weighted plastic tube (Lund, 1949) lowered to a depth of 3 m. The subsurface temperature was measured on each occasion with a mercury in glass thermometer mounted in a Ruttner bottle.

Samples of rotifers were prepared for counting by narcotising with .04% procaine hydrochloride ($NH_2.C_6H_4.COO.CH_2.CH_2.N(C_2H_5)_2.HCL$) and preserving in 4% formaldehyde. The rotifers were then concentrated by sedimentation and counted with an inverted microscope. Live rotifers were examined for species determination on return to the laboratory. The density of algae (Bailey-Watts, unpub. data) was estimated for each sample.

Notholca squamula was cultured on a diet of *Asterionella formosa* Hass. in modified Chu's medium number 10 (Chu, 1942; Lund *et al.*, 1975) diluted 1 : 4 with glass distilled water. The alga was originally isolated from Loch Leven and subsequently grown in modified Chu 10. It was added to the rotifer medium as a dense suspension in the exponential phase of growth. The rotifers were kept at 10°C and illuminated on a 12-12 (LD) photoperiod by white fluorescent light (Ascot Ltd, England) at a light intensity of 2,200 lux.

Results

Fig. 2 shows the seasonal occurrence of *N. squamula* in Loch Leven in relation to temperature and the appearance of the diatom *Asterionella formosa*. *N. squamula* was seldom found when the water temperature was above 10°C, although it regularly occurred in the plankton at lower temperatures. Numbers of *N. squamula* were particularly high early in 1977 following a dense bloom of *A. formosa*. In 1978 *N. squamula* appeared shortly after *A. formosa*. Both remained at low density until the end of March. The density of *A. formosa* then increased rapidly, but the expected rise in the numbers of *N. squamula* did not follow. It

Hydrobiologia 73, 177-180 (1980). 0018-8158/80/0732-0177$00.80.
© Dr. W. Junk b.v. Publishers, The Hague.

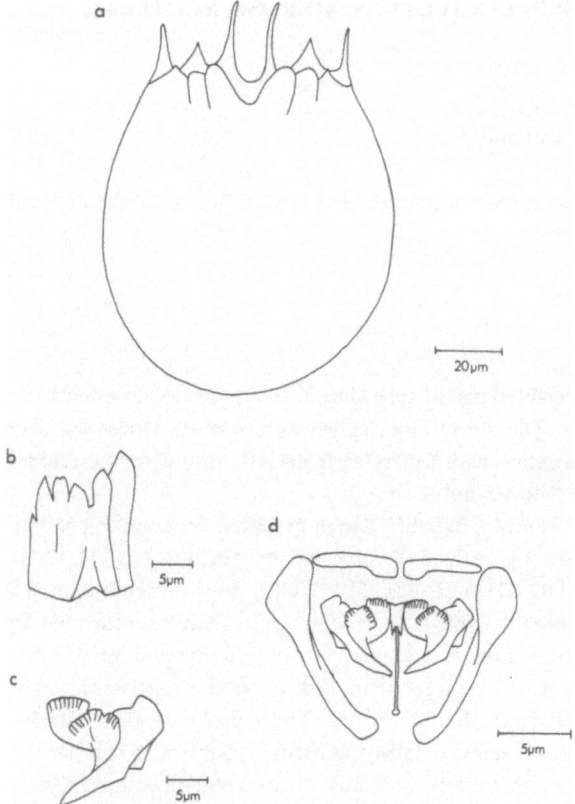

Fig. 1. *Notholca squamula* a) lorica ventral b) uncus, c) ramus, d) trophi.

and their guts empty, indicating that death had occurred shortly after hatching. In contrast, when animals were incubated at 10°C, 63% of the eggs hatched and there were no offspring mortalities within the first few days of hatching.

In samples of live plankton, specimens of *Notholca squamula* were seen to feed on colonies of *Asterionella formosa* whilst rejecting cells of *Microcystis* sp., *Chlorella* sp. and small flagellates. On approaching a colony of *A. formosa,* the animal appeared to select a suitable cell and pull the free end into its mastax. The end of the cell was then broken (Fig. 3) by violent crushing movements of the trophi and the cell contents removed. The trophi of *N. squamula* are very robust and seem to be well-suited to this purpose (Fig. 1b, c, d). The rejected piece of *A. formosa* frustule, which remained attached to the colony, was usually 1/4 – 2/3 of its original length. Occasionally the frus-

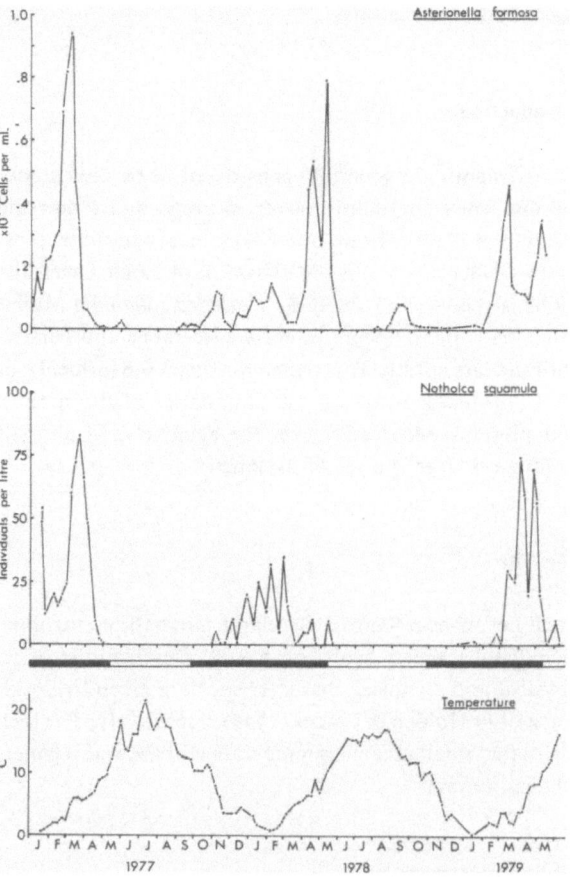

Fig. 2. Seasonal variation in the population of *Notholca squamula* in relation to temperature and the density of *Asterionella formosa*. The shaded areas indicate the periods during which the temperature was less than 10°C.

seems likely that the reproductive capacity of the rotifer was reduced by the sharp increase in temperature which occurred early in April, so that it was unable to take advantage of the abundance of food. *N. squamula* died out soon afterwards when the water temperature rose above 10°C. In 1979 numbers of *N. squamula* increased rapidly in mid March following a high density of *A. formosa*. A second increase in the numbers of *N. squamula* occurred towards the end of April when *A. formosa* numbers were still relatively low, but high numbers of small centric diatoms were present in the plankton. Despite the presence of abundant food, the numbers of *N. squamula* decreased rapidly in early May when the water temperature again increased beyond 10°C.

In the laboratory cultures of *N. squamula* could be kept only if the incubation temperature did not exceed 10°C. When the incubation temperature was raised to 15°C most of the animals died and only 29% of the amictic eggs laid by the survivors hatched. However, none of the newly hatched young survived; their loricas remained crumpled tule remained complete, except for a large hole close to its free end. Only rarely did any cell contents remain in the broken frustule. The detached piece of frustule was taken into the stomach along with the cell contents and was later discharged in the faeces. Pieces of *A. formosa* frustules were identifiable amongst the gut contents of animals taken from the loch, even at times when the density of *A. formosa* was comparatively low ($\simeq$ 100 cells ml.$^{-1}$). On one occasion *N. squamula* was seen to feed on a filament of *Melosira italica* subsp. *subarctica* Müller. One end of the filament was taken into the mastax and, when later discarded, the terminal cell, previously intact, was empty. A large hole was evident in one side of the frustule.

Discussion

Whilst the most important factor affecting the distribution of *Notholca squamula* appears to be temperature, detailed information on the subject is scarce. For the most part, the animal is referred to merely as a cold water species (Beach, 1960; Stemberger, 1974; Nogrady, 1976; Dartnall, 1977). Observations on the plankton of Loch Leven suggest that the species is successful only when the water temperature does not exceed 10°C. Data from Carlin (1943) and Pejler (1957) for ponds and lakes in Sweden seems to support this hypothesis. In the laboratory the species is unable to maintain its population at temperatures above 10°C, although individuals may survive at higher temperatures for some time. This could explain why specimens of *N. squamula* are occasionally found at unusually high temperatures (Gillard, 1948; Zankai & Ponyi, 1971; Stemberger, 1976).

It has been suggested that *N. squamula* feeds on the microscopic algae associated with decaying plant material (Pejler, 1962b), and on small nannoplankters (Naumann, 1923). However, this has not been confirmed, and an attempt to culture them on *Chlorella* sp. was unsuccessful (De Beauchamp, 1938).

Diatom frustules have been found amongst the stomach contents of *Notholca* spp. (Pourriot, 1965; Pejler, 1962a, 1977b; Thane-Fenchel, 1968) but have so far been considered as unlikely to form the major part of their natural diet (Pourriot, 1965; Pejler, 1977a). In Loch Leven *N. squamula* appears to feed almost exclusively on *A. formosa*. When the water temperature is favourable the availability of this food species appears to be an important factor in limiting its population size.

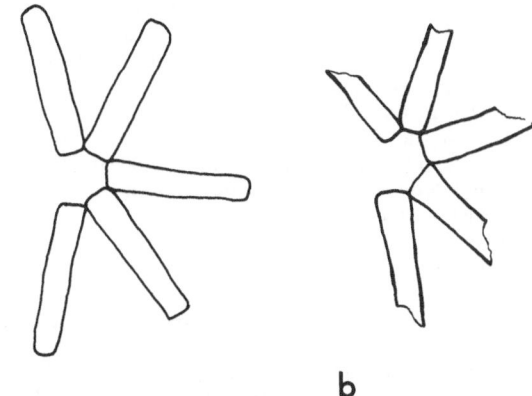

Fig. 3. a) A colony of *Asterionella formosa* as found in cultures which have not been subjected to grazing. b) Broken, empty frustules of a colony of the diatom which has been grazed by *Notholca squamula*. (Traced from photograph).

Acknowledgements

I thank Dr D. G. George, Dr A. E. Bailey-Watts, Dr P. S. Maitland, Mr J. D. Hamilton and Miss E. Rutkowski for reviewing the draft of this paper, and Mrs M. Wilson for typing the manuscript. I am particularly indebted to Dr A. E. Bailey-Watts for allowing me to use his data on the algae of Loch Leven.

This work was supported by a Research Training Grant from the Natural Environment Research Council.

References

Beach, N. W. 1960. A study of the planktonic rotifers of the Ocqueoc River system, Presque Isle County, Michigan. Ecol. Monogr. 30: 339-358.

Carlin, B. 1943. The planktonic Rotifera of Motalastrom: on the taxonomy and ecology of planktonic Rotifera. Meddn. Lunds. Univ. limnol. Instn. 5, 255 pp.

Chu, S. P. 1942. The influence of the mineral composition of the medium on the growth of planktonic algae. Part 1. Methods and culture media. J. Ecol. 30: 284-325.

Dartnall, H. J. G. 1977. Antarctic freshwater rotifers. Arch. Hydrobiol. Beih. 8: 240-242.

De Beauchamp, P. 1938. Les cultures de Rotifères sur Chlorelles. Trav. Stn. Zool. Wimereux. 13: 27-38.

Gillard, A. A. M. 1948. De Brachionidae (Rotatoria) van België met Beschouwingen over de Taxonomie van de Familie. Natuurw. Tijdschr. 30: 159-218.

Holden, A. V. & Caines, L. A. 1974. Nutrient chemistry of Loch Leven. Proc. R. Soc. Edinb. 74: 101-121.

Lund, J. W. G. 1949. Studies on Asterionella I The origin and nature of the cells producing seasonal maxima. J. Ecol. 37: 389-419.

Lund, J. W. G., Jaworski, G. H. M. & Butterwick, C. 1975. Algal bioassay of water from Blelham Tarn, English Lake District and the growth of planktonic diatoms.Arch. Hydrobiol. Suppl. 49: 49-69.

Naumann, E. 1923. Spezielle Untersuchungen über die Ernährungsbiologie des tierischen Limnoplanktons II über den Nahrungserwerb und die natürliche Nahrung der Copepoden und der Rotiferen des Limnoplanktons. Acta. Univ. Lund nov. ser. avd. 2, 19 no. 6 17 pp.

Nogrady, T. 1976. Canadian rotifers I Lac Echo, Quebec. Naturaliste can. 103: 425-436.

Pejler, B. 1957. Taxonomical and ecological studies on the plankton Rotatoria from central Sweden. Svenska. Vetensch. Handl. 6: 1-52.

Pejler, B. 1962a. Morphological studies on the genera Notholca, Kellicottia, and Keratella (Rotatoria). Zool. Bidr. Upps. 33: 295-309.

Pejler, B. 1962b. On the taxonomy and ecology of benthic and periphytic Rotatoria. Zool. Bidr. Upps. 33: 328-415.

Pejler, B. 1977a. General problems of rotifer taxonomy and global distribution. Arch. Hydrobiol. Beih. 8: 212-220.

Pejler, B. 1977b. On the global distribution of the family Brachionidae (Rotatoria). Arch. Hydrobiol. Suppl. 53: 255-306.

Pourriot, R. 1965. Recherches sur l'ecologie des Rotiferes. Vie et Milieu, Suppl. 21, 224 pp.

Smith, I. R. 1974. The structure and physical environment of Loch Leven, Scotland. Proc. R. Soc. Edinb. 74: 81-100.

Stemberger, R. S. 1974. Temporal and spatial distributions of the planktonic rotifers in Milwaukee Harbor and adjacent Lake Michigan. Proc. 17th Conf. Great Lakes Res. Internat. Assoc. Great Lakes Res. pp. 120-134.

Stemberger, R. S. 1976. Notholca laurentiae and N. michiganensis new rotifers from the Laurentian Great Lakes region. J. Fish. Res. Bd. Can. 33: 2814-2818.

Thane-Fenchel, A. 1968. Distribution and ecology of non-planktonic brackish-water rotifers from Scandinavian waters. Ophelia 5: 273-297.

Zankai, N. P. & Ponyi, J. E. 1971. The horizontal distribution of the Rotiferan plankton in Lake Balaton. Annal. Biol. Tihany. 38: 285-304.

ECOLOGY OF SESSILE ROTIFERS

Robert L. WALLACE

Biology Department, Ripon College, Ripon, Wisconsin, U.S.A. 54971

Keywords: Ecology, feeding, invertebrate behavior, larval biology, Rotifer, sessile rotifer, substrate-dependent survivorship, substrate selection

Introduction

The sessile rotifers are taxonomically defined as all the individuals belonging to the families Flosculariidae, Conochilidae (Order Flosculariaceae), and Collothecidae (Order Collothecaceae) (Edmondson, 1940, 1944). Free-swimming species are, however, found in each family. The Conochilidae are totally planktonic. Although most of my remarks in this review will concern those species which as adults are permanently attached to surfaces, complete understanding of the evolution of this group is not possible without considering the planktonic forms.

Research on sessile rotifers has been mainly limited to use of natural populations because with available techniques few species have been reliably maintained in culture. Tiefenbacher (1972) and Wallace (unpubl. exp.) have maintained reproducing cultures of *Floscularia* for up to three months. Sometimes a species may bloom in hay infusions but when this happens it usually dies out relatively quickly (Edmondson, pers. comm.). Cori (1925) apparently holds the record for longest continuous culture by maintaining a population of *Cupelopagis vorax* for over one year. Some workers have taken advantage of situations in which a species is particularly plentiful in an aquarium or greenhouse pool (Koste, 1974; Surface, 1906; Vasisht & Dawar, 1969). Fortunately, many questions concerning the ecology of sessile rotifers can be considered without culturing a single species. Nevertheless, investigators must be opportunists because under natural conditions each species tends to develop to peak densities for a few weeks and then essentially disappears. Thus one takes up a new problem when suitable material appears; during one season progress is usually made on several problems rather than the investigator concentrating his full attention on just one.

Although the first serious study on the ecology of sessile rotifers was published about 35 years ago (Edmondson, 1944, 1945) and several contributions have been made since then, progress in answering some fundamental questions has been slow. For example, substrate selection and elucidation of its adaptive value has been conclusively demonstrated for only one species (*F. conifera*, Edmondson, 1945), although many species are so restricted in their distribution on surfaces that substrate selection seems probable (Wallace, 1977c). This slow progress is due to many reasons, not the least of which is the difficulty of working in the littoral zone with sessile organisms (Pejler, 1962). Sessile rotifers are fairly common on lakes and rivers in Europe and North America, but their particular distribution patterns appear dependent on the presence of specific substrates and on local chemistry of the habitat (see Edmondson, 1944; Koste, 1970a; Tiefenbacher, 1972; and Wallace, 1977a for detailed accounts).

One of the keys to understanding the ecology of a sessile rotifer no doubt lies in comprehending the relationship it has with its substrate(s). In any locality there are many surfaces on which larvae can settle, but each may provide a different potential for reproductive success. Since discrimination by larvae among surfaces is potentially crucial to post-metamorphic survival and reproduction, understanding the biology of the larval stage is critically important if we are to fully comprehend the life history and evolution of sessile rotifers. Thus I will begin this review by considering the larval stage.

Before proceeding with this discussion it is necessary to clear up a matter of language. Investigators often report their research on larval attachment by using terms such as choice, preference, or decision. They do not mean to

181

Hydrobiologia 73, 181-193 (1980).   0018-8158/80/0732-0181$02.60.
© *Dr. W. Junk b.v. Publishers, The Hague.*

imply consciousness on the part of the larva, nor do they wish to convey the idea that a larva remembers past encounters so that two potential substrates are cognitively compared. Such language is merely a convenient literary device to explain this biological phenomenon.

Larval Biology

Edmondson (1944) and Ruttner-Kolisko (1974) suggest that the young motile stage is not a true larva, but several workers have used the term (Cori, 1925; Koste, 1972; Vasisht & Dawar, 1969; Wallace, 1975; Wright, 1959). It does seem appropriate to do so, since the organism goes through a fairly extensive reorganization in body plan after permanent attachment. There is also a conceptual parallel with sessile marine invertebrates which prompts the use of the term.

Except for a few papers, larval biology has been ignored. Apparently no one has undertaken a study of sessile rotifers with a comprehensive review of larval biology as one of the goals. Consequently, the information that can be gathered about the larval stage is not at all complete. Here is an excellent area for research.

Egg hatching

Little information is available concerning egg hatching in sessile rotifers. Surface (1906) suggested that the frequent contractions of *Sinantherina socialis* embryos eventually burst the egg membrane. Similar contractions can be seen in embryos of many species but the actual hatching mechanism may be more sophisticated, perhaps involving imbibition of water and/or action of enzymes. Development time is believed to be controlled solely by temperature (Edmondson, 1945, 1971) and parental age (Edmondson, 1945; Ruttner-Kolisko, 1974). Nutritional history of the parent is thought to have little influence on egg development (Edmondson, 1971; Seitz & Halbach, 1973; cited in Hoffman, 1977). Total development time has been found to be on the order of one to three days. In two experiments Edmondson (1945) determined the development time of *F. conifera* eggs to be 2.4 and 3.8 days. Champ & Pourriot (1977b) found development time in *S. socialis* to be approximately 1.0 to 2.3 days (depending on the time of laying). Development time for *Lacinularia flosculosa* is on the order of two days (Pourriot *et al.*, 1972). These rates are slower than those reported for some planktonic rotifers at approximately the same tempera-

ture (20°C) (e.g. 0.5 to 1.0 days in *Brachionus urceolaris*, Ruttner-Kolisko, 1974). No systematic inquiry of egg development time in sessile rotifers has been undertaken.

There have been only two published accounts of the influence of photoperiod on hatching (both on *S. socialis*) and they disagree. Surface (1906) states that, ... 'the colonies usually hatch out during the morning hours, the young ball breaking away about noon or afterwards.' That is, the oldest eggs of the colony all hatch at about the same time and the larvae form into a planktonic colony (see *Larval escape*). He never observed, 'under natural conditions,' the formation of a larval colony in the evening hours. Champ & Pourriot (1977a) reported, 'The rhythm of hatching shows a synchronization with photoperiod...'. However, they stated that, 'hatching of the eggs during July occurs daily in the laboratory, as in nature, at about 3:00 o'clock in the afternoon' (see also Champ & Pourriot, 1977b). I have recently had opportunity to investigate egg hatching and larval colony release in a population of *S. socialis* collected from Buffalo Lake (Marquette Co., Wisconsin). In this strain hatching preceded larval colony release by only one or two hours, not the three to four which Surface reported, and peak larval colony release occurred between 0800 and 1100 hours (Fig. 1). Perhaps this conflict in time of egg hatching arises because of differences in the geographic-varieties. Surface's *Sinantherina* were originally isolated, 'some years before from a small pond...' and were currently growing in an aquarium. In lentic habitats (i.e., Surface's and mine) egg hatching and subsequent larval colony release early in the day will provide for a wide distribution since the photopositive larvae do not settle until after dark. In the Loire river habitat where Champ & Pourriot collected their specimens, a long planktonic existence would not be advantageous, since colonies might drift downstream out of the favorable habitat.

Champ & Pourriot (1977a, b) hypothesize that hatching in their population of *S. socialis* is controlled by a photolabile hatching-inhibitor produced in the dark. They base their hypothesis on the observation that groups of eggs held in photoperiods with progressively longer dark intervals (i.e., L/D of 16 : 8, 14 : 10, 12 : 12) require longer periods of light before hatching occurs.

Any species which uses light as an aid to larval navigation (e.g., *S. socialis*, see *Substrate selection behavior*) or requires light to make the choice of the optimal substrate (e.g., see following) should have egg hatching synchronized to photoperiod. Hatching soon after sunrise, in the first instance, would insure a wide dispersal of larvae, whereas in the second case larvae would have a full day to search for their optimal substrate. An example of this second case may be a strain of *Collotheca gracilipes* collected from a small pond in the University of Washington Arboretum (King Co., Washington). This strain prefers the abapical surface of *Elodea canadensis* leaves in the light but fails to discriminate between the upper and lower surfaces in the dark. Adults attached to the undersurface grow faster and produce more offspring than those on the upper surface (Wallace & Edmondson, in prep.).

Larval escape

Most sessile rotifers are oviparous and house their eggs within their tubes until the eggs hatch. Escape after hatching, therefore, can be quite different depending on tube construction. In many Flosculariidae the tube is a hollow concretion of jelly-like material secreted by the adult (e.g. *F. melicerta*) or formed by a cement gland into a more-or-less impenetrable structure (e.g., *Limnias melicerta*). The tube may or may not be formed with addition of detritus (e.g., *F. conifera*). In these tubes eggs are free and rest below the main body of the adult. Larvae escape by squeezing past the adult which can block the opening. I have never seen evidence which would suggest that larvae of this group escape by burrowing through the tube wall.

Larval escape by the colonial species, *S. socialis*, is unusual and worth noting. This sessile rotifer does not have a tube but carries its eggs attached to a specialized structure on the adult called the oviferon (Edmondson, 1959). According to Surface (1906) after the egg is expelled from the cloaca the animal bends dorsally so that the egg touches the slightly posterior adhesive gland of the oviferon to which it adheres. Upon hatching a larva is immediately free-swimming, but remains attached to its parent by a thin filament. Larvae apparently hatch at approximately the same time so that the threads of several (usually more than 25) larvae become entangled producing a larval colony which remains attached to the parental one. Finally, all the larval threads break, and the photopositive larval colony swims away.

Escape after hatching appears to be quite different in those species which produce a gelatinous tube (some members of all three families). In these species, besides the ability to exit the tube at the top opening, larvae can burrow through the jelly. In *C. gracilipes* burrowing is not

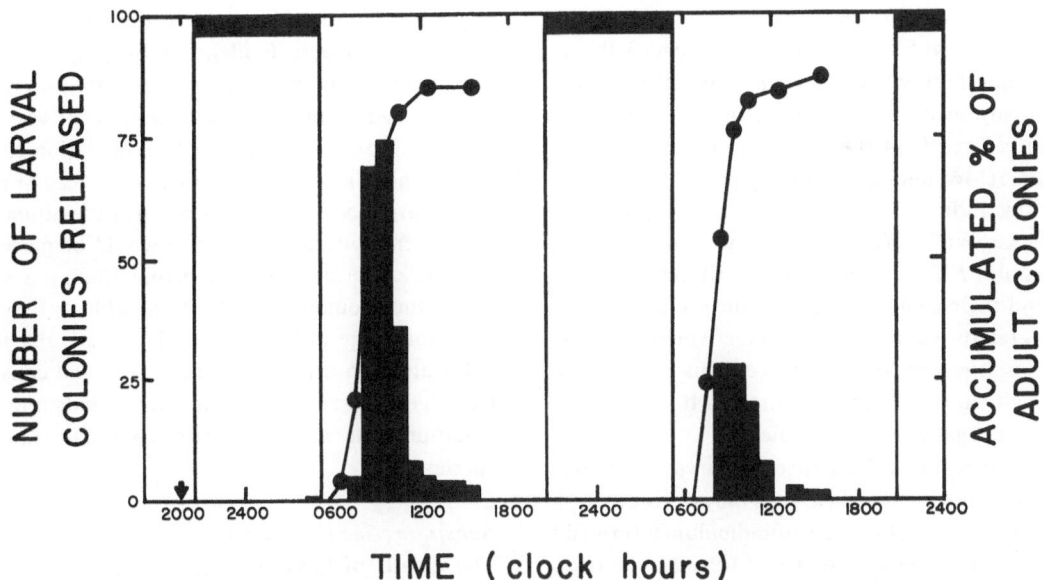

Fig. 1. Diel pattern of egg hatching and larval colony release in *S. socialis*. Histogram – number of larval colonies released; Solid circles – Accumulated percent of adult colonies which exhibited egg hatching. Material was collected from Buffalo Lake (Marquette Co., Wisconsin) (26.5° C) on 12 July 1979 between 1000 and 1200; 112 colonies were placed in 1-ml pond water-filled plastic depressions and held at 27 ± 1° C in an ≃ 16:9 L/D photoperiod which approximated the natural one. At 2000 of 12 July (arrow) all residual larval colonies were removed. At 0430 of 13 July, during the first observation period, a single larval colony was removed. (I believe it was inadvertently left from the evening before.) Observations during the dark period were made by using a Kodak 1A dark red filter.

always at right angles to the axis of the tube (the shortest route possible). The direction taken by a larva depends on the egg's position in the tube at time of hatching. When *C. gracilipes* larvae burrow they twist, bend, and contract at irregular intervals. By these movements and with the aid of their ciliated corona they propel themselves through the jelly-tube wall. One individual, whose complete hatching/escape sequence I observed, moved 180 μm in 6 minutes, a rate of 30 μm per minute (Wallace, unpubl. obs.).

Feeding

As far as I can determine, little work has been done concerning whether larvae are capable of feeding. One might assume that Flosculariidae larvae are suspension feeders since their ciliary apparatus is not that much unlike the adult corona. In fact, there is evidence which suggests that the larvae of two species do feed. *Ptygura beauchampi* larvae double in size during their first 4.5 hours of life and their mastaces are active throughout larval life (Wallace, 1975). (I have been unable to confirm feeding in larvae by direct observation using carmine or carbon black particles. However, adult *P. beauchampi* did not feed on these materials either. These particles are probably too large for both larvae and adults.) Larval *S. socialis* will feed on carmine (Wallace, unpubl. obs.).

Even assuming that Flosculariidae larvae do feed, there remains the question, what is their net energy balance. Larvae of many sessile marine invertebrates are known to feed and can reside in the plankton for considerable lengths of time (Meadows & Campbell, 1972; Crisp, 1974). Obviously this indicates a positive or, at worst, neutral energy balance. Wallace (1975) showed that in senescent (> 10 hours old) *P. beauchampi* larvae, swimming speed was slow and their ability to settle on the preferred substrate was greatly reduced. Most larvae which delayed metamorphosis beyond 20 hours died. This may mean that, although the larvae grew at first, in the long run there was a net negative energy balance.

Before metamorphosis, Collothecidae larvae resemble those of the family Flosculariidae – it is only during metamorphosis that the funnel-shaped infundibulum is formed (Wright, 1959). (To see the similarity between the two families one can compare larvae of *C. campanulata longicaudata* (Wright, 1959, Fig. 1; Koste, 1972, Fig. 2) and *C. trilobata* (Koste, 1970b, Fig. 1d, e) with those of *P. beauchampi* (Donner, 1964, Fig. 36a) and *P. pectinifera* (Koste, 1974, Fig. 2). Because Collothecids are raptorial, one would not expect the larvae to feed until after settling

and metamorphosis has taken place. Larvae may be suspension feeders before metamorphosis and adults raptorial, but that does not seem likely since the larval trophi would be maladapted to processing small particles. Koste (1970b) has reported that *C. trilobata* larvae do not feed until after metamorphosis. Collothecidae larvae, therefore, probably function solely on stored food and must find a suitable substrate before that reserve is depleted.

The problem of a larval energy budget is not merely an academic one. Edmondson (1944) found that the Collothecidae are comparatively rare in 'large localities'. These were 'areas with relatively little vegetation and large surface area'. Intermediate and small localities have, according to Edmondson, smaller surface area and more aquatic vegetation. The mean number of species he found in large localities was 0.5 for Collothecidae and 3.4 for Flosculariidae. Calculations of the G-statistic for heterogeneity (Sokal & Rohlf, 1969), using Edmondson's (1944) data from Table 14, demonstrated that the numbers of Flosculariidae species in the three locality-types were homogeneous ($0.9 > P > 0.5$). However, the numbers of Collothecidae species in the three locality-types were not all homogeneous ($P < 0.005$). Edmondson suggested that the explanation for the paucity of Collothecid rotifers in large localities was in the adult feeding type. The Flosculariidae he stated are probably better able to feed in places with water currents. Evidence supporting this hypothesis must come from feeding rate studies done on representatives of both families under conditions of varying water velocity. However, an alternate explanation can be proposed which is based on whether larvae feed or not. Since in the large locality class there was little aquatic vegetation, then Collothecidae larvae would be more likely to become exhausted and die before finding a surface to which they could attach. By being able to remain in the plankton longer, Flosculariidae larvae are able to locate a substrate even if plants are rare. As I will consider later (see *Age of larvae*) the larval energy budget may in part determine the specific substrate-search strategy for a species.

Substrate selection behavior

The details of larval substrate selection behavior have been described for only two species, *P. beauchampi* (Wallace, 1975) and *S. socialis* (Surface, 1906).

P. beauchampi has an unusual distribution. In any given locality it is limited to one or more of the following four substrates: 1) the left side of planorbid snails, 2) dead, brown grasses, 3) trap doors of the largest of three morphological distinct traps of the carnivorous hydrophyte *U. vulgaris* and, 4) the colonial bluegreen algae, *Gloeotrichia* (Edmondson, 1944; Wallace, 1977b). I have observed substrate selection behavior of *P. beauchampi* in response to two of its substrates, *U. vulgaris* and *Gloeotrichia echinulata* (Wallace, 1975, and unpubl. obs., respectively) and found them to be quite similar. Following is a summary of this behavior.

Once a larva has passed through a refractory period, during which time it will not settle, it will then react to surfaces in distinct ways. When encountering a surface other than its preferred one, a larva may pause for a few seconds with its corona in contact with the surface, but usually it swims away. The behavior of presenting the corona towards surfaces probably indicates that the larva's sensory apparatus is present there. When a larva encounters its preferred substrate, new behaviors can be seen. If, for example, a *P. beauchampi* larva of the strain that normally settles exclusively on *Gloeotrichia* encounters a *Gloeotrichia* colony, it will stop and begin swimming in among the vegetative filaments of the algae. Eventually the larva will attach to a vegetative filament and begin characteristic bending and twisting movements. Similar movements have been reported for the *Utricularia*-strain (Wallace, 1975). Larvae have been seen to detach from one site, swim around the *Gloeotrichia* colony, and reattach to another site several times before permanent attachment is made. During each attachment phase they undergo the characteristic twisting and bending movements. In the *Gloeotrichia*-strain I failed to see one particular behavior which was exhibited by the *Utricularia*-strain. In the latter, a dominant behavior was one in which the larva would attach its foot to the surface and bend over, touch the substrate with its corona, straighten up and repeat the process over and over, advancing to the left or right – eventually circumscribing a circle. It has been suggested that this behavior may establish a minimum distance between individuals to prevent crowding (Wallace, 1975). It is not known whether this movement is missing from the behavioral repertoire of the *Gloeotrichia*-strain or whether it cannot be accomplished because of interference by algal trichomes. Any time before final attachment is accomplished, larvae may swim away. In both strains nearly 100% of the larvae die within 48 hours if their preferred substrate is not available; very few ever metamorphose on another surface or while free-swimming (Wallace, 1977b).

Surface (1906) investigated the substrate selection behavior of *S. socialis* which apparently settled exclusively on *Myriophyllum* in aquaria at the University of Pennsylvania. According to Surface, once a photopositive larval colony is placed in darkness it will begin its settling behavior within 30 minutes if the proper substrate is available. The colony begins to swim at random and when coming into contact with an object, larvae orient their coronae towards its surface. This behavior against suggests that the larval sensory apparatus is located somewhere on the coronal field, but in this case the availability of food and not a chemical cue, as in *P. beauchampi* (Wallace, 1978), is responsible for the behavior. Surface reports that permanent attachment can occur if the colony remains at one site for a period of time. Apparently this happens if there is a sufficient quantity of debris on which the larvae feed. While feeding is going on a few larvae will venture out away from the others in short jerky movements, return, and then begin again. Each time a new sortie is undertaken a larva moves further away from the colony. Eventually a large number of individuals are doing this. Larvae advance slowly over previously unexplored surfaces and often return to areas where other larvae are found. This no doubt aids is maintaining the integrity of the colony until permanent attachment occurs. In areas where previous explorers have been, larvae move more quickly – presumably because food material is no longer present. Larvae may keep track of one another by the adhesive threads which once held them together, and which they apparently secrete continuously. Eventually a few individuals attach permanently and others join the colony which is usually completed within 90 minutes after onset of darkness.

In larvae of sessile marine invertebrates, planktonic navigation and substrate selection behavior are in part controlled by larval photosensitivity. Perhaps the most characteristic behavior of marine invertebrate larvae is a positive phototaxis (Crisp, 1974). Unfortunately little is known about larval response to light in sessile rotifers. As far as I can determine from reviewing descriptions and drawings in the literature, all sessile rotifer larvae have eye spots. Some species possess eyes as larvae and adults, while others have them only in the larval stage. If larvae are photoreceptive, how does that function in the life history of the organism? Perpetually photopositive larvae (such as *S. socialis*) would provide for greater and more rapid dispersion in the habitat, regulated of course, by the periodicity of egg hatching (see *Egg hatching* above). However, not all species can have perpetually photopositive larvae since settling would be delayed until after dark. This would be maladaptive for the strain of *C. gracilipes* in the Arboretum pond, since its larvae cannot distinguish between optimal and suboptimal substrates in darkness. Obviously more work needs to be done on this question.

Age of larvae

It is well known that as larvae of sessile marine invertebrates grow older, their physical and behavioral characteristics change. An interesting aspect to aging is that as a larva ages it changes the way it reacts to a particular substrate type (Knight-Jones, 1953; Crisp & Meadows, 1963). Sessile rotifer larvae age in the same way. *P. beauchampi* larvae, 0-120 minutes old, are rapid swimmers (Fig. 2), and are refractory to settling even when offered their preferred substrate (Wallace, 1975). Settling usually begins after about 120 minutes although a few may settle earlier than that. The most intense settling activity occurs in larvae which are approximately 200 minutes old (Wallace, 1975). After about 5 hours larval swimming speed decreases and continues to decline until death (Fig. 2). If larvae are deprived of their preferred substrate for 20 hours or more, they begin to lose their ability to settle. This loss is probably due to the larvae becoming physically weakened (net negative energy budget) so that eventually they are incapable of settling on any surface (see also Knight-Jones, 1953).

Not all sessile rotifers react like *P. beauchampi* when deprived of their preferred substrate; *C. gracilipes, F. conifera* and *S. socialis* will within 24 hours abandon searching for their preferred substrate and will settle anywhere, even on the bottom of plastic petri dishes or sometimes on the air-water interface (Surface, 1906; Wallace, unpubl. obs.). When considering why some species would abandon searching but not others, we must explore two points. One, I propose that only those species whose larvae feed can afford the luxury of a protracted search. If correct in the assumption that larvae of the family Collothecidae do not feed, then I would predict that they would abandon searching behavior relatively quickly, be-

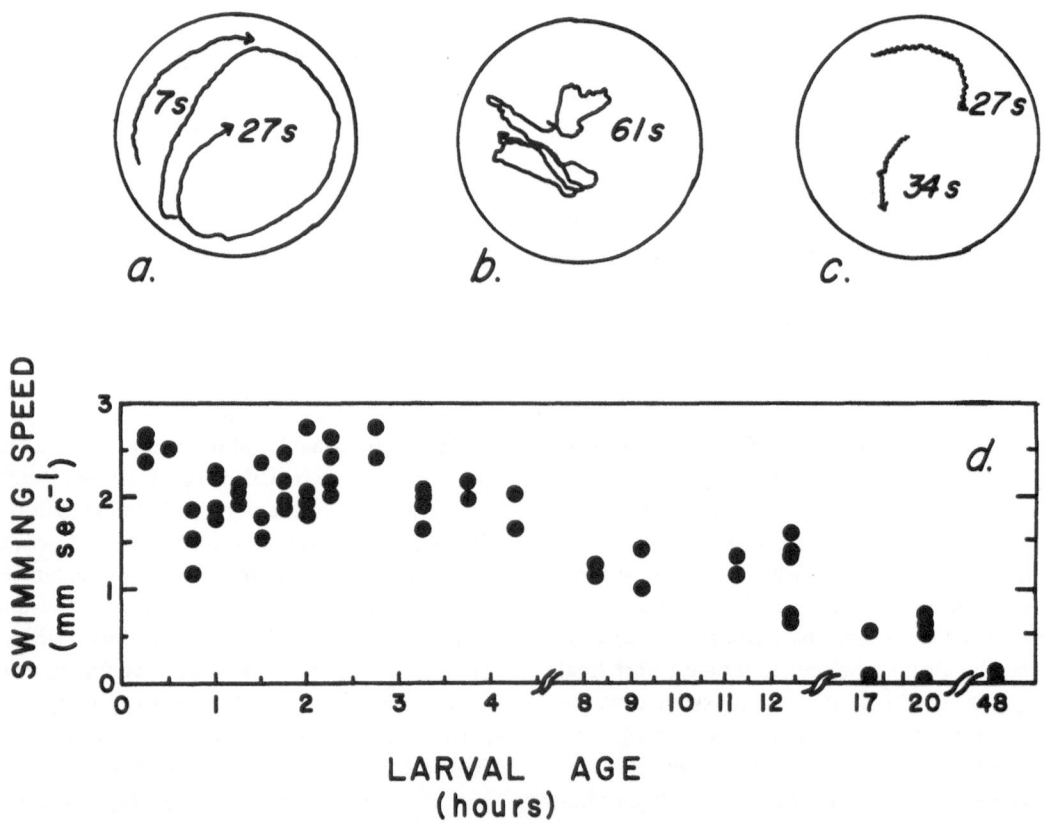

Fig. 2. *P. beauchampi* larval swimming behavior and speed. a, b, c–camera lucida tracings of swimming larvae, ≃ 15 mins (a), ≃ 8 hrs (b), and ≃ 25 hrs (c) old. The length of time and direction that each larva swam is indicated. d –swimming speeds of larvae at various ages.

fore their energy supply was exhausted. Only larvae which feed (perhaps only Flosculariidae) may prolong searching for their preferred substrate, but they do not always do so. This is the second point which is considered below.

To abandon substrate searching behavior or not probably indicates important relative differences in the adaptive values of those substrates to adults. If a species is unable to accept alternative substrates (e.g., *P. beauchampi*) it indicates that the preferred substrate provides adults with a habitat which is extraordinarily superior to all others. If the search is abandoned after a certain amount of time it indicates that although the preferred substrate is better than alternate ones, those substrates also provide an adequate habitat.

Jennings and Lynch (1928) showed that parental age at time of egg deposition is important to subsequent survivorship and reproduction of offspring in the planktonic rotifer *Proales sordida*. Presumably the same thing takes place in sessile rotifers but this has not been rigorously investigated (see Edmondson, 1945). It would also be of interest to know if parental age has an effect on larvae. That is, are larvae from older adults in some way less competent in substrate selection than those from younger adults?

Metamorphosis
Initiation and control of metamorphosis is not understood, but once a larva permanently attaches to a substrate the process proceeds until the rotifer achieves adult form. Wright's (1959) description of metamorphosis in *C. companulata longicaudata* is summarized below (see also Koste, 1972).

Once the larva is permanently attached it contracts so that it is reduced to about one-half its previous size. In one to two minutes the body rises off the substrate but the foot remains attached. The foot then detaches from the substrate and the whole animal lifts off the substrate but remains attached to the surface by a developing peduncle. The body remains contracted as the peduncle continues to elongate to approximately 90 μm or more. After the peduncle is completed, activity ceases for about 20 minutes during which time Wright believes the corona to be developing. Within one hour of the original attachment the corona emerges and the foot has begun to elongate further. At this point the metamorphosis is essentially complete. A less complete description of larval metamorphosis in *Limnais melicerta* was reported by Wright (1954).

Distribution on substrates

The distribution of sessile rotifers on substrates in the field will be determined by two factors: 1) substrate selection by larvae and 2) substrate-dependent survivorship of adults. Components of the first factor are the larva's ability to discriminate between substrates and on the probability of encountering each substrate. The second factor involves how well the individual does once permanent attachment has occurred.

Evidence of substrate selection
In the course of their investigations researchers often list all substrates on which a species was found. Sometimes species have been discovered only on a particular substrate type. Although it is tempting to suggest that such highly restricted distributions represent larval choice, it is only circumstantial evidence, and may actually indicate an extreme example of substrate-dependent survivorship. Before substrate selection may be attributed to them it must be statistically proven that the larvae actually prefer one substrate over all others present in the system. This has been done very few times in relation to the number of restricted distributions described in the literature (Wallace, 1977a). For example, Wallace (1977b) reported that most *P. beauchampi* larvae died ($>$95%) after 48 hours if their preferred substrate (trap doors of the large trapping organs of *U. vulgaris*) was not available. *P. brevis* is a second species whose distribution is restricted. Its larvae select the forks of highly dissected leaves (Edmondson, 1944).

Other species, while having restricted distributions in nature, fail to discriminate among several naturally occurring substrates in experiments. For example, with very few exceptions, Edmondson (1939, 1944) found *C. gracilipes* attached to *U. vulgaris*. However, in laboratory experiments larvae did not discriminate between *U. vulgaris* and two other macrophytes, *Myriophyllum* sp. and *Potamogeton pusillus*.

Hierarchical preferences for an array of substrates exist as well. *F. conifera* larvae appear to have a clear preference for tubes of conspecifics but will also colonize macrophytes (Edmondson, 1945; Wallace, 1977a). When Edmondson (1944) offered them a choice of three plants, 65% chose *U. vulgaris*, 26% *Chara*, and 9% *Potamogeton pusillus*. *Collotheca gracilipes* from the Arboretum pond also selected its substrates in a certain order: *Lemna* sp. $>$ *Elodea canadensis* $>$ *Myriophyllum spicatum* $>$ *Nymphea* sp. $>$ plastic petri dishes (Wallace and Edmondson, in prep.). Presumably such ranking of substrates indicates an individual substrate attractiveness based on the presence of factors (settling factors or cues) which induce larvae to settle, and which vary in their quantity and/or quality on each substrate. This hypothesis will be taken up again in the following section (see *Substrate selection mechanisms*).

Habitat loyalty (Doyle, 1976), perhaps more properly termed substrate loyalty in this case, is a phenomenon which apparently occurs in sessile rotifers. Substrate loyalty in Doyle's terms is, 'a property of populations which are distributed over two or more habitat [substrate] types' and which 'is exhibited when the probability that offspring will return to the parental [substrate] type is higher than the mean probability of choosing that [substrate], averaged over the whole population'. No sessile rotifer populations have been found to exhibit substrate loyalty in a single lake. However, two species are known to exhibit a type of substrate loyalty. In Green Lake (King Co., Washington) *P. beauchampi* larvae settle on *Gloeotrichia echinulata* but did not accept *U. vulgaris* when offered. However, in Mud Pond (Grafton Co., N.H.) larvae colonized *U. vulgaris* but were never found in colonies of *Gloeotrichia* sp. The second species which appears to exhibit substrate loyalty is *C. gracilipes*. Larvae of that species collected from the Arboretum Pond preferred *Elodea* to *Utricularia* but Edmondson (1944) found the opposite preference in a population collected from Bird Preserve Pond (New Haven Co., Conn.).

True generalist species, in which no substrate preferences at all are seen, may exist (Tiefenbacher, 1972; Wallace, 1977a) but this is impossible to determine without detailed larval choice experiments. If they exist, it would indicate that no single substrate imparts a sufficiently large or consistent competitive, and therefore selective, advantage to the adult.

Substrate selection mechanisms

The mechanism whereby a larva selects a surface on which it will attach will eventually be described in sensory physiology terms. Presently we can only speculate as to where the sensory receptors are located (probably on the coronal field) and how they function. The specific settling factors which evoke sensory receptor response are just beginning to be discovered.

During the life of a larva it may encounter several surfaces, therefore, the real biological question is, how does a larva process the information it receives about a substrate to 'know' if it is a preferred one? Perhaps sessile rotifer

larvae operate in a manner similar to the way which has been suggested for larvae of marine invertebrates. Müller, Wicker & Eiben (1976) propose that larvae exhibit 'appetitive or searching behavior which continues until the larva is presented with a specific releasing stimulus triggering fixation (attachment) and metamorphosis'. In other words, the larval cue-sensing apparatus probably functions in a simple stimulus/reponse manner. Only if the settling factor is present in detectable levels will larvae begin to settle. The way larvae respond to a surface changes as they grow older (usually by showing less discrimination). Therefore, one would expect that the threshold level of stimulus required to elicit a response would decrease with larval age. In terms of Levins & MacArthur (1969) I propose that a larva's substrate choice 'is based on simple sensory coding of... physical and/or chemical aspects of a surface; thus, although the sensory system may respond differently to each (substrate) in the environment the processing system is limited to very simple decision rules' (behavioral mechanisms). Therefore,

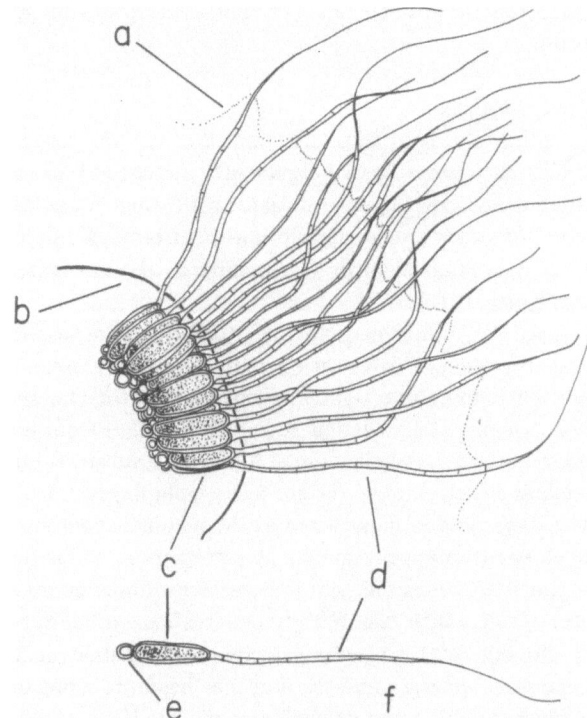

Fig. 3. A section through a colony of *Gloeotrichia echinulata* showing its construction. a–outer limit of the water soluble settling factor which can be seen when stained with crystal violet. b–central core of firm mucilage. c–akinete. d–vegetative filament. e–heterocyst. f–single trichome of *Gloeotrichia*.

in an experimental universe in which each substrate type has a specific amount of settling factor associated with it, I would predict that, given equal substrate areas, larvae should settle on the substrates in proportion to the amount of detecable settling factors associated with each. The strain of *P. beauchampi* which settles exclusively on *Gloeotrichia* may illustrate this idea (Wallace, unpubl. exp.). This rotifer settles in among trichomes of *G. echinulata* in response to a water soluble settling factor held between the solid jelly core and the outer tips of the trichomes (Fig. 3). This material, whose properties have not yet been identified, can be isolated by crushing several *Gloeotrichia* colonies in a ground glass mortar and pestle. If this crude preparation is presented to larvae they will usually begin settling activities within 30 minutes. When a serial dilution of the preparation is presented to several batches of larvae, a decreasing percentage of larvae will settle within 24 hours (Fig. 4). This mechanism may be responsible for the hierarchical preference for substrates exhibited by *C. gracilipes* (*Elodea*-strain) and *F. conifera* (Edmondson, 1944).

Certain marine invertebrate larvae will respond to more than one stimulus. Barnacles, for example, will react to a chemical (arthropodin) and/or the physical texture of the surface (Crisp, 1974). *F. conifera* may be an example of a sessile rotifer which uses multiple stimuli: perhaps one for conspecific tubes and other for macrophytes. Stimuli which prevent settling may also exist (e.g., the hypothetical spacing behavior used by *P. beauchampi* larvae, see section substrate selection behavior).

Some information is available as to the nature of cues which larvae use in choosing their substrates. Both chemical and physical cues are known.

There is no single cue used by each species: apparently they can vary on a geographic basis. I have suggested that *P. beauchampi* larvae select *U. vulgaris* because of the presence of a unique chemical associated with glandular trichomes present on the large traps of that plant (Wallace, 1978). However, the strain that selects *Gloeotrichia* apparently does so because of a water soluble material produced by the alga (see above). These two factors are probably not identical, since I have discovered the strain which settles on *Gloeotrichia* will not colonize *Utricularia*. The *P. melicerta* group (three species) and *C. algicola* also attach to *Gloeotrichia* (Edmondson, 1940) and may use the same cue as *P. beauchampi*. The predatory rotifer *Acyclus inquietus* is restricted to colonies of its prey *Sinantherina* and probably uses a cue specifically related to that genus (Edmondson, 1940). Larval *F. conifera* no doubt use cues which reside in the tubes of conspecifics. Tiefenbacher's report (1972) that he rarely saw colony formation in this species is probably, as he pointed out, due to the small population sizes with which he worked. *Collotheca epizootica* apparently prefers Cladoceran carapaces (Monard, 1922). *Ptygura brevis* and *Cupelopagis vorax* both use physical shape of the substrate as a cue. *Ptygura brevis* has a preference for the forks of highly dissected leaves such as *Myriophyllum*. *Cupelopagis vorax* seems to prefer broad flat surfaces such as water lilies (Edmondson, 1944; Wallace, 1977a).

Significance of substrate selection
Since adult sessile rotifers are not locomotory, the key to their survival rests in larval substrate choice. Natural selection will eliminate any individual which, in the larval stage, does not choose an optimal substrate, that is, one with a balance of the following variables: abundant food, limited predation, and favorable physical/chemical conditions; Wallace, 1978). Consequently, as we have considered above, it is no surprise to discover that larvae use more-or-less specific cues in making their decision. The adaptive significance of substrate selection has been suggested for several species (Edmondson, 1944, 1945; Wallace, 1977a, b, 1978) and two general strategies have so far become apparent: enhancement of feeding and predator avoidance.

To date Edmondson's work on *F. conifera* is the only one which has conclusively demonstrated an adaptive significance for substrate selection (Edmondson, 1945).

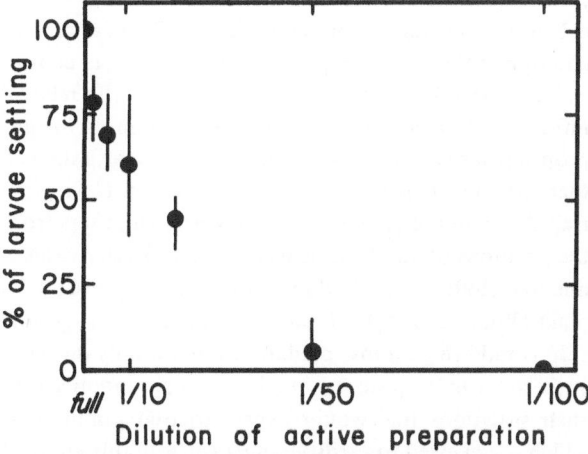

Fig. 4. Settling of *P. beauchampi* larvae in the presence of different dilutions of *Gloeotrichia* settling-factor. Means with ranges are shown. At least four experiments of 25 larvae each were used at the respective dilutions of the active preparation.

Colonial adults live longer and produce more offspring then solitary individuals. Presumably two or more animals which are close together can reinforce each other's filtering ability in some nonlinear manner, so that energy intake is greater than that of a solitary rotifer of the same age. Feeding enhancement may be important to all colonial rotifers whether they form intra- or interspecific colonies (see Tiefenbacher, 1972; Wallace, 1977). Being in a colony may also provide protection from predators. Surface (1906) states that those *S. socialis* larvae which do not form a larval colony eventually settle down near the adult one, but they 'do not long survive the attacks of their enemies…' Solitary *Conochilus dossuarius* and *C. unicornis,* which have been removed from their colony by using pronase to hydrolyze the jelly matrix, are very vulnerable to predation by *Asplanchna,* but generally colonies are not (Gilbert, 1980). However, in incidental laboratory observations I have seen whole colonies ingested by Sticklebacks. Perhaps predation by fish can account for their behavior of remaining in the shadows of lily pads (Edmondson, 1959).

Acyclus inguietus and *Cupelopagis vorax,* it is believed, also gain a feeding advantage by substrate selection. *Acylus* is found in colonies of its prey, *Sinantherina,* on whose larvae it feeds (Edmondson, 1940). *Cupelopagis* is said to prefer broad flat surfaces over narrow ones because flat surfaces aid in prey capture (Edmondson, 1944; Wallace, 1977a). Some plants with highly dissected leaves are, however, adequate for *Cupelopagis.* I have seen a substantial population develop on *Ceratophyllum* in an aquarium.

Several species, *P. beauchampi,* the *P. melicata* group, and *C. algicola,* probably avoid predation by selecting the colonial bluegreen algae, *Gloeotrichia,* as a substrate. This substrate may provide a refuge from aufwuchs consumers such as insect larvae and snails which will graze upon sessile rotifers. For example, Tiefenbacher (1972) reports that insect larvae (chironomids) can tear off pieces of *F. ringens* tubes. I have discovered tube fragments of *L. melicerta* in fecal matter from snails which were offered *E. canadensis* leaves colonized by that sessile rotifer. *P. beauchampi* probably avoids predation by colonizing the trap doors of *U. vulgaris,* as well as *Gloeotrichia* (Wallace, 1978). In this case most predators will themselves fall prey to the carnivorous plant before they can consume *P. beauchampi.* Three species are believed to live in cryptic places which predators cannot reach. *Ptygura brevis* lives in forks of leaves; *P. velata* lives within the cavities of highly curled *Sphagnum erythrocalyx* leaves (Edmond-

son, 1944); and *C. ambigua* apparently also lives within the cavities of highly curles *Spagnum* leaves (Hudson & Gosse, 1886).

Recent progress I have made towards discovering the cue used by *P. beauchampi* larvae to select *Gloeotrichia* will help test the general applicability of the predator avoidance hypothesis. A locality is required in which there is a substantial population of *P. beauchampi* colonizing *Gloeotrichia.* Hundreds of larvae may be isolated by periodically removing them from concentrated samples of *Gloeotrichia.* Since the *Gloeotrichia* settling factor can be isolated, *P. beauchampi* larvae can be induced to settle on natural or artificial substrates by adding the settling-factor preparation to a vessel containing larvae and substrates. After the larvae have settled, their position on each substrate can be determined, and predators added for a certain period of time. The survivorship on the different substrates can then be statistically compared. It would be particularly interesting to use this technique to determine whether *P. beauchampi* attached to the trap doors of *U. vulgaris* are less subject to predation than those induced to attach elsewhere.

Substrate-dependent survivorship
Substrate-dependent survivorship in marine organisms has been well documented, e.g., barnacles (Connell, 1961) and serpulid polychaetes (Knight-Jones *et al.,* 1971). In substrate-dependent survivorship, larval recruitment on all substrates may or may not be equal, but larval survivorship is definitely not. In all likelihood, except for Edmondson (1945), no one has bothered to look for it in the sessile rotifers. Cases in which larvae colonize several substrates, but survive on only a few, will appear as a restriction of the species to those substrates. This, of course, will resemble larval substrate selection but the relative influence of each can be determined by running the appropriate larval choice experiments. Doing that, I showed that although *P. beauchampi* larvae settle on *U. vulgaris* trap doors in a certain pattern, this pattern differs from the positions of adults in field collections. I believe this is due to dislodgement of adults during prey capture by the plant (Wallace, 1977b). If substrate-dependent survivorship is mild (e.g., a low predation rate on only one substrate used by a species) it might easily go unnoticed. In such situations one would expect to find populations which are skewed towards younger (measurably shorter) individuals to develop on suboptimal substrates. In the spring of 1976 I had an opportunity to look at the size frequence distribution of *F. ringens* and *Stephanocerous*

fimbriatus on four macrophytes in the Aboretum Pond. Some interesting results were found (Table 1). Pairwise comparisons of the four *F. ringens* populations using the Mann-Whitney U-statistic (Sokal & Rohlf, 1969), demonstrated that the population on *Nymphaea* was shorter than on any other substrate ($P > 0.99$). The populations on *Lemna, Myriophyllum,* and *Elodea* were statistically indistinguishable (Kruskal-Wallis statistic, $0.2 > P > 0.1$, Sokal & Rohlf, 1969). Some of the tubes appeared to be damaged in the same way as Tiefenbacher (1972) described, but unfortunately I did not correlate damage with substrate type. The *S. fimbriatus* populations on three of the four macrophytes were significantly different from one another (Kruskal-Wallis, $P < 0.025$). I discounted the possibility that these results were an artifact of a rough collection procedure since care was taken in handling the plant material and no free *Floscularia* or *Stephanocerous* tubes were found in the vessels used for transport and temporary storage. The shorter populations may have resulted from predators foraging on *Nymphaea* or from abrasion by submerged objects on the undersurface of the floating lily pads.

One may ask why natural selection would permit a species to continue colonization of an obviously suboptimal substrate (i.e., *Nymphaea*); why have the behavioral mechanisms not evolved to eliminate settling there? I suggest that the mechanisms currently in use continue simply because they are adequate to maintain a viable population. Eventually a new one may be adopted which is more suitable for prevailing conditions (*Nymphaea* might no longer be colonized), but natural selection processes are very slow.

Food and feeding habits

Feeding in sessile rotifers is an area in need of much research. Although there have been some interesting reports on feeding behaviors, many researchers have had only enough material or time to gather incidental observations on the diets of a few species.

Members of the sessile families can be separated easily by their feeding type. Flosculariidae and Conochilidae are suspension feeders which gather small particles from the water by generating feeding currents with their ciliated coronae. Diatoms have been reported to be an important component of the food in *F. ringens* (Tiefenbacher, 1972), *L. flosculosa* (Pourriot *et al.,* 1972), and *S. socialis* (Champ & Pourriot, 1977b). Members of the genus *Ptygura* are thought to feed mainly on detritus (Koste, 1970a), but probably take any small particles. The *Conochilus unicornis-hippocrepis* group also feeds on bacteria and detritus (Pourriot, 1977). *Conochilus* (= *Conochiloides*) *natans* has been reported to feed on diatoms (Centrales) and Chrysomonadales (Pourriot, 1977).

Collothecidae are raptorial. They have a corona which is modified into a lobed funnel-shaped structure (infundibulum) the margins of which are extended by lobes and which usually possess supple setae (however, members of the genera *Acyclus, Atrochus,* and *Cupelopagis* have no coronal setae). Edmondson (1959) suggests that these setae guide thigmotactic prey towards the corona. Prey are captured when the lobes of the corona fold in quickly forcing them through the mouth. Prey capture can be aided by whip-like action from the setae (Wright, 1958; Edmondson, 1959). Prey are first stored in the vestibulum

Table 1. Average length (±1SE) of F. ringens and S. fimbriatus on four macrophytes in the Arboretum Pond on 17 May 1976.

| Rotifer species | Macrophyte species | | | |
	Lemna sp.	Elodea canadensis	Myriophyllum spicatum	Nymphaea sp.
F. ringens	787±94um (n=38)	1021±124 (n=22)	909±59 (n=40)	240±41 (n=40)
S. fimbriatus	847±65 (n=29)	------ (n=4)	833±68 (n=35)	669±44 (n=70)

before being passed on through the pharynegeal tube into the proventriculus where they are macerated by the trophi. Collothecids generally feed on diatoms, green and colorless flagellates, ciliates, and small rotifers (Berzins, 1951, 1952). For example, *C. h. heptabrachiata* feeds on *Gymnodinum* and colorless flagellates (Sladecek, 1962), *C. trilobata* feeds on diatoms, *Euglena,* dinoflagellates, and small rotifers (Koste, 1970b), *Cupelopagis vorax* consumes ciliates and small rotifers (Koste, 1973), *Stephanocerous* prefers *Euglena* and *Chilomonas* (Valerio, 1975), and *Acyclus inquietus* is somewhat specialized, feeding on the larvae of *Sinantherina* (Beauchamp, 1912; Edmondson, 1959).

Several interesting feeding phenomena have been observed or proposed which need to be examined further. Obviously Edmondson's hypothesis that colonial rotifers have higher feeding rates than solitary individuals needs to be confirmed by appropriate experiments. Wright (1950) describes a behavior used by *F. ringens* whereby individuals rid themselves of unwanted material which has reached the mastax. Basically the trophi seize particles and push them back into the feeding current which has been reversed. He observed ciliary reversal in *P. pilula* as well, and suggested that it is a 'generic strategem'. Valerio (1975) describes a selection mechanism used by *Stephanocerus* sp. Selection for *Euglena* and *Chilomonas* is reported to take place in the vestibule where potential prey are temporarily held. Periodically other organisms and particulate matter are released and then the preferred prey swallowed. This is not unlike what Lucks (1929, cited in Berzins, 1952) has described: prey are held in the vestibule and seem to be evaluated before consumption. Berzins (1952) has hypothesized the release of an invisible chemical prey-lure from the 'mouth cavity' of several Collothecidae. The material is said to be projected into the surrounding water by 'vomiting' and small 'wave-like motions' of the setae. This lure supposedly acts selectively on prey guiding them along the chemical gradient. The lure from *C. h. heptabranchiata* 'exerts no influence on *Peridina* species, but stimulates a marked reaction on some... species of Chlorophyta'. Wright (1958) reported that he could not confirm the existence of this hypothetical attractant.

Conclusions

Clearly more research on all aspects of sessile rotifers needs to be done. For example detailed studies should be undertaken which correlate reproductive rate with food type and availability, as Edmondson (1965) did for the planktonic rotifers (e.g., see Champ, 1978). Also missing from the literature is sufficient information on sexuality and the occurrence of amphoteric reproduction (e.g., see Champ & Pourriot, 1977b). It would also be interesting to know how males go about searching for mictic females (e.g., is it by random swimming or do they use the same substrate cues as the larvae?).

At the present time, developing an evolutionary scenario for the sessile rotifers would be a difficult but interesting problem. In so doing, several questions must be kept in mind. These would include the following: 1) How did the larval stage evolve? 2) How did the use of cues of recognize a particular substrate develop, and how does natural selection modify existing substrate selection mechanism when an optimal substrate becomes inadequate? 3) How did coloniality develop? 4) What pressures led to the abandonment of sessile life by the planktonic species?

Acknowledgments

Preparation of this paper was supported in part by a Dartmouth Teaching Fellowship, a Grant-in-Aid of Research from the Society of the Sigma Xi, an N.S.F. grant to W. T. Edmondson and R. L. Wallace (DEB 75-03107A01), and a grant from the Faculty Development Fund of Ripon College. I wish to thank W. Brooks and L. Wallace who read and improved this manuscript, and L. Toussaint who typed it.

References

Beauchamp, P. de. 1912. Rotifères communiques par MM, H. K. Harring et C. F. Rousselet: Contribution a l'étude des Atrochidés. Bull. Soc. zool. France 37: 182-187.

Berzins, B. 1951. On the Collothecacean Rotatoria. Aktiv. Zool. I (37): 565-592.

Berzins, B. 1952. Notes on the feeding of some Rotifera. J. Quekett. micr. Cl., Ser. 4, 3: 334-336.

Champ, P. 1978. Dynamique d'une population d'un Rotifère épiphyte thermophile (Sinantherina socialis) en présence de pollution thermique. Arch. Hydrobiol. 83: 213-231.

Champ, P. & Pourriot, R. 1977a. Reproductive cycle in Sinantherina socialis. Arch. Hydrobiol. Beih. 8: 184-186.

Champ, P. & Pourriot, R. 1977b. Particularties biologiques et écologiques du Rotifère Sinantherina socialis (Linne). Hydrobiologia 55: 55-64.

Connell, J. H. 1961. The influence of interspecific competition and other factors on the distribution of the barnacle Chthamalus stellatus. Ecology 42: 710-723.

Cori, C. I. 1925. Zur Morphologie und Biologie von Apsilus vorax Leidy. Zeits für Wiss. Zool. 125: 557-584.

Crisp, D. J. 1974. Factors influencing the settlement of marine invertebrate larvae. In: P. T. Grand & A. M. Mackie (eds.) Chemoreception in Marine Organisms. pp. 177-265, Academic Press, N.Y.

Crisp, D. J. & Meadows, P. S. 1963. Absorbed layers: The stimulus to settlement in barnacles. Proc. R. Soc. B. 158: 364-387.

Donner, J. 1964. Die Rotatorien-Synusien submerser Makrophyten der Donau bei Wien und mehrerer Alpenbäche. Arch. Hydrobiol. Suppl. 27: 227-324.

Doyle, R. W. 1976. Analysis of habital loyalty and habitat preference in the settlement behavior of planktonic marine larvae. Am. Nat. 110: 719-730.

Edmondson, W. T. 1939. New species of Rotatoria, with notes on heterogonic growth. Trans. Am. microsc. Soc. 58: 459-472.

Edmondson, W. T. 1940. The sessile Rotatoria of Wisconsin. Trans. Am. microsc. Soc. 59: 433-459.

Edmondson, W. T. 1944. Ecological studies of sessile Rotatoria. Part I. Factors affecting distribution. Ecol. Monogr. 14: 31-66.

Edmondson, W. T. 1945. Ecological studies of sessile Rotatoria. Part II. Dynamics of populations and social structure. Ecol. Monogr. 15: 141-172.

Edmondson, W. T. 1959. Rotifera. In: W. T. Edmondson (ed.) Freshwater Biology 2nd Ed., John Wiley and Sons, Inc., N.Y. pp. 420-494.

Edmondson, W. T. 1965. Reproductive rate of planktonic rotifers as related to food and temperature in nature. Ecol. Monogr. 35: 61-111.

Edmondson, W. T. 1971. Reproductive rate determined indirectly from egg ratio. In: W. T. Edmondson and G. G. Winberg (eds.). A manual on methods for the assessment of secondary productivity in fresh waters. pp. 165-169, IBP Handbook No. 17.

Gilbert, J. J. 1980. Observations on the susceptibility of some protists and rotifers to predation by Asplanchna girodi. In this volume, pp. 87-91.

Hoffman, W. 1977. The influence of abiotic environmental factors on population dynamics in planktonic rotifers. Arch. Hydrobiol. Beih. 8: 77-83.

Hudson, C. T. & Gosse, P. H. 1886. The Rotifera or wheel animalcules. Vol. I and II, Longmans, Green, and Co., London.

Jennings, H. S. & Lynch, R. S. 1928. Age, mortality, fertility, and individual diversities in the rotifer Proales sordida Gosse. I. Effect of age of the parent on characteristics of the offspring, J. exp. Zool. 50: 345-407.

Knight-Jones, E. W. 1953. Decreased discrimination during settling after prolonged planktonic life in larvae of Spirorbis borealis (Serpulidae). J. mar. biol. Ass. U.K. 32: 337-345.

Knight-Jones, E. W., Bailey, J. H. & Isaac, M. J. 1971. Choice of algae by larvae of Spirorbis particularly of Spirorbis spirorbis. In: D. J. Crisp (ed.), 4th European Marine Biology Symp. pp. 89-104. Cambridge Univ. Press, Cambridge.

Koste, W. 1970a. Über die sessilen Rotatorien einer Moorblänke in Nordwestdeutschland. Arch. Hydrobiol. 68: 96-125.

Koste, W. 1970b. Das Radertier-Porträt Collotheca trilobata, ein seltenes sessile Radertier. Mikrokosmos 2: 195-200.

Koste, W. 1972. Das Radertier-Porträt Collotheca canpanulata longicaudata. Mikrokosmos 4: 97-99.

Koste, W. 1973. Das Rädertier-Porträt, ein merkwürdiges festsitzendes Rädertier: Cupelopagis vorax. Mikrokosmos 4: 101-106.

Koste, W. 1974. Das Rädertier-Porträt Ptygura pectinifera die 'Kammträgerin', eine Einwanderin in Warmwasseraquarien. Mikrokosmos 6: 182-187.

Levins, R. & MacArthur, R. 1969. An hypothesis to explain the incidence of monophagy. Ecology 50: 910-911.

Meadows, P. S. & Campbell, J. I. 1972. Habitat selection by aquatic invertebrates. Adv. Mar. Biol. 10: 271-382.

Monard, A. 1922. Une nouvelle espèce de Rotateur: Floscularia epizootica, nov. spec. Bull. Soc. Neuchâtel. Sci. nat. 46: 66-68.

Müller, W. A., Wicker, F. & Eiben, R. 1976. Larval adhesion, releasing stimuli and metamorphosis. In: G. O. Mackie (ed.) Coelenterate Ecology and Behavior. pp. 336-346, Plenum Press.

Pejler, B. 1962. On the taxonomy and ecology of benthic and periphytic Rotatoria. Zool. Bidr. Uppsala 33: 327-422.

Pourriot, R. 1977. Food and beeding habits of Rotifera. Arch. Hydrobiol. Beih. 8: 243-260.

Pourriot, R., Royer, G. & Peltier, M. 1972. Proliferation de rotifères épiphytes et pollution thermique dans la Loire. Bull. fr. Pisc. 244: 111-118.

Ruttner-Kolisko, A. 1974. Plankton Rotifers. Biology and Taxonomy. Binnengewässer (Suppl. Ed.) 26 (part 1): 1-146.

Sládeček, V. 1963. Notes on the ecology of two sessile rotifers from acid waters. Tech. Water 7: 563-568.

Sokal, R. R. & Rohlf, F. J. 1969. Biometry. W. H. Freeman and Co., San Francisco, p. 776.

Surface, F. M. 1906. The formation of new colonies of the rotifer, Megalotrocha alboflavicans. Ehr. Biol. Bull. 11: 182-192.

Tiefenbacher, L. 1972. Beiträge zur Biologie und Ökologie sessiler Rotatorien unter besonderer Berücksichtigung des Gehäusebaues und der Regenerations-fähigkeit. Arch. Hydrobiol 71: 31-78.

Vasisht, H. S. & Dawar, B. L. 1969. Anatomy and histology of the Rotifer Cupelopagis vorax Leidy. Res. Bull. (N.S.) Panjab Univ. 20: 207-221.

Valerio, C. E. 1975. Mechanismos para la captura y selection del alimento en Rotiferos especializados. Bremesia 5: 39-45.

Wallace, R. L. 1975. Larval behavior of the sessile rotifer Ptygura beauchamp: Edmondson. Verh. int. Ver. Limnol. 19: 2811-2815.

Wallace, R. L. 1977a. Distribution of sessile rotifers in an acid bog pond. Arch. Hydrobiol. 79: 478-505.

Wallace, R. L. 1977b. Substrate discrimination by larvae of the sessile rotifer Ptygura beauchampi Edmondson. Freshw. Biol. 7: 301-309.

Wallace, R. L. 1977c. Adaptive advantages of substrate selection by sessile rotifers. Arch. Hydrobiol. Beih. 8: 53-55.

Wallace, R. L. 1978. Substrate selection by larvae of the sessile rotifer Ptygura beauchampi. Ecology 59: 221-227.

Wallace, R. L. & Edmondson, W. T., in prep. Mechanism and adaptive significance of substrate selection by the sessile rotifer Collotheca gracilipes Edmondson.

Wright, H. G. S. 1950. A contribution to the study of Floscularia ringens. J. Queckett microsc. Cl., Ser. 4, 3: 103-116.

Wright, H. G. S. 1954. The ringed tube of Limnias melicerta Weisse. Microscope 10: 13-19.

Wright, H. G. S. 1958. Capture of food by Collothecid Rotatoria. J. Quekett microsc. Cl., Ser. 4, 5: 36-40.

Wright, H. G. S. 1959. Development of the peduncle in a sessile rotifer. J. Quekett microsc. Cl., Ser. 4, 5: 231-234.

RESPIRATION OF DIFFERENT STAGES AND ENERGY BUDGETS OF JUVENILE BRACHIONUS CALYCIFLORUS

Norbert LEIMEROTH

Johann Wolfgang Goethe-Universität, Frankfurt/M., G.F.R.

Keywords: *Brachionus calyciflorus,* respiration, energy budget, assimilation efficiency, K_1, K_2.

Abstract

Respiration data for different stages of *Brachionus calyciflorus,* fed with three concentrations of *Kirchneriella lunaris* at 20°C, are presented. Increasing oxygen consumption from 4.1 to 4.6 .10^{-3} $\mu l/h$ x ind. with food decreasing from 5.10^6 to 10^6 and 4.10^5 cells/ml has been found for adult females with one egg, but other age groups showed divergent results. Based on the respiration data for age groups 0 to 12 and 12 to 24 h old and some other results and calculations – e.g. dry weight and caloric content of eggs and females, ingestion rates/h for the different concentrations of food – energy budgets for juvenile, growing *B. calyciflorus* are presented.

Introduction

The fresh-water rotifer *B. calyciflorus* is a useful experimental organism for determining population parameters, because it is easy to cultivate, has a short life cycle, and a high reproductive potential. Energy budgets for adult, egg-producing females also have been determined (Galkovskaya, 1963). Respiration data used for these investigations were based on large numbers of individuals (see also Pourriot & Deluzarches, 1970). By using Cartesian divers it is possible to determine oxygen consumption of individuals of different age and physiological constitution. Such work has been done in *B. plicatilis* (Doohan, 1973) and *B. rubens* (Pilarska, 1977), including energy budgets for different age groups. The present paper gives complete data for repiration of *B. calyciflorus* and energy budgets for juveniles of this species.

Material and methods

B. calyciflorus were derived from batch cultures at the Zoologisches Institut in Frankfurt. The animals for experimental work were cultured in slowly rotating vessels to prevent sedimentation of the food algae. Vessels with different contents were used, from 2.5 ml for single females up to 100 ml for large numbers. Three different food concentrations of *Kirchneriella lunaris* have been used: 4.10^5, 10^6 and 5.10^6 cells/ml, that is 9.2, 23 and 115 μg dry weight/ml. Temperature for the cultures and all experiments was 20°C, and the rotifers were transferred into new Medium (Stammlösung A, Halbach, 1974) and algae in 24 h intervals.

Oxygen consumption was determined with stoppered Cartesian divers after Klekowski, 1971, with a gas phase from 0.8 to 2 μl, following the instructions given by Doohan, *loc. cit.* Respiration rates of adults were measured using one animal per diver. Two juveniles were put together in larger divers, but only a single one was used in small divers. Manometer readings were taken at intervals of 20 min. and for not longer than two to three hours to prevent starvation effects and low oxygen stress. Ivlev's (1945) oxycaloric coefficient was used to convert oxygen consumption to calories.

For the energy budgets the respiration data of the two juvenile age groups have been integrated over the whole period of growth (i.e. time from hatching until the first egg is attached). This period was derived from single-animal cultures. Production data have been calculated from measurements of dry weight on a Cahn electrobalance. After washing three times with distilled water, groups of about 30 animals (adult females with and without eggs) were put into small containers and dried for one hour at 105°C. Also containers with only distilled water were dried and their weight subtracted from the dry weight of the rotifers plus distilled water. Production is the difference dry weight of a female without egg minus dry weight of an egg, expressed in calories (for caloric content of *B. calyciflorus* see Störkel, 1977). Assimilation is respiration plus production. Consumption is calculated, using data from 1.10^6 cells/ml. It is possible to do this, because there is a close

Hydrobiologia 73, 195-197 (1980).   0018-8158/80/0733-0195$00.60.
© *Dr. W. Junk b.v. Publishers, The Hague.*

relationship between concentration of food and ingestion rate for small food particles like *Kirchneriella* (Starkweather & Gilbert, 1977).

Results and discussion

Table 1 shows the results of the diver experiments in $\mu l.10^{-3}$/h x ind. and cal x 10^{-6}/h x ind. In females with one egg there is an increase in oxygen consumption with decreasing food concentration. In opposition to this respiration in females without eggs is nearly the same in 5.10^6 and 10^6 cells/ml and decreases slightly in 4.10^5 cells/ml. This is in accordance with the dry weight measurements. The dry weight of females plus eggs increased with decreasing food from about 0.4 μg to 0.55 μg/ind., the dry weight of females without eggs from 0.32 to 0.39. So it seems probable that the main reason for the increase of respiration in females plus eggs is the enlargement of the eggs (dry weight of eggs: 0.08 μg in 5.10^6 cells/ml, 0.13 in 10^6, and 0.15 in 4.10^5). The reason for this enlargement of eggs is not known.

Pilarska (1977) did not find such variations of egg weight for *B. rubens,* but she also found the largest animals in her intermediate food concentrations from 4.10^5 to 10^6 cells/ml of *Chlorella vulgaris.* In this investigation 4.10^5 cells/ml of *Kirchneriella* seemed to be the best concentration for the rotifers, while higher concentrations seemed to have the negative influences, as described by Halbach & Halbach-Keup (1974). Another point of interest in Table 1 is the higher respiration rate of juveniles aged 12 to 24 h, compared to that of females without eggs. Pilarska's work about *B. rubens* does not provide data for adult females without eggs, so her results cannot be compared with others. The high oxygen consumption may be due to the fact that the metabolism of growing *B. calyciflorus* is higher than the metabolism of postovigerous or not yet egg-producing adults.

Table 2 shows the elements of the energy budgets for the three food concentrations. The data are not directly comparable with those of Doohan and Pilarska because their results are based on other concentrations and/or age groups. Nevertheless the calculated efficiencies are comparable with those found by other authors. All coefficients are within the range given by Galkovskaya for *B. calyciflorus.* She gives assimilation efficiencies from 0.78 to 0.21,

Table 1. Respiration of different stages of *Brachionus calyciflorus*

Groups of animals	Concentration of food		
	$5 \cdot 10^6$	$1 \cdot 10^6$	$4 \cdot 10^5$
females without eggs	3.0428 ± 0.41485	3.0604 ± 0.5461	2.7241 ± 0.5777
	14.70 ± 2.00	14.78 ± 2.64	13.16 ± 2.79
	n = 12	n = 8	n = 8
females with one egg	4.1115 ± 0.7313	4.5699 ± 0.7049	4.8064 ± 0.5176
	19.86 ± 3.53	22.07 ± 3.40	23.215 ± 2.50
	n = 12	n = 10	n = 10
females with two eggs	6.7834 23.76	6.5914 31.84	6.4883 31.34
	6.5386 31.58	6.5021 31.41	6.4837 31.32
	5.3632 25.90	6.2346 30.11	5.7323 27.69
juveniles 0 – 12 h	2.1141 ± 0.5234	2.5552 ± 0.7108	1.8521 ± 0.3121
	10.21 ± 2.53	12.34 ± 3.43	8.95 ± 1.51
	n = 7	n = 8	n = 8
juveniles 12 – 24 h	3.6172 ± 0.3724	3.3493 ± 0.2041	3.2948 ± 0.8333
	17.47 ± 1.80	16.18 ± 0.99	15.91 ± 4.025
	n = 7	n = 7	n = 6

First line: Mean oxygen consumption in $\mu l \times 10^{-3}$/h x ind. ± S.D.
Second line: Respiration in cal $\times 10^{-6}$/h x ind. ± S.D.
Third line: Number of animals.

Table 2. Elements of an energy budget for growing *Brachionus calyciflorus*

Parameters	Concentration of food		
	$5 \cdot 10^6$	$1 \cdot 10^6$	$4 \cdot 10^5$
Period of growth (h)	24.50 ± 1.25	31.28 ± 7.83	32.62 ± 6.63
Consumption	7.644 ± 0.39	4.504 ± 1.128	2.349 ± 0.477
Production	1.1955	0.989	0.672
Respiration	0.343 ± 0.009	0.483 ± 0.091	0.487 ± 0.049
Assimilation (P + R)	1.5385	1.4715	1.159
Efficiencies A/C	0.20	0.33	0.49
P/C = K_1	0.16	0.22	0.29
P/A = K_2	0.78	0.67	0.58

Period of growth: Time from hatching until the first egg is attached, in hours. Efficiencies in absolute numbers. All other data in cal $\times 10^{-3}$/ind. for period of growth.

the higher efficiencies occurring at the lower food concentrations. This seems to be in accordance with the coefficients of the present paper reaching from 0.49 to 0.20 in intermediate and high food concentration. Also the gross production efficiencies from 0.29 to 0.16 are within the range given by Galkovskaya (0.36-0.04). The net production efficiencies seem to be very high, especially for $5 \cdot 10^6$ cells/ml (0.78), compared with data given by Galkovskaya (0.20-0.69) and Pilarska (0.5 for high food concentrations). But Galkovskaya's coefficients were calculated for adult females, her production values are egg production, and Pilarska found the highest values for K_2 in juvenile *B. rubens*. So it seems possible that my very high values for K_2 in juvenile *B. calyciflorus* correspond to reality.

References

Doohan, M. 1973. An energy budget for adult Brachionus plicatilis Muller (Rotatoria). Oecologia 13: 351-362.

Galkovskaya, G. A. 1963. Utilisation of food for growth and conditions for maximum production of the rotifer Brachionus calyciflorus Pallas. Zool. Zh. 42: 506-512.

Halbach, U. & Halbach-Keup, G. 1974. Quantitative Beziehungen zwischen Phytoplankton und der Populationsdynamik des Rotators Brachionus calyciflorus Pallas. Befunde aus Laboratoriumsexperimenten und Freilanduntersuchungen. Arch. Hydrobiol. 73: 273-309.

Ivlev, V. S. 1945. The biological productivity of waters. Usp. sovrem. Biol. 19: 98-120.

Klekowski, R. Z. 1971. Cartesian diver microrespirometry for aquatic animals. Pol. Arch. Hydrobiol. 18: 93-114.

Pilarska, J. 1977. Eco-physiological studies on Brachionus rubens Ehrbg. (Rotatoria). I. Food selectivity and feeding rate. II.

Production and respiration. III. Energy balances. Pol. Arch. Hydrobiol. 24: 319-354.

Pourriot, R. & Deluzarches, M. 1970. Sur la consommation d'oxygene par les Rotifers. Ann. Limnol. 6: 229-248.

Starkweather, P. L. & Gilbert, J. J. 1977. Feeding in the rotifer Brachionus calyciflorus. II. Effect of food density on feeding rates using Euglena gracilis and Rhodotorula glutinis. Oecologia 28: 133-139.

Störkel, K. U. 1977. Kalorimetrische Untersuchungen an Brachionus calyciflorus Pallas und Kirchneriella lunaris Möbius. Thesis, University of Frankfurt/Main.

FIELD EXPERIMENTS ON THE OPTICAL ORIENTATION OF PELAGIC ROTIFERS

Kurt PREISSLER[1]

Biologische Fakultät der Universität München, GFR

[1] This publication is dedicated to Pater Dr. Josef Donner on the occasion of his 70th birthday

Keywords: Pelagic Rotifers, 'avoidance of shore', optical orientation

Abstract

'Avoidance of shore' by pelagic rotifers is considered to be the result of an optical orientation. Field experiments show that the spatial light distribution in the shore region determines the preferred direction of migration. The behaviour of *Eudiaptomus gracilis* was tested in comparison to that of rotifers.

Introduction

Pelagic and littoral rotifers show not only morphological and physiological adaptations to their biotope, but also a specific behaviour related to the nature of their habitat. When released inshore the pelagic *Asplanchna priodonta* migrates away from the shore, whereas the littoral *Euchlanis dilatata* does not prefer any direction (Preissler, 1977a). Because other pelagic species react like *Asplanchna* when tested in various lakes, the phenomenon of avoidance of the shore can probably be generalized. The question then arises, whether this orientation of pelagic rotifers is governed by the same optical clue as in pelagic crustaceans (Siebeck, 1968), i.e. by the spatial light distribution close to the shore.

Material and methods

In order to test the foregoing hypothesis, the circular perspex cylinder ('arena') developed by Siebeck (1968) for field experiments on the orientation of planktonic crustaceans was modified. The arena makes it possible to draw conclusions about the animals, behaviour from changes in their horizontal distribution in an area separated from the lake. After the rotifers are transferred from, the pelagic to the littoral zone they are released in the centre of the arena and allowed to disperse. At the end of the experiment all 18 radial chambers at the periphery are closed simultaneously by means of a conical perspex ring. The position of the chamber in which most of the animals are found then indicates the preferred direction of migration (for more technical details see Preissler, 1977b). The arena was posted at a distance of 1 m from the shore, so that the rotifers were exposed simultaneously to light incident from above, from the side (e.g. plants), and from below (e.g. sea chalk, stones reflecting light).

The idea behind this analysis is to test whether it is possible to disturb the behaviour observed in arena experiments by eliminating the light patterns from any or all of the 3 directions mentioned above. If so, this would provide evidence for optical orientation. Moreover, it might be possible to show which optical cues control the behaviour in question.

Results and discussion

To determine the effect of light, 3 series of experiments (diffuse light, cloudiness 0/10) were performed near the western shore of Bansee (Seeon, Bavaria) at the same spot where some work (Preissler, 1977a, b) had been done previously. First, the arena was placed on a black disk (Fig. 1, A); second, a black 'collar' was added to eliminate the influence of light from the sides (Fig. 1, B). Under both conditions the maximum numbers of individuals of

199

Hydrobiologia 73, 199-203 (1980).   *0018-8158/80/0733-0199$01.00.*
© *Dr. W. Junk b.v. Publishers, The Hague.*

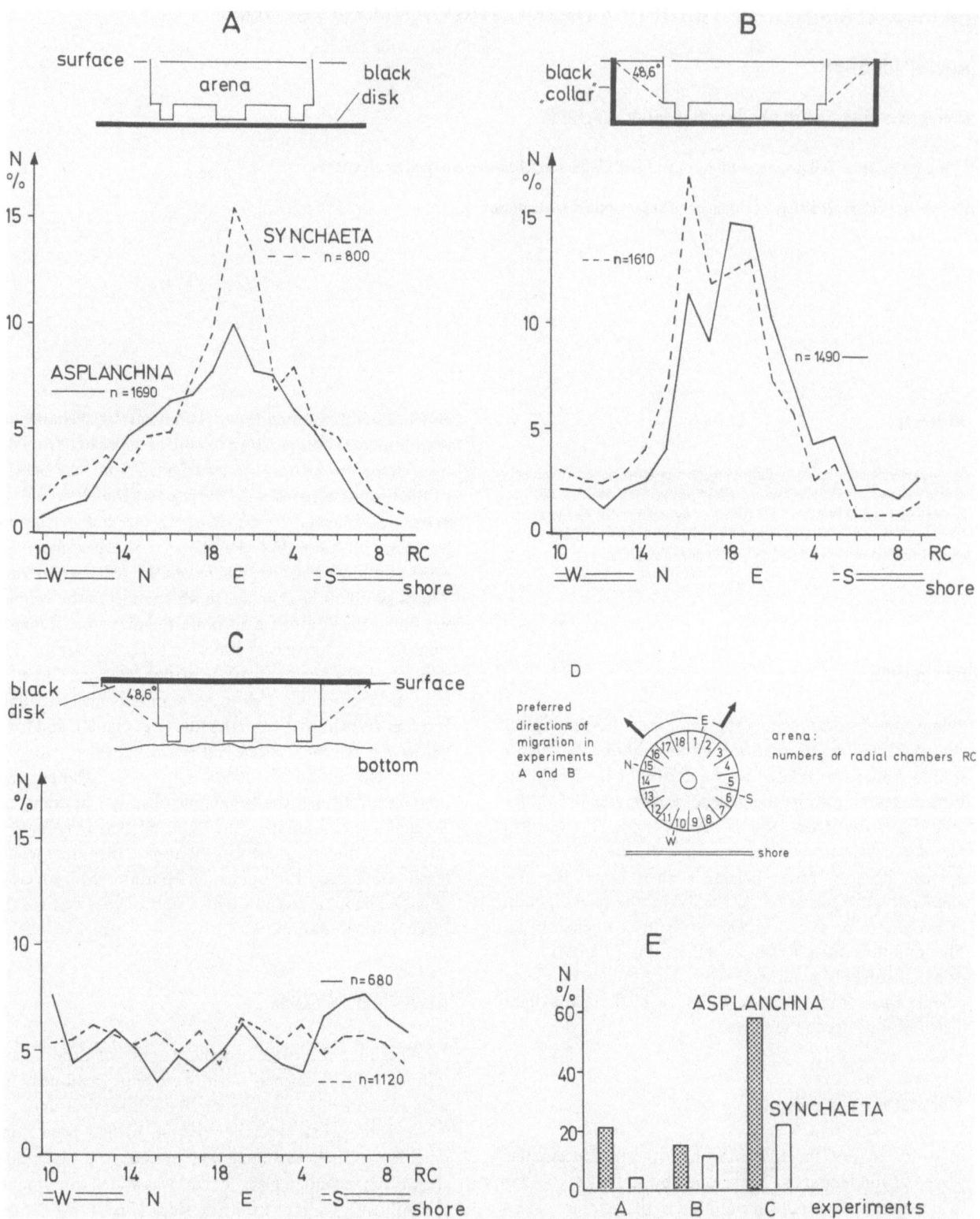

Fig. 1. Three experiments to determine the effect of light from below (A), from the sides (B) and from above (C). N = % of individuals found in all radial chambers. D: arena from above. E: N = % of animals released which remained in the centre of the arena.

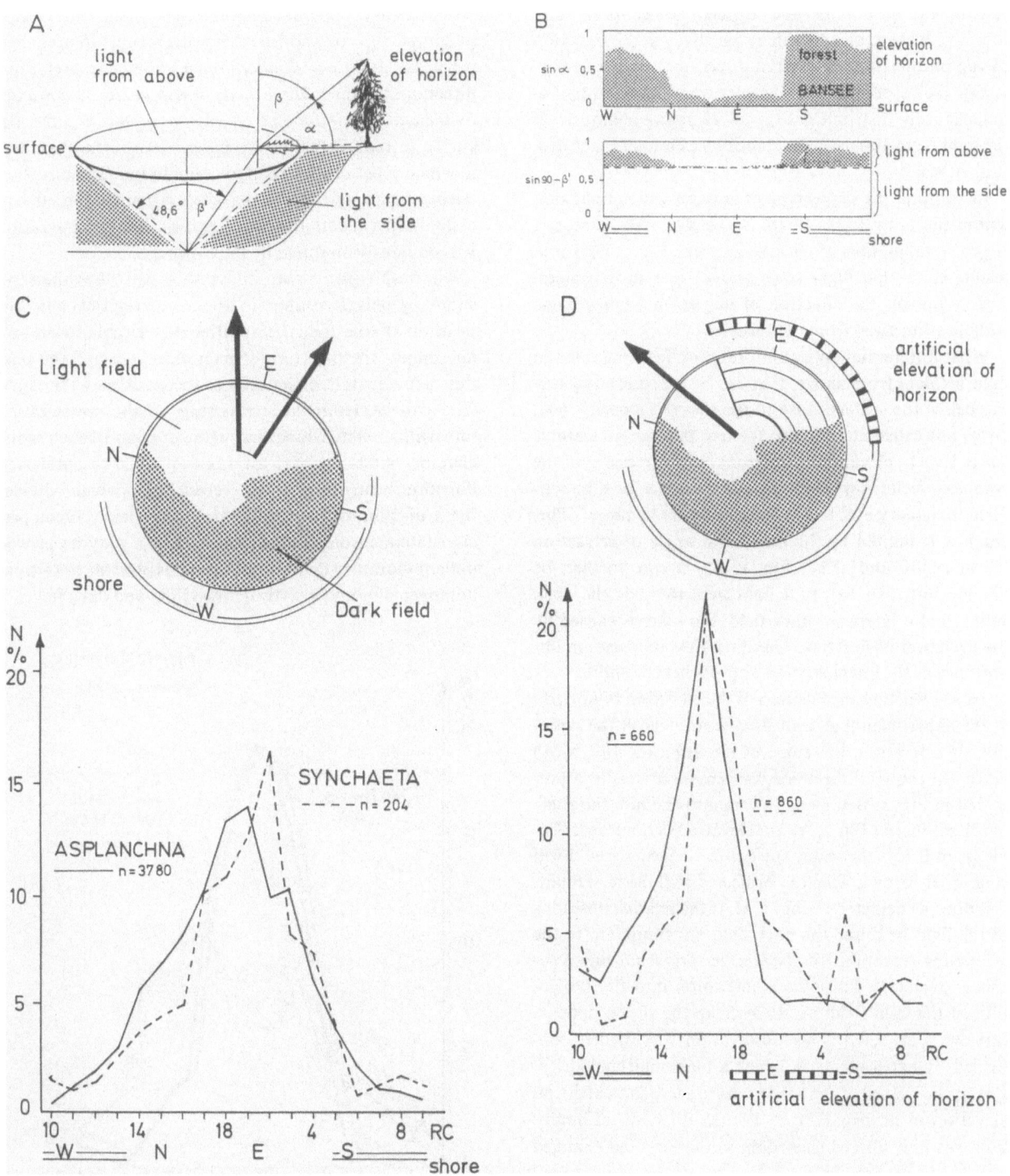

Fig. 2. A: Special light conditions within the circular 'window' as seen from below, angle of elevation: α, of incidence: β, of refraction: β' and of maximum refraction: 48.6°. B: representation of the elevation of horizon at the experimental site in Bansee above water surface (top) and how it is seen after refraction (below). C: results of experiment to show the influence of light field and dark field (top) on the preferred directions of migration (= arrows). D: Results of experiment with an artificial elevation of horizon.

Asplanchna priodonta and *Synchaeta pectinata* were found in those radial chambers which lay in the direction of the pelagic zone (preferred directions: arrows in Fig. 1, D). However, the same species were found to be homogeneously distributed over all 18 radial chambers, if the light from above was excluded by a circular black disk (Fig. 1, C).

In addition the percentage of animals which remained within the central area of the arena distinctly increased (100%: total number of released animals, Fig. 1, E). These results show that 'light from above', in a most general sense, controls the direction of migration when pelagic rotifers swim away from the shore.

What differential optical information lies concealed in light incident from above, that can be recognized by any eye below the surface close to the shore? Siebeck (1968, 1979) has called attention to the fact that the area above water level (= elevation of horizon) in all directions of the compass, seen from below the water surface by a punctiform imaginary eye, falls within a circular 'window'. This window is limited by the maximum angle of refraction (0° at zenith and 48.6°, Fig. 2, A). It can, further, be divided into two halves: a light field towards the open water, and a landward dark field; the latter is caused by the elevation of horizon. This dark field increases in importance as the imaginary eye approaches the shore.

To test whether the position of the elevation of horizon at the experimental area in Bansee influenced the direction of migration, the angle of elevation (α, Fig. 2, A) above the centre of the arena was measured (results represented in Fig. 2, B above). Taking into account the angle of refraction (β', Fig. 2, A), the elevation of horizon is visible from below the water surface in a 'compressed' form (Fig. 2, B below), which is projected within the circular 'window' as depicted in Fig. 2, C. If the relative positions of the light field and the dark field are compared to the swimming directions of *Asplanchna priodonta* and *Synchaeta pectinata*, an obvious migration into the projection of the light field, i.e. away from the shore, appears (arrows, Fig. 2, C). This observation was further confirmed by experiments in which a black half-cylindrical screen was placed around the arena to serve as an artificial elevation of horizon (Fig. 2, D): the direction of migration was now altered according to the new and changed position of the light field.

It can therefore safely be said that the migration of pelagic rotifers away from the shore is the consequence of an optical orientation governed by the position of the light field.

The pelagic copepod *Eudiaptomus gracilis* was tested in comparison to and together with rotifers in arena experiments in Bansee: *Eudiaptomus* was seen to prefer one direction, but the highest totals of *Asplanchna priodonta*, *Synchaeta pectinata* and *Keratella quadrata* now occurred in radial chambers right and left of the chambers in which most of the copepods were found. The direction selected by pelagic rotifers basically remained the same as in the experiments described earlier, but the migration was obviously modified by the copepods.

The results dealt with above show that 'avoidance of shore' by pelagic rotifers is based on an optical clue: the position of the light field within the circular 'window' determines the direction of migration. Though the relation between this direction and the position of the light field can be described at this stage of the investigation only with a relative inaccuracy (the opening of each radial chamber stretches over an angle of 20°), the preferred direction seems to lie in the vertical plane which divides the projection of the light field symmetrically. Such precise statement of position, however, can only be proved under laboratory conditions; there it is possible to arrange homogeneous and exactly defined light and dark fields.

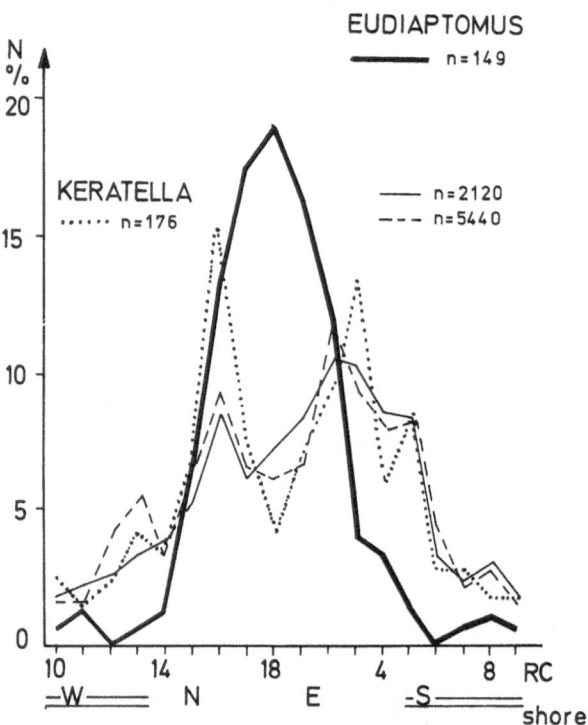

Fig. 3. Results of experiments with Eudiaptomus gracilis; symbols of Asplanchna and Synchaeta like in Fig. 2.

Some preliminary experiments appear to confirm this suggestion (Preissler, in preparation). This, then, would mean that rotifers 'utilize' the spatial light distribution in the shore region in a way comparable to pelagic crustaceans (Siebeck, 1968, 1979). *Eudiaptomus gracilis,* when tested together with rotifers, actually prefers the same direction, which is normally chosen by rotifers. If one considers the high numbers of animals (n in Fig. 1 and 2), which are released at the centre of arena, it should not be excluded, that the behaviour of rotifers could be influenced by interspecific competition or crowding effects. Dumont (1972) has discussed these problems in detail. The results of arena experiments, however, show, that the basic direction of the rotifers under these conditions is not changed, but that migration seems to be disturbed by the presence of the copepods.

References

Dumont, H. J. 1972. A competition-based approach of the reverse vertical migration in zooplankton and its implications, chiefly based on a study of the interactions of the rotifer Asplanchna priodonta (Gosse) with several Crustacea Entomostraca. Int. Rev. ges. Hydrobiol., 57: 1-38.

Preissler, K. 1977a. Horizontal distribution and 'avoidance of shore' by rotifers. Arch. Hydrobiol. Beih., 8: 43-46.

Preissler, K. 1977b. Do rotifers show 'avoidance of the shore'? Oecologia, 27: 253-260.

Siebeck, O. 1968. Uferflucht und optische Orientierung pelagischer Crustaceen. Arch. Hydrobiol. Suppl. 25: 1-118.

Siebeck, O. 1979. The importance of the spatial orientation of pelagic Copepods for their horizontal distribution. Naturwissenschaften, 66: 266.

WORKSHOP ON TAXONOMY AND BIOGEOGRAPHY

H. J. DUMONT

Zoological Institute, Univ. Gent, Belgium

Keywords: species concept, taxonomy, global distribution, passive dispersal

Introduction

The species concept in Rotifers, as in many other animal groups, is applied by various investigations in a rather lax way. Separation of species is often based on ad hoc criteria, such as the structure of the lorica; there is, as yet, insufficient insight into the contribution of morphology to reproduce isolation, or to avoidance of interspecific competition. An exception is the structure of the mastax; if habitat partitioning through selective feeding is occurring within rotifers, one should indeed expect this structure to show high morphological specificity. However, the same holds true for the corona, which has been little studied, and also for the various receptors found on the head region, and sometimes elsewhere on the body.

As to geographical distribution, it is still customary to consider all rotifers as potentially cosmopolitan animals. One cannot deny that the reproductive strategies of the rotifers are indeed very strongly oriented towards achieving maximum dispersal: parthenogenesis (a single female can start a new population); drought and digestion resisting stages (transportation by various means over long distances outside water possible); rapid maturation following emergence from the egg and short generation time (probability of elimination of egg propagules as well as of young populations through predation low). Yet, this strategy is partly counterbalanced by the non-availability of suitable niches to species being dispersed passively to 'new terrorities'.

Nothing is known as yet about the biological niche, but recent information on the environmental niche shows that tropical-subtropical species rapidly appear in temperate climatic belts if tropical conditions temporarily occur here, as during the long and hot summer of 1976 in Europe (Coussement et al., 1976), or in the heated effluents of nuclear power plants (Lair, this symposium). The idea that relatively stable tropical-subtropical species and species-associations exist, and also some arctic-subarctic associations, has only recently gained considerable support (Green, 1967; Ruttner-Kolisko, 1972; Koste, 1978 and, especially, Pejler, 1977a, b). There also appears to exist at least one typical high-mountain rotifer species, *Hexarthra bulgarica* (Dumont et al., 1978). However, the existence of a set of endemic psammophilic species of the genus *Notholca* in lake Baikal (Kutikova, 1970), and of other genera (Kutikova, pers. comm.) as well, suggests that in ancient lakes strictly local speciation can proceed in rotifers as in bisexual groups. The question as to why these endemics do not spread outside Baikal has apparently not been asked, yet might be of considerable importance to a better understanding of rotifer chorology. Would these species for example have lost the faculty of producing resting eggs, or would their resting eggs not be drought-resistant?

A warning should, finally, be given against too rapid generalizations. Our knowledge about rotifers distribution patterns, on a world scale, still presents numerous and enormous gaps. This can be exemplified by the European situation, doubtlessly the best documented one in the world. Berzins (1978) lists 1330 species for this continent, and calls 150 among these true cosmopolitans. However, upon analysing the ranges of the remainder, it appears that 445 have been found only one time. Such a situation should at least inspire caution in making statements about the geographical behaviour of the group as a whole.

Hydrobiologia 73, 205-206 (1980).   0018-8158/80/0733-0205$00.40.
© Dr. W. Junk b.v. Publishers, The Hague.

Discussion

A. Ruttner-Kolisko stated that the cosmopolitism of the rotifers certainly does not need to involve all species. At least, the statement should be mitigated to a potential cosmopolitism.

The discussion then focussed on species concepts and evolution. C. King, drawing from his genetical work, stressed the profound and rapid changes in gene frequencies that can occur in 'continuous' populations that look morphologically identical. He found that some genetically different but morphologically similar *Asplanchna* would not even mate. Taking into account the rapid reproduction of some rotifers he also used standard estimates of mutation rates and ratios of deleterious to advantageous mutations to obtain an idea on the speed of evolution in rotifers. This should be high, especially in Bdelloids, which have eliminated sex completely. Limnologists should, so to speak, carry not only plankton nets with them, but also equipment for electrophoresis.

C. Ricci remarked that she found it impossible to apply species concepts to Bdelloids. J. Donner admitted that all species names in Bdelloids should be regarded as preliminary, but that investigators should state very clearly on what principle(s) they founded their taxonomy.

L. Kutikova then said that, for the time being and probably for some time to come, we will remain forced to rely on a typological concept of the species in rotifers. A. Herzig protested against pure typology, calling attention to the ecological dimension of species. He cited the example of the species couple *Arctodiaptomus bacillifer/ A. salinus* (copepoda calanoida). Both are morphologically very similar, but the first one is confined to alpine lakes, while the second one occurs in the saline lakes and ponds of the pannonian plain. H. Dumont agreed, and remarked that similar phenomena are seen in the rotiferan genus *Hexarthra*. W. Koste then complained about the lack of care and thoroughness of many taxonomists, and concluded the discussion on the statement that we are today witnessing the stone age of rotifer taxonomy.

References

Berzins, B. 1978. Rotatoria. In: Limnofauna Europaea, 2nd. ed. (ed. J. Illies). Fischer, Stuttgart, 54-91.

Coussement, M., Henau, A. M. de & Dumont, H. J. 1976. Brachionus variabilis Hempel and Asplanchna girodi De Guerne, two rotifer species new to Europe and Belgium, respectively. Biol. Jaarb. Dodonaea, 44: 118-122.

Dumont, H. J., Coussement, M. & Anderson, R. S. 1978. An examination of some Hexarthra species (Rotatoria)from Western Canada and Nepal. Can. J. Zool., 56: 440-445.

Green, J. 1967. Associations of Rotifers in the zooplankton of the lake sources of the White Nile. J. Zool., 151: 243-278.

Koste, W. 1978. Rotatoria. 2 Vols. Borntraeger Berlin. 673 pp., 234 plates.

Kutikova, L. A. 1970. Kolovratki Fauna SSSR. In: Fauna USSR, vol. 104. Akademia Nauk, Leningrad, 744 pp.

Kutikova, L. A. 1979. Personal communication.

Pejler, B. 1977a. General problems of rotifer taxonomy and global distribution. Arch. Hydrobiol. Beih., 8: 212-220.

Pejler, B. 1977b. On the global distribution of the family Brachionidae (Rotatoria). Arch. Hydrobiol. Suppl. 53: 255-307.

Ruttner-Kolisko, A. 1972. Rotatoria. Binnengewässer, 26: 99-234.

VARIATION IN THE GENUS KERATELLA

Birger PEJLER

Limnological Institute, Uppsala, Sweden

Keywords: Rotatoria, Keratella, seasonal variation, local variation

Abstract

The literature on variation in *Keratella* is reviewed. The old idea of a thorough endogenous control has to be rejected, but internal factors ought to play a certain role beside influences from current and previous environment. In certain cases there is probably a succession of genetically different clones during the course of the year (cf. King, 1972, 1977), but the seasonal variation in lake populations of, e.g., *K. cochlearis* ought to be mainly non-genetical. There is some evidence that temperature and food exert an influence on the morphology, via rate of growth, but probably other abiotic and biotic factors are at work as well. The existence of allometric relationships is clearly demonstrated for several species. The variation in spine length has been suspected by some authors to consititute just the function of size variation which is thus considered primary. Some of the variation found is obviously non-adaptive. An attempt is made at explaining the existence of discontinuous variation within a single lake. Implications on taxonomy and speciation are briefly discussed.

The topic of this paper represents a classical research object within limnology. Studies began about at the same time as those on seasonal variation in the genus *Daphnia*. However, the work made on *Keratella* has mainly concerned the specialists. The investigations on daphnids have been cited much more in text-books and popular publicatons, probably because such fascinating theories have been connected with them.

The pioneer work on variation in *Keratella* (and on rotifers in general) was done by Lauterborn (1898, 1900, 1904) on material from the Rhine and some neighbouring localities. In these studies he found a continuous morphological variation in *K. cochlearis* (Gosse) during the course of the year. A winter form, f. *macracantha,* with a relatively uniform appearance (a long posterior spine etc) was succeeded by three different series, each of them being more and more pronounced morphologically during the course of the summer. A reduction in spine length also occurred in each of the series. In addition there exists a complex of forms, deviating morphologically from the three series and not undergoing reduction in spine length, thus not representing a series. This complex was called f.

robusta and the three series *tecta*-series, *hispida*-series and *irregularis*-series. F. *robusta* is said to be restricted to ponds with a heavy growth of macrophytes, whereas the forms of the three series only occur in waters with a more or less large area devoid of such vegetation. The latter were found in ponds as well as in lakes (e.g. L. Constance) and slowly flowing rivers. According to Gillard (1948, 1949) f. *robusta* can be regarded as an 'ecological race' and is designated with his nomenclature as '*K. cochlearis* OE robusta'.

Lauterborn presents a really comprehensive and convincing material of data. On the other hand he is very restrictive concerning interpretations of his results, quite contrary to some other earlier workers. Thus Krätzschmar (1908, 1913) founded a theory of 'cyclomorphosis' based on his studies of the seasonal variation in the *K. quadrata* (Müll.) complex. His scheme of the life cycle (condensed in Krätzschmar, 1908, Fig. 20) has been quoted in many text-books because of its perspicuity. The following pattern is described: From the resting eggs long-spined amictic females are hatched, which produce a sequence of other amictic individuals, in which the posterior spines get successively shorter for each generation. After a certain number of amictic generations, mictic females appear and, arisen from these, males and resting eggs, which after a resting period will form the starting-point of a new cycle. Krätzschmar speaks of a 'successively decreasing vitality of the parthenogenetic females' and a 'degenerative process', which finally causes sexual propagation.

Krätzschmar based his view on experimental work: He cultured his animals at different temperatures, light intensity, amount of food, concentration of chloride etc, which factors, however, did not apparently influence the morphology. Therefore, Krätzschmar concludes that endogenous factors alone are decisive for the seasonal variation.

Hartmann (1918) adheres to Krätzschmar's opinion, though he believes that external factors may modify the extent of the variation (which is to some extent also admitted by Krätzschmar). Apparently Sudzuki (1964,

207

Hydrobiologia 73, 207-213 (1980). *0018-8158/80/0733-0207$01.40.*
© *Dr. W. Junk b.v. Publishers, The Hague.*

pp. 28 and 32) shares these ideas as well. However, Ruttner-Kolisko (1949) presents convincing evidence that Krätzschmar partly based his view on misinterpretations. She worked with material from the same lake as the senior author, L. Lunzer Obersee, and found that two species of the *K. quadrata*-complex existed there (three species in later decades) with different spine length. Both these species (*K. quadrata* s. str. and *K. hiemalis* Carlin) were incorporated into Krätzschmar's scheme. The individuals of *K. quadrata* s. str. cultured by Ruttner-Kolisko produced reductional forms as in Krätzschmar's experiments, but she interpretes this result in another way. As Krätzschmar's ideas are thus rejected, Ruttner-Kolisko also proposes that the term 'cyclomorphosis' not be used any more, as it has been connected with an obligatory relation between morphological variation and sexual cycle.

As early as 1911 Dieffenbach & Sachse obtained experimental results which contradicted Krätzschmar's theory, and more recently very clear evidence against endogenous control has been put forward by Rauh (1963) and Halbach (1970), based on studies of species of the related genus *Brachionus*. The original aim of Rauh's study was in fact to analyse the variation of *Keratella cochlearis,* but on account of the difficulty in cultivating this species, *Brachionus* was chosen instead. Regarding the evidence from field studies against an endogenous periodicity, reference is made to Ruttner-Kolisko, op. cit. and the extensive discussions by Buchner, Mulzer & Rauh (1957) and Buchner & Mulzer (1961).

The form hatching from the resting egg is a crucial point in this argumentation. However, it appears that a successful hatching has very rarely occurred. Dieffenbach & Sachse (op. cit.) mention such a result, in which the resulting forms were long-spined (thus in accordance with Krätzschmar's scheme). Likewise Sudzuki (op. cit., p. 28) says in connection with *K. cochlearis* that he has 'verified the fact that the specimens with the longest spine hatch out from the dormant eggs'.

Nipkow (1961, pp. 417-419) succeeded in hatching resting eggs of *K. quadrata* (as well as of several other rotifers). The individuals appearing from such eggs had evidently spines of intermediate length. When further cultivated they produced offspring of reductional forms, similar to those of Ruttner-Kolisko. The mictic females found in the lake (L. Zürich) had relatively long spines.

Other authors call attention to the difficulty of obtaining offspring from resting eggs in the laboratory, and, thus, they conform to an indirect argumentation. E.g., Amrén (1964b) mentions that many of the ponds and puddles he investigated on Spitsbergen freeze to the bottom during winter and that the first appearing individuals in the spring therefore have to take their origin from resting eggs. This first generation is characterized by very short or non-existent posterior spines, quite contrary to the cases reported above. The offspring of these females are equipped with somewhat longer spines (verified in cultures) and they give to a generation with still longer spines, a sequence which could be followed through some generations.

As stated above, an overwhelming evidence against endogenous control has now been cumulated. However, this must not mean that such forces are never at work. Nobody now ought to question the nice results obtained by Nipkow (1952), showing that the first generation of *Polyarthra* is devoid of fins. Almost as good evidence is obtained by Amrén (1964a), regarding *K. quadrata,* in favour of internal factors determining the appearance of the generation hatching from the resting eggs, as well as of those following next. Possibly this is a widespread phenomenon within Rotatoria.

The seasonal variation analysed by Amrén had no apparent connection with either temperature or food. In most cases, however, a very obvious correlation to temperature exists, long appendages being found at low temperatures, short or none at high. For *K. cochlearis* especially, many univocal studies were made, showing this connection in a variety of lakes and ponds: Lauterborn, 1900 and 1904 (excl. f. *robusta*); Züscher, 1912; Ammann, 1913 and 1923; Schreyer, 1921; Schneider, 1922; Vialli, 1924; Robert, 1925; Wesenberg-Lund, 1930; Varga, 1941; Carlin, 1943; Entz & Sebestyén, 1946; Buchner, Mulzer & Rauh, 1957; Parise, 1960; Buchner & Mulzer,1961 and Hillbricht-Ilkowska, 1972. Experimental evidence proving the influence of temperature has been put forward by Pourriot (1964) and Lindström & Pejler (1975). Regarding *K. quadrata* f *frenzeli,* a form typical of larger lakes, the same conditions seem to occur according to Carlin, 1943. Klement (1957) reports a similar cycle for a form within the '*quadrata*-series' living an a pond. For the eulimnoplanktic *Kellicottia longispina* (Kellicott), belonging to the same subfamily, the indication of a similar influence of temperature is very strong, to judge from Ammann, 1913; Schreyer, 1921; Vialli, 1924; Robert, 1925;Varga, 1941; Carlin, 1943 and Hakkari, 1969. Such is the case for the related genus *Notholca* as well, not only for the true plankters, but also for benthic and periphytic forms, e.g. those occurring in rockpools (see Björklund, 1972). This comprehensive material should be enough to refute the general applicability of the well-known buoyancy theory suggested by Wesenberg-Lund and Ostwald, which is based upon quite reverse conditions existing in certain cladocerans. In spite of this, the mentioned theory is often reported in text-books still to-day as the probable explanation for the seasonal variation of all plankters!

However, there are exceptions from the trend discussed above. The pond-living f. *robusta* of *K. cochlearis* has already been mentioned as an example. Gallagher (1955 and 1957) reports a reverse course of variation for a form of *K. cochlearis* found in an artificial pond. Likewise, the pond-living forms of *K. quadrata* do not follow the regular pattern described, their variation being rather erratic: in some cases non-existent, in some cases correlated to temperature in one way or the other (see especially Ruttner-Kolisko, 1948 and Buchner & Mulzer, 1961). Even in true lakes *K. quadrata* sometimes does not show any pronounced seasonal variation (Ruttner-Kolisko, 1949; Parise, 1969). An interesting deviation from the close correlation with temperature is constituted by the 'spring peak' discussed by Carlin (op. cit.): Though the temperature is constant or somewhat increasing, the spines get obviously longer during April and May in *Keratella cochlearis*, *K. quadrata* and *Kellicottia longispina*. Carlin ascribes this peak to the improved food conditions during these months. A similar peak develops in *Notholca caudata*, which is cold stenothermal and disappears in summer, thus not being capable of demonstrating any variation related to temperature.

Keratella hiemalis, which is a cold-water form, normally does not show any seasonal variation (Ruttner-Kolisko. 1949), but Hutchinson (1967, pp. 891-892) attempts to trace a tendency, basing his argument upon the limited material of Pejler (1957). In arctic lakes no obvious seasonal variation seems to have been reported, but in smaller water bodies the pattern discussed by Amrén (cf. above) may occur, whereby temperature is apparently not involved.

Going to the other extreme, seasonal variation has in some cases been shown to exist in tropical waters. The species studied are *Keratella tropica* (Apstein), *Brachionus calyciflorus* Pallas and *B. caudatus* Barrois & Daday (see Green, 1960 and 1977; Nayar, 1965 and Arora, 1966). In these cases temperature cannot be considered responsible for the morphological changes.

As temperature is evidently not the only factor lying behind seasonal variation, other agents have to be sought for. As much more experimental evidence is obtained concerning some species of *Brachionus*, sime hints could

be expected from the studies of this genus. Thereby it is interesting to find agreeing conclusions in Rauh (1963) and Halbach (1970) on the basis of their very elaborate investigations. Both authors talk of temperature and food as important factors, which influence developmental rate: At low temperatures and low concentration of food particles the development is slow, whereby the length of the spines increases.

Beside temperature and food Halbach (op. cit.) analyses another factor, the so-called '*Asplanchna*-substance', now a central topic within rotifer research and summarized in several other papers (see, e.g., Gilbert, 1966 and Halbach, 1971a). The *Asplanchna*-substance acts even more strongly upon the morphology of *Brachionus* than temperature and food.

Now it remains to be considered if the mentioned results from the *Brachionus* experiments can be applied upon seasonal variation in *Keratella* as well. The influence of temperature has already been discussed. Also, it was reported above that Carlin regarded food to be active concerning the 'spring peak' of some species. However, the effect was here quite reverse to that expected from Rauh's and Halbach's investigations. On the other hand, the studies on *local* variation in *K. cochlearis* made by Pejler (1962) are quite consistent with the mentioned research on *Brachionus*. In fact, Pejler suggested the same idea as Rauh and Halbach regarding the influence of food and temperature, referring to Edmondson (1960), who had found that birth-rate is positively correlated with temperature, as well as with quantity of phytoplankton. (These ideas are still more developed in Edmondson, 1965). Pejler found in Swedish lakes during the summer, at roughly equal temperatures, a very strong correlation between trophic degree and spine length of *K. cochlearis:* In oligotrophic lakes only more or less long-spined individuals occurred all through the summer, while forms with short spines or without spines dominated in the eutrophic. F. *tecta*, devoid of spines, was even shown to be one of the best indicators of eutrophy. Quite similar results from Polish lakes were obtained by Hillbricht-Ilkowska (1972).

The third factor stated to be active by Rauh and Halbach, the *Asplanchna*-substance, has not been considered regarding *Keratella*. However, it is probably of less importance to the true lake plankters, on account of the dilution effect.

On the other hand, recent ecological investigations (also in rotifers) have shown that the abiotic interrelations have been too often overestimated and the biotic ones neglected. Many possibilities exist concerning biotic effects. Thus, Halbach (1970, pp. 311-312) refers to several earlier papers concerning a direct or indirect effect of food quality. The role of competition is discussed by Snell (1977), that of selective predation by Halbach & Jacobs (1971), Nilsson & Pejler (1973) and Green (1977).

However, abiotic factors other than temperature may also be conceived as agents. Edmondson (1948), for example, points at a possible effect of calcium, Green (1960) at the connections with floods in tropical lakes. The influence of turbulence on cladocerans has been discussed very much, and it is interesting to find that some authors connect rotifer variation with morphometric conditions of the lake. Thus, Green (1977) says that 'dwarfing' occurs in lakes with a low ratio of drainage area to surface area. Ruttner-Kolisko (1972, p. 143) maintains that *hispida* forms of *K. cochlearis* mainly occur in strongly turbulent shallow and small water bodies. Finally, Berzinš (1958) and Hillbricht-Ilkowska (1972) describe a vertical stratification concerning the morphology of *Kellicottia longispina* and *Keratella cochlearis,* respectively. In both cases, more short-spined forms are found, on an average, in the superficial layers than farther down.

The two last examples show that ecoclinal variation can be of local as well as temporal character. Another local ecocline was reported by Wermel (1930), who showed that the morpholgy of *Keratella serrulata* (Ehrbg) changed successively within a boggy pool parallel to a gradient of pH. Similarly, Pejler (1957, 1958) found that the length of the posterior spines of *K. hiemalis* decreased concomitantly with the annual heat budget of the water body (i.e. on the whole with rising height above sea-level).

Concerning the adaptive value, it is quite easy to understand the meaning of the predator *Asplanchna* eliciting longer spines in its prey *Brachionus.* As Halbach (1970, 1971 b) has pointed out, this arrangement is of advantage to both predator and prey. It is much more difficult to understand which benefit could be connected with a variation induced by temperature or food. Several authors have speculated regarding this matter. Hartmann (1918), for example, mainly considers the mechanism of locomotion, while Carlin (1943) believes that the appendages have the function of 'catching' the turbulent currents and utilizing them for floating.

It has been noted by some authors that the variation in spine length may be just a function of a varying size of the body, the last being the primary phenomenon (see Green 1960; Pejler, 1962; Hutchinson, 1967, p. 877; Ruttner-Kolisko, 1972, pp. 115 and 126). In fact it is possible to

discern this way of thinking in as early a work as Lauterborn (1904, p. 612). Now several studies (on different species) have been made showing allometric relations between spines and size of the body: Margalef, 1947; Green, 1960; Magis 1962; Pejler, 1962; Fergg, 1963; Amrén, 1964a; Hutchinson, 1967; Halbach, 1970; Björklund, 1972; Guiset, 1977; Nauwerck, 1978. Regarding the paper by Halbach on *Brachionus calyciflorus,* it ought to be noted that the growth of separate individuals was also followed.

This leaves the background of size variation to be discussed, a great topic treated in diverse general expositions (see, e.g., Margalef 1955). Here, only the relationship between metabolism and body size should be briefly touched upon. A short survey of this problem is given by Odum (1971, pp. 77-79). He does not discuss the applications on planktic organisms, which, however, is done by Brooks & Dodson (1965) and Brooks (1968). The well-known 'size-efficiency hypothesis', put forward by these authors, implies that larger zooplankters have a better metabolic economy than smaller ones, which stands in agreement with Winberg's law. According to Nilsson & Pejler (1973, pp. 69-71) a large body must, therefore, be an advantage in an environment poor in food, i.e. in oligotrophic lakes, and it is also shown that larger species and larger infraspecific forms are found in such lakes. However, what is here true in comparison of different lakes has also to be true for seasonal comparisons. It then appears as a striking fact that the conditions during winter are generally more oligotrophic than those of the summer. Thus, the winter forms should be larger, which is also the normal case. This may be conceived to form at least part of the explanation of the varying body size.

Of course the spines may be partly regarded as an adaptation for escaping predation (cf. above concerning *Asplanchna-Brachionus),* but a correlation between spine length and predation pressure has not been demonstrated for any *Keratella* or *Kellicottia* species (see also Nilsson & Pejler, loc. cit.)

However, surely not all variation is adaptive. Some examples where this is apparently not the case were mentioned by Pejler (1957, p. 41). Pure deformities are sometimes reported: Milković, 1934; Klement, 1955, 1957, 1959; Thomasson, 1957. Such forms are predominantly found in smaller waters and it seems probable that genetic drift is at work in these cases. It may also be mentioned here that students of brachionids in ponds and pools often talk of an erratic variation (see, e.g., Buchner & Mulzer, 1961), which should be compared with the relatively uniform conditions in real lakes.

As 'microgeographical isolation' ought to be one reason for the local form variation, the question may be raised whether new species could originate in this way. However, the possibilities of dispersal are probably in general strong enough to counteract isolation, and no evidence of such a microgeographical speciation seems to have been demonstrated. On the other hand, speciation based on 'macrogeographical isolation' surely occurs, and Pejler (1977, pp. 275-276) mentions some examples of probably recent evolution of new species or subspecies.

For some species a discontinuous variation within single waters has been found. Regarding *Keratella cochlearis*, this phenomenon has been demonstrated by Carlin (1943, pp. 56-58); Pejler (1957, pp. 6-13 and 40-41, 1962); Parise (1960, pp. 31-34, 1961, p. 123); Fergg (1963); Hutchinson (1967, pp. 879-880) and Nauwerck (1978, p. 277). Fig. 4 in Hakkari (1969) seems to show an indication of a similar variation in *Kellicottia longispina*. Several explanations are conceivable – camouflaged sibling species, polymorphism connected with apomixis etc. (see Pejler, 1957). One fact to be stressed is that the lakes containing two or three separate forms of *K. cochlearis* are deep enough to be stratified during the summer, and possibly the long-spined forms have developed in the hypolimnion (also if they later on can be encountered in the epilimnion as well). This hypothesis suggested by Pejler (1962, p. 12) is supported by Hillbricht-Ilkowska (1972), who found the spine-less form (f. *tecta*) chiefly in the epilimnion of a Polish lake. Three separate forms of *K. cochlearis* were encountered only in two of the investigated Swedish lakes. Both these lakes form parts of lake chains (and, in addition, are stratified). It appears possible that one of the forms has developed in an adjacent water and then been brought to the investigated lake. This view is supported by comparisons of samples collected in different years (see further Pejler, 1962, pp. 12-13).

It is obvious that the knowledge or ignorance of variation has influenced taxonomy to a very great extent. At the time of the typological species concept a multitude of species was described within the variable form complexes now treated. However, during the first years after the appearance of the pioneer works on variation, a reverse tendency is often traced. Thus, the genus *Polyarthra* is treated as a single species by, e.g., Lauterborn (1904) and Hartmann (1918). Wesenberg-Lund (1900) takes offence at the 'non-scientific species making' and recognizes, e.g., only two species of *Synchaeta,* all the other being considered seasonal variants. Von Daday (1897, p. 132) even sus-

pects that *K. cochlearis* and *K. quadrata* ('*Anuraea aculeata'*) should belong to the same species, basing his hypothesis upon studies in Lake Balaton. However, by biometric methods it has been possible to distinguish infraspecific variation from interspecific and to establish, on a firmer basis, new species within varying form complexes (e.g. by Carlin, 1943). Surely much more can be done within this field, especially if more modern methods are applied.

Different opinions have been expressed as well regarding the genetical background of the variation. The earlier writers were influenced by ideas of their time. Krätzschmar's reasoning, apparently inspired by August Weismann, has already been mentioned. Purely lamarckistic elements are incorporated into the explanations given by Hartmann, who writes, e.g., in 1918, p. 288, that some characters may be 'in certain cases acquired hereditarily due to external influence' (translated from German).

Using as a base the current scientific thinking, however, it is appropriate to ask to what extent the variation is genetically founded. This problem has been approached, e.g., by culturing forms from different ponds under equal conditions (see Buchner, Mulzer & Rauh, 1957; Buchner & Mulzer, 1961; Rauh, 1963 and Halbach, 1970). Thereby it was shown that the different clones derived in this way reacted similarly, though not identically, to the environmental factors to which they were exposed. Buchner & Mulzer (op. cit.) discuss three cooperating factor complexes: internal factors (evidently hereditary to their character), current environment and previous environment (with a subsequent effect).

Strictly genetical analysis has been performed by King (1972, 1977), who considers seasonal variation to be 'largely, but not entirely, non-genetic in origin'. Samples containing *Euchlanis dilatata* Ehrbg, or species of *Asplanchna* were collected at biweekly or weekly intervals and clones reared from them. These clones were found to differ in diverse physiological characters, and a genetic change through time was thus demonstrated (in two different rotifer families). King discusses two alternative hypotheses, those of 'incomplete' and 'complete genetic discontinuity', respectively. Most evidence is in favour of the model of complete discontinuity, which presupposes a low gene flow and a high competition between genotypes adapted to different environmental conditions. This model is also supported by the results of Snell (1977), who presents evidence of a succession of genetically distinct populations, each one developing from resting eggs at different times of the year. Though no corresponding investigations have been made regarding brachionids it seems quite probable that a similar pattern can exist there as well. However, concerning the lake-dwelling populations of *Keratella* and *Kellicottia*, there are some circumstances contradicting this idea. Though very intense investigations were performed over 6 years by Carlin (1943, pp. 103-104 and 143) a sexual period was never recorded for the common species *Keratella cochlearis* and *Kellicottia longispina* and only once for *Keratella quadrata*. Similar results were obtained by Ruttner-Kolisko (1949, pp. 443 and 460) and Pejler (1957, p. 43[1]). If no resting eggs are formed, and consequently not hatched, the mentioned model of King & Snell cannot be applied. Out of the three models discussed by King (1972) then only one remains, viz. that of a purely physiological adaptation. In other words, the *seasonal* variation in *lake* populations of, e.g., *Keratella cochlearis* ought to be regarded as a mainly non-genetic phenomenon. On the other hand, genetic factors are probably largely responsible for the different pattern of variation shown in different lakes, which stands in agreement with the argumentation held by Buchner and his co-workers, as well as by Snell (1977).

The material and understanding of rotifer variation has indisputably increased since the days of Lauterborn and Krätzschmar. In spite of this, a simple universal solution has never seemed more remote than today. Probably, such a general solution does not exist. The problem may be compared with that of mictic-female production (see the review by Gilbert 1977), different conditions occurring in different species. Nature certainly does not always provide simple solutions for inquirers of truth. This should not, of course, discourage us from trying to clarify things as far as is possible.

References

Ammann, H. 1913. Temporalvariationen einiger Planktonten in oberbayerischen Seen. 1910-1912. Arch. Hydrobiol. 9: 127-146.

Ammann, H. 1923. Temporalvariationen einiger Planktonwesen. Beitr. Nat DenkmPflege 8: 297-317.

Amrén, H. 1964 a. Temporal variation of the rotifer Keratella quadrata (Müll.) in some ponds on Spitsbergen. Zool. Bidr. Uppsala 36: 193-208.

An unfortunate printing error was discovered on this page: Formations of resting eggs were observed in *K. longispina* in Lapland only in tarns, not in lakes (not the reverse, which was written).

Amrén, H. 1964 b. Ecological and taxonomical studies on zooplankton from Spitsbergen. Ibid. 36: 209-276.

Arora, H. C. 1966. Cyclomorphosis (form variations) in some species of Indian planktonic Rotatoria. Int. Rev ges. Hydrobiol. 51: 623-632.

Berzinš, B. 1958. Ein planktologisches Querprofil. Rep. Inst. Freshw. Res. Drottningholm 39: 5-22.

Björklund, B. G. 1972. Taxonomical and ecological studies of species of Notholca (Rotatoria) found in sea- and brackish water, with description of a new species. Sarsia 51: 25-65.

Brooks, J. L. 1968. The effects of prey size selection by lake planktivores. Syst. Zool. 17: 273-291.

Brooks, J. L. & Dodson, S. I. 1965. Predation, body size and composition of plankton. Science 150: 28-35.

Buchner, H. & Mulzer, F. 1961. Untersuchungen über die Variabilität der Rädertiere. II. Der Ablauf der Variation im Freien. Z. Morph. Ökol. Tiere 50: 330-374.

Buchner, H., Mulzer, F. & Rauh, F. 1957. Untersuchungen über die Variabilität der Rädertiere. I. Problemstellung und vorläufige Mitteilung über die Ergebnisse. Biol. Zbl. 76: 289-315.

Carlin, B. 1943. Die Planktonrotatorien des Motalaström. Zur Taxonomie und Ökologie der Planktonrotatorien. Medd. Lunds Univ. Limnol. Inst. 5: 256 pp.

Daday, E. von, 1897. Rotatorien. In: G. Entz, Resultate der wissenschaftlichen Erforschung des Balatonsees 2, 1: 121-133.

Dieffenbach, H. & Sachse, R. 1911. Biologische Untersuchungen an Rädertieren in Teichgewässern. Int. Rev. Ges. Hydrobiol. Hydrogr., Biol. Suppl. ser. 3. 93 pp.

Edmondson, W. T. 1948. Ecological applications of Lansing's physiological work on longevity in Rotatoria. Science 108: 123-126.

Edmondson, W. T. 1960. Reproductive rates of rotifers in natural populations. Mem. Ist. ital. Idrobiol. 12: 21-77.

Edmondson, W. T. 1965. Reproductive rate of planktonic rotifers as related to food and temperature in nature. Ecol. Monogr. 35: 61-112.

Entz, G. & Sebestyén, O. 1946. Das Leben des Balaton-Sees. Arb. ung. biol. Forsch. Inst. 16: 179-411.

Fergg, I. 1963. Untersuchungen über die Variabilität der Rädertiere. IV. Vergleichende biometrische Untersuchungen an Keratella cochlearis und K. quadrata. Verh. dt. Zool. Ges. München 1963: 253-268.

Gallagher, J. J. 1955. Cyclomorphosis in the rotifer Keratella cochlearis. Diss. 13, 388, Ann Arbor, Mich.

Gallagher, J. J. 1957. Cyclomorphosis in the rotifer Keratella cochlearis (Gosse). Trans. Am. micr. Soc. 76: 197-203.

Gilbert, J. J. 1966. Rotifer ecology and embryolog cal induction. Science 151: 1234-1237.

Gilbert, J. J. 1977. Mictic-female production in monogonont rotifers. Arch. Hydrobiol. Beih. 8: 142-155.

Gillard, A. A. M. 1948. De Brachionidae (Rotatoria) van België met Beschouwingen over de Taxonomie van de Familie. Natuurwet. Tijdschr. 30: 159-218.

Gillard, A. A. M. 1949. On the systematical aspect of the study of animal microorganisms. Biol. Jaarb. 16: 206-214.

Green, J. 1960. Zooplankton of the River Sokoto. The Rotifera. Proc. zool. Soc. Lond. 135: 491-523.

Green, J. 1977. Dwarfing of rotifers in tropical crater lakes. Arch. Hydrobiol. Beih. 8: 232-236.

Guiset, A. 1977. Some data on variation in three planktonic genera. Ibid. 8: 237-239.

Hakkari, L. 1969. Zooplankton studies in the lake Längelmävesi.

south Finland. Ann. Zool. Fenn 6: 313-326.

Halbach, U. 1970. Die Ursachen der Temporalvariation von Brachionus calyciflorus Pallas. Oecologia 4: 262-318.

Halbach, U 1971 a. Das Rädertier Asplanchna – ein ideales Untersuchungsobjekt. III. Ein Räuber liefert seiner Beute die Abwehrwaffen. Mikrokosmos 1971: 360-365.

Halbach, U. 1971 b. Zum Adaptivwert der zyklomorphen Dornenbildung von Brachionus calyciflorus Pallas (Rotatoria). I. Räuber-Beute-Beziehung in Kurzzeit-Versuchen. Oecologia 6: 267-288.

Halbach, U. & Jacobs, J. 1971. Seasonal selection as a factor in rotifer cyclomorphosis. Naturwiss. 57: 326.

Hartmann, O. 1918. Studien über den Polymorphismus der Rotatorien. Arch. Hydrobiol. 12: 209-310.

Hillbricht-Ilkowska, A. 1972. Morphological variation of Keratella cochlearis (Gosse) (Rotatoria) in several Masurian lakes of different trophic level. Pol. Arch. Hydrobiol. 19: 253-264.

Hutchinson, G. E. 1967. A treatise on limnology. Vol. 2. Introduction to lake biology and the limnoplankton. New York, London and Sidney. 1115 pp.

King, C. E. 1972. Adaptation of rotifers to seasonal variation. Ecology 53: 408-418.

King, C. E. 1977. Genetics of reproduction, variation, and adaptation in rotifers. Arch. Hydrobiol. Beih. 8: 187-201.

Klement, V. 1955. Uber eine Missbildung bei dem Rädertier Keratella cochlearis und über eine neue Form von Keratella quadrata. Zool. Anz. 155: 321-324.

Klement, V. 1957. Zur Rotatorienfauna des Monrepos-Teiches bei Ludwigsburg. Jh. Ver. vaterl. Naturk. Württemberg 112: 238-263.

Klement, V. 1959. Zur Rotatorienfauna des Monrepos-Teiches bei Ludwigsburg. 2. Beitrag. Ibid. 114: 193-221.

Krätzschmar, H. 1908. Über den Polymorphismus von Anuraea aculeata Ehrbg. Int. Rev. Ges. Hydrobiol. Hydrogr. 1: 623-675.

Krätzschmar, H. 1913. Neue Untersuchungen über den Polymorphismus von Anuraea aculeata Ehrbg. Ibid. 6: 44-49.

Lauterborn, R. 1898. Vorläufige Mitteilungen über den Variationskreis von Anuraea cochlearis Gosse. Zool. Anz. 21: 597-604

Lauterborn, R. 1900. Der Formenkreis von Anuraea cochlearis. I. Morphologische Gliederung des Formenkreises. Verh. naturh.-med. Ver. Heidelb., N. F., 6: 412-448.

Lauterborn, R. 1904. Der Formenkreis von Anuraea cochlearis. II. Die cyklische oder temporale Variation von Anuraea cochlearis. Ibid. 7: 529-621.

Lindström, K. & Pejler, B. 1975. Experimental studies on the seasonal variation of the rotifer Keratella cochlearis (Gosse). Hydrobiologia 46: 191-197.

Magis, N. 1962. Croissance allometrique chez Brachionus falcatus Zacharias. Ann. Soc. r. Zool. Belg. 92: 153-169.

Margalef, R. 1947. Notas sobre algunos Rotiferos. Publ. Inst. Biol. apl. Barcelona 4: 135-148.

Margalef, R. 1955. Temperatura, dimensiones y evolución. Ibid. 19: 13-94.

Milkovic, Z. 1934. Degenerationserscheinungen bei Anuraea cochlearis Gosse. Zool. Anz. 106: 252-255.

Nauwerck, A. 1978. Notes on the planktonic rotifers of Lake Ontario. Arch. Hydrobiol. 84: 269-301.

Nayar, C. K. G. 1965. Cyclomorphosis of Brachionus calyciflorus Pallas. Hydrobiologia 25: 538-544.

Nilsson, N.-A. & Pejler, B. 1973. On the relation between fish fauna ' and zooplankton composition in North Swedish lakes. Rep. Inst. Freshw. Res. Drottningholm 53: 51-77.

Nipkow, F. 1952. Die Gattung Polyarthra Ehrenberg im Plankton des Zürichsees und einiger anderer Schweizer Seen. Schweiz. Z. Hydrol. 14: 135-181.

Nipkow, F. 1961. Die Rädertiere im Plankton des Zürichsees und ihre Entwicklungsphasen. Ibid. 23: 398-461.

Odum, E. P. 1971. Fundamentals of ecology. 3rd ed. Philadelphia, London and Toronto. 574 pp.

Parise, A. 1960. I Rotiferi del Lago di Nemi. Arch. Oceanogr. Limnol. 12: 1-93.

Parise, A. 1961. Sur les genres Keratella, Synchaeta, Polyarthra et Filinia d'un lac italien. Hydrobiologia 18: 121-135.

Pejler, B. 1957. On variation and evolution in planktonic Rotatoria. Zool. Bidr. Uppsala 32: 1-66.

Pejler, B. 1958. Taxonomic Studies on planktonic Rotatoria. Uppsala Univ. Arsskr. 1958: 240-243.

Pejler, B. 1962. On the variation of the rotifer Keratella cochlearis (Gosse). Zool. Bidr. Uppsala 35: 1-17.

Pejler, B. 1977. On the global distribution of the family Brachionidae (Rotatoria). Arch. Hydrobiol., Suppl. 53: 255-306.

Pourriot, R. 1964. Etude expérimentale de variations morphologiques chez certaines espèces de Rotifères. Bull. Soc. zool. France 89: 555-561.

Rauh, F. 1963. Untersuchungen über die Variabilität der Rädertiere. III. Die experimentelle Beeinflussung der Variation von Brachionus calyciflorus und Brachionus capsuliflorus. Z. Morph. Ökol. Tiere 53: 61-106.

Robert, H. 1925. Sur la variabilité de quelques espèces planctoniques du lac de Neuchâtel. Ann. Biol. lac. 14: 5-38.

Ruttner-Kolisko, A. 1949. Zum Formwechsel- und Artproblem von Anuraea aculeata (Keratella quadrata). Hydrobiologia 425-468.

Ruttner-Kolisko, A 1972. Rotatoria. Binnengewässer 26: 99-234.

Schneider, G. 1922. Das Zooplankton der Eifelmaare, insbesondere die Cyclomorphose von Anuraea cochlearis und Notholca longispina. Verh. naturh. Ver. preuss. Rheinl. 77: 7-34.

Schreyer, O. 1921. Die Rotatorien der Umgebung von Bern. Int. Rev. Hydrobiol. 9: 311-370 and 491-537.

Snell, T. 1977. Clonal selection: Competition among clones. Arch. Hydrobiol. Beih. 8: 202-204.

Sudzuki, M. 1964. New systematical approach to the Japanese planktonic Rotatoria. Hydrobiologia 23: 1-24.

Thomasson, K. 1957. Über eine Missbildung bei dem Rädertier Keratella cochlearis. Zool. Anz. 158: 31.

Varga, L. 1941. Über die Zyklomorphose einiger Planktontiere des Balaton-Sees. Math. term. Ert. 60: 546-582. (Hungarian, with German summary.).

Vialli, M. 1924. Richerche sui Rotiferi pelagici del Plancton lariano. In: R. Monti (ed.), La Limnologia del Lario: 217-282.

Wermel, J. 1930. Über die Variabilität der Anuraea aculeata v. serrulata Ehrbg. und Arcella vulgaris Ehrbg. in einem Moortümpel. Int. Rev. Ges. Hydrobiol. Hydrogr. 24: 140-146.

Wesenberg-Lund, C. 1900. Von dem Abhängigkeitsverhältnis zeischen dem Bau der Planktonorganismen und dem spezifischen Gewicht des Süsswassers. Biol. Zbl. 20: 606-619, 644-656.

Wesenberg-Lund, C. 1930. Contributions to the biology of the Rotifera. II. The periodicity and sexual periods. K. danske vidensk. Selsk., nat.-math. Afd. 9, 2, 1: 1-230.

Züscher, M. 1912. Das Plankton des Schlossgrabens und des Schlossteiches zu Münster in Westfalen unter besonderer Berücksichtigung der Temporalvariationen von Anuraea cochlearis und Ceratium hirundinella. Trier (Inaug.-Diss. Münster). 50 pp.

ON THE EVOLUTIONARY PATHWAYS OF SPECIATION IN THE GENUS NOTHOLCA

L. A. KUTIKOVA

Zoological Institute Academy of Sciences, Leningrad, 199164, U.S.S.R.

Keywords: Rotatoria, taxonomy, geographical distribution

Abstract

An attempt is undertaken to determine criteria and limits of taxonomic rank in the genus *Notholca* proceeding from the standpoint of level-values of the characters evolution. Level-values of characters were established using Vavilov's principle of homologic series of hereditary variability. As a result of taxonomic revision such terms as group of species, species and subspecies are recommended. Analysis of the scheme of relations between representatives of the genus and data on their ranges gives a clue to relations between phylogeny and geographic distribution of the group over the vast territory of the Asian continent which may have been its centre of speciation.

Introduction

The modern species concept generally accepted in current taxonomy can without doubt successfully clarify various intricate problems of rotifer taxonomy. These problems are concerned primarily with rotifer polymorphism and polyphenism and are particularly obscure where the law of homologous series in hereditary variability is involved (Vavilov, 1922; Mayr, 1965). To avoid inexact conclusions in establishing relations between various taxonomic categories and to correctly describe new rotifer species, Ruttner-Kolisko (1963) has suggested two ways: statistical analysis of characters in natural populations and cultivation of genetically identical clones in the laboratory. Thus, apart from morphological characteristics, diagnosis of the species described should include ecological and physiological ones, as well as reproductive isolation. Starobogatov (1977), assessing the results of using biological and typological species concepts distinguished two stages in systematical revisions at the species level. The first step i.e. a morphotypical approach has revealed differing forms, distribution areas and their ecological specificity. The second step, a biological one, establishes the final taxonomic status of the primarily distinguished forms using the data obtained from statistical comparisons of characteristics of individuals found together, experimental hybridization, caryological analysis and other methods which convincingly show the existence of various populations.

Unfortunately, these suggestions are not currently used in practice and the taxonomic conclusions are based mainly on the typological approach or, to be more exact, on the morphological approach, and descriptions of new species are based on such characters. It is only in recent years that the method of variation statistics has been used in the procedure.

Material and method

The species of the genus *Notholca* are a typical example of polytypical species, though from the standpoint of modern concepts it is impossible at present to provide strong evidence in favour of their segregation. The morpho-typological approach to the representatives of this genus allowed us to reveal that interspecific as well as intraspecific characters can be easily arranged into Vavilov's homological series of hereditary variability. Using the principle of the homological series a scheme of relationships in the genus *Notholca* was constructed (Table 1).

Being objective to some extent the method, however, involves certain difficulties in establishing the level-value of characters since it is here that subjective attitude and intuition of a taxonomist would be expected.

The table columns present all the recently known representatives of the genus *Notholca* and lines present all their characters: morphological ones (structure of the lorica and its details), some ecological ones (salinity) and a few geographical ones as well.

Hydrobiologia 73, 215-220 (1980).   0018-8158/80/0733-0215$01.20.
© *Dr. W. Junk b.v. Publishers, The Hague.*

Table I. Variability of species of the genus *Notholca*.

Characters. Morphological characters	*squamula* (Müll.)	–*michiganensis* Stemberger	–*balina* Focke	*müllerii* Focke	–*frigida* Yaohnov	*jugosa* Gosse	*laurentiae* Stemberger	*kostel* Chengalath	*cristata* Grese	*lapponica* Ruttner-Kolisko	*verae* Kutikova	*acuminata* Ehrb.+*extensa* Olofsson	*marina* Focke	*labis* Gosse +*limnetica*	*labtstyla* (Olofsson) Levander	*complexa* Kutikova	*caudata* Carlin	*intermedia* Voronkov	*haueri* Thomasson	*cinctura* Skorikov	*grandis* Voronkov	*psammarina* Buchholz et Ruhmann	*lamellifera* Vassiljeva et Kutikova	*determinata* Vassiljeva et Kutikova	*rectospina* Kutikova	*kozhovi* Vassiljeva et Kutikova	*gataglast* Kutikova	*lyrata* Tikhomrov	*foliacea* (Ehrb.)	*japonica* (Marukawa)+*kisseleui* Kutikova	*angulata* Daday	*striata* (Müll.)	–*bipalium* (Müll.)	–*liepettesent* Bjorklund	*thraththoides* Skorikov	*olchonensis* Tikhomrov+*deviata* Vas. Kut.	*jasontekii* Tikhomrov	*baicalensis* Yaohnov	*cornuta* Carlin
I. lateral spines of the lorica																																	+	+	+	+	+	+	+
dorsal spines	+	+	+	+	+	+	+	+	+	+	+	+	+	+	+	+	+	+	+	+	+	+	+	+	+	+	+	+	+	+	+	+	+	+	+	+	+	+	+
without caudal spines	+	+	+	+	+	+	+	+	+																														
with caudal spines										+	+	+	+	+	+	+	+	+	+	+	+	+	+	+	+	+	+	+	+	+	+	+	+	+	+	+	+	+	+
6 anterior dorsal spines	+	+	+	+	+	+	+	+	+	+	+																												
4 anterior dorsal spines								+	+																				+										
2. shape of the lorica																																							
broad ovate, ratio length/width<1.5			+									+						+	+			+							+			+	+		+	+	+	+	+
elongated ovate, ratio 1/w <2	+						+			+			+		+	+	+	+	+	+	+	+	+	+	+	+	+	+	+	+			+	+	+	+	+	+	+
elongated, ratio 1/w>2	+	+				+	+																								+			+			+	+	+
bell-shaped				+				+	+																											+	+	+	+
diamond-shaped					+	+																	+		+	+					+				+				
Size of the lorica: large>350µm																				+	+	+	+		+	+	+		+	+	+						+	+	+
medium 150-350 µm	+	+	+	+	+	+	+	+	+	+	+	+	+	+	+	+	+	+	+	+	+	+	+	+	+	+	+	+	+	+	+	+	+	+	+				
small ≤140 µm											+	+	+		+	+	+																						
Markings on the lorica, longitudinal stripes																			+	+	+	+	+	+	+	+	+	+	+					+		+			
stripes inconspicuous, smooth	+	+	+																												+		+	+	+	+	+	+	+
dotty																										+				+		+				+			
granular					+	+			+																														
with wavy ridges								+	+	+	+																		+	+									
Length and shape of the dorsal spines																																							
long (>30 µm)	+	+	+	+	+	+	+	+	+	+	+	+	+	+	+	+	+	+	+	+	+	+	+	+	+	+	+	+	+	+	+	+	+	+	+	+	+	+	+
short (≤30 µm)	+	+	+	+	+	+	+	+	+	+	+	+	+	+			+											+											
straight or slightly curved down	+	+	+	+	+	+	+	+	+	+	+	+	+	+		+										+		+	+	+	+	+	+	+	+	+	+	+	+
curved															+		+	+	+	+	+	+	+	+	+	+	+		+	+									
the median one longest																									+		+					+	+	+	+	+	+	+	+
the lateral one longest																										+				+	+	+	+	+	+	+		+	+
the lateral one shortest	+	+																															+				+	+	+
the intermediate one longest																																				+		+	+

the intermediate one shortest
Posterior margin of the lorica
    cone-shaped
    oval
    rectangular-rounded
    with a segment
Posterior margin of the ventral
    plate lamellate
    slightly protrusible
Margins of the cloacal membrane
    nearly trapezium-shaped
    wing-shaped
    nearly semicircular
Dorsal plate convex or keeled

3. Length and shape of the lateral
    spines: long (>80 μm)
    short (usually ≤60 μm)
    wing-shaped
    lash-shaped
    extended distally
Caudal spine: variable
    needle shaped
    broadly spade shaped
    extended distally
    angular
    forked
Spine of the dorsal plate mobile
Spine of the dorsal plate immobile
2 caudal spines
Spine of the ventral plate mobile

Ecological-geographical characters.
  Habitat
    salt and brackish water only
    fresh water only
    salt and fresh water
    cold stenothermous
    widespread
    rare
    endemic to lake Baical

Result and discussion

Grouping characters in the genus *Notholca* according to level values allow us to represent continuous and discontinuous series in their evolution. Such characters as the presence of caudal, dorsal and lateral spines and a number of anterior dorsal spines belong to the first level. These particular characters show similarity in the series of hereditary variability between genera of the family Brachionidae (e.g. *Keratella, Brachionus* in the number of anterior dorsal spines) and between rather distant families in whose genera *(Macrochaetus, Lecane, Squatinella)* the number of spines on the lorica changes. On the second level similarity in the form of lorica, its marking, size, structure of dorsal spines and cloacal membrane shows that more definite and therefore more close relations exist in this group. And finally the third level unites rather variable characters, usually of a polyphenic type, which form continuous series of parallel variability (e.g. the shape of caudal and lateral spines).

To determine the taxonomic rank of representatives of the genus *Notholca* one can use the level-value of characters which show the degree of relationship between them. The first level of characters corresponds to the subgenetic or superspecific category, and for such species as *N. squamula, N. acuminata, N. striata, N. caudata, N. cinetura, N. lamellifera,* and *N. triarthroides* the most suitable term is 'group of species' (Cain, 1954: Ruttner-Kolisko, 1974). Characters of the second level correspond to the rank 'species'. Characters of the third level correspond to the rank 'subspecies'. The term 'Artgruppe' and 'Formenkreis' used by Koste (1978) seem of little use, since various meanings have been attributed to them at different times (Kleinschmidt, 1900; Rensch, 1929). The use of the above mentioned names makes the problem of limits and criteria of rotifer species still more urgent.

To study relations in genus *Notholca* one can use the scheme presented in Fig. 1. This scheme allows us to trace three separate complexes of species: the 1st- without posterior spines on the lorica; 2nd with caudal spines developed to some extent; 3rd with well-developed posterior spines*. The parallelism in morphological variability allows us to distinguish the characters typical of species complexes, as well as of separate species. Thus *N. angulata* with its diamond-shaped lorica and lateral angles is a

typical transitional form between species of complex II and complex III. The dotted sculpture of the lorica in *N. japonica, N. foliacea, N. kostei,* and *N. olchonensis* is similar in complexes I, II and III. On the level of intraspecific parallel variability *N. complexa* provides a link between the species *N. acuminata* and *N. caudata.* Parallelism is more obvious in interspecific and particularly intraspecific variability. Specific characters which are used in diagnosis are usually mosaic and are relatively stable whereas some intraspecific ones (e.g. changes of caudal spines) are surprisingly similar in the direction of their variability. A distinct tendency towards parallel variability is observed in the group of endemic species of Lake Baikal (of the dorsal spines the lateral ones are the longest). In establishing taxonomic rank therefore one should proceed from the level-value of the characters (Table 1).

The scheme presented here provides a clue for a study of the evolution of species of the genus *Notholca* from the point of view of geographical speciation. In Pejler's opinion (Pejler, 1977), research into distribution of rotifers of the family Brachionidae, genus *Notholca* included, can shed light on the origin of several genera. However, he finds no obvious relation between phylogeny and distribution.

Species of the genus *Notholca* include a great number of Baikalian endemics (11 endemic species out of a total of 39 species in the genus). As is known, the fauna of Lake Baikal consists of two complexes: one is usually termed palaeolimnic, being a more ancient complex closely associated with the inhabitants of oligotrophic waterbodies of the Holarctic, while the second one is called mesolimnic, and originated later from remnants of the faunas of no longer-existing waterbodies, originally located in the Central Asian and Mongolian territories in the upper Cretaceous (Martinson, 1961; Kozkov, 1973). One can assume that the formation of the mesolimnic rotifer complex of Lake Baikal started in the Palaeocene in the territory of the Baikal area, which was a centre of speciation for the genus *Notholca,* like North America was for the genus *Keratella* (Berzins, 1954). Evolution of this complex continued until the Late Pliocene – Early Pleistocene, i.e. until the intensive cooling which ultimately led to the Pleistocene glaciations. As glaciers did retreat, a number of species *(N. jugosa, N. triarthroides)* could penetrate into the arctic and as well as into temperate latitudes. Some of them diverged and evolved into new species *(N. intermedia, N. caudata, ? N. grandis, N. cinetura).* The series of endemic species includes the species groups *N. triarthroides* which originated by way of adaptive radiation. It is also

* The scheme of the species relation presented is almost identical to that obtained by using a computer (Kutikova & Menshutkin, in press).

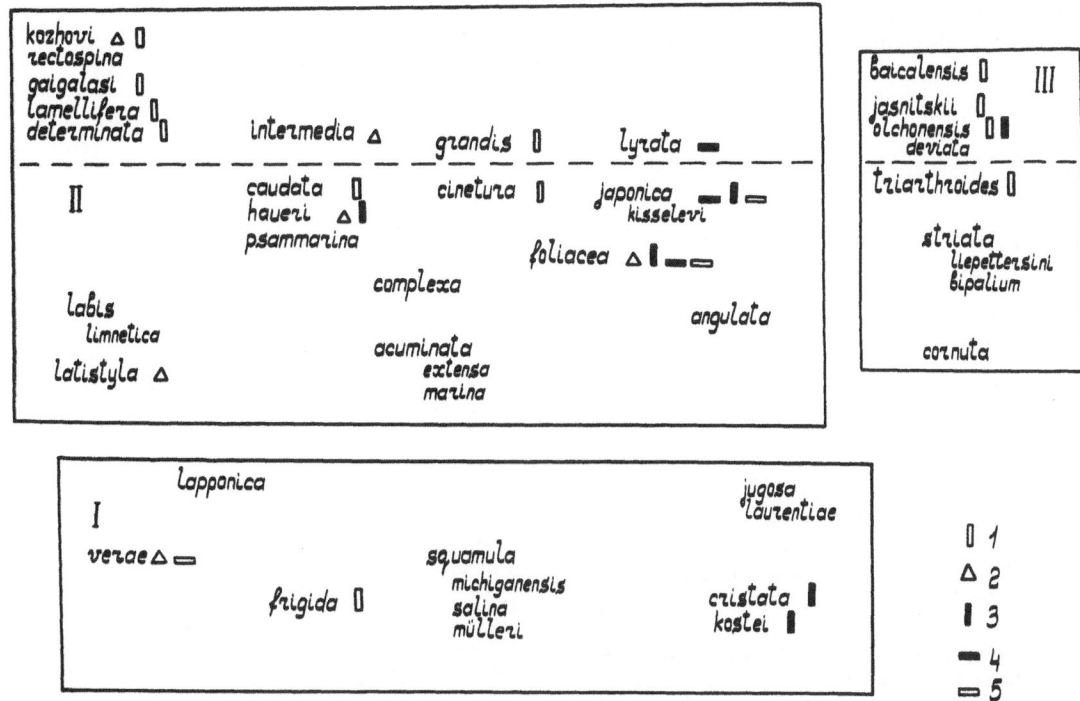

Fig. 1. Similar characters in the species of different complexes.
1. Large size; 2. Posterior margin of ventral plate protrusible; 3. Dorsal plate with dotty marking; 4. Caudal spine forked; 5. Lorica with wavy ridges.
Above the interrupted line: endemics of Lake Baikal.

interesting to mention here the occurrence of *N. japonica* in the sea of Japan (apart from other seas), *N. japonica kisselevi* in South-Kuril shallow waters, and the variable species *N. complexa* in Hanka Lake and *N. angulata* in Mongolia (the latter found also in Lake Baikal) i.e. in the border regions of those large basins of the upper Cretaceous which spread from Japan and South Korea to Soviet Central Asia and were in some connection with the eastern seas.

Though they are relatively euryhaline and cold stenothermous it would definitely be wrong to think that all the species of the genus *Notholca* have originated in these areas. Apparently as a result of a different rate of speciation palaeolimnic species such as *N. squamula, N. acuminata, N. labis* are more ancient and forms closely related to them may have given origin to new species and subspecies in other areas. These species possess the most extensive polymorphism, as we can see in numerous geographical races (in *N. squamula* – subsp. *michiganensis, salina, mülleri*; in *N. acuminata* – *marina*) and also in related species: *N. cristata, N. kostei, N. lapponica, N. latistyla*. Geographic isolation has in all probability led to the emergence

of *N. striata (subsp. bipalium, liepetterseni), N. lapponica, N. psammarina, N. haueri*. Some characters in *N. verae* found in Antarctica, can be treated as indirect evidence that it is a palaeolimnic type of the *N. squamula* species group.

Conclusion

Species of the genus *Notholca* fall into two groups. The more ancient one *(N. squamula, N. acuminata, N. labis)* is cosmopolitan in its distribution and displays high polymorphism. The other, larger group of species of this genus shows well-defined ranges with the centre of speciation in the vast territory of Central Asia, from which the dispersal of younger species *(N. caudata, N. triarthroides, N. cinetura)* started. This distribution is probably associated with the glacial period (Pejler, 1962). A number of endemic species is recorded from the border region of Central Asia *(N. complexa, N. angulata)*. The greatest number of endemics inhabits Lake Baikal where intense speciation continues and morphological variability is usually observed with poor stability of characters. The above-mentioned levels of characters also show different rates of evolution in various groups of the genus *Notholca*.

References

Berzins, B. 1954. A new Rotifer, Keratella canadensis. J. Quekett. micr. Cl., ser. 4: 112-115.

Cain, A. I. 1954. Animal species and their evolution. Hutchinson's Univ. Libr. London, 190 pp.

Kleinschmidt, O. 1900. Arten oder Formenkreise? J. Ornithol. 48: 134-139.

Koste, W. 1978. Rotatoria. Die Rädertiere Mitteleuropas. Gebrüder Borntraeger, Berlin-Stuttgart, 2 Vols., 673 pp., 234 plates.

Kozhov, M. M. 1973. Formation and evolutionary pathways of the fauna of Lake Baikal. In: Problems of Evolution, Vol. 3: 5-30, Publ. House 'Nauka', Novosibirsk (In Russian).

Martinson, G. G. 1961. Mesosoic and Cenozoic molluscs in continental deposits of the Siberian platform of Transbaikalia and Mongolia. Ed. Acad. Sci. USSR. Moskow-Leningrad, 332 pp. (in Russian).

Mayr, E. 1965. Animal species and evolution. Harvard Univ. Press, Cambridge. Mass., 797 pp.

Pejler, B. 1962. Notholca caudata Carlin (Rotatoria), a new-presumal glacial relict. Zool. Bidr. Uppsala 33: 453-457.

Pejler, B. 1977. On the global distribution of the family Brachionidae (Rotatoria). Arch. Hydrobiol. 53, 2: 255-306.

Rensch, B. 1929. Das Prinzip der geographischen Rassenkreise und das Problem der Artbildung. Borntraeger, Berlin, 206 pp.

Ruttner-Kolisko, A. 1963. The interrelationships of the Rotifera. In: The lower Metazoa. Univ. Calif. Press: 263-272.

Ruttner-Kolisko, A. 1972. Rotatoria. Binnengewässer 26: 99-234.

Starobogatov, J. I. 1977. On relationships between biological and typological species concepts. Zh. obshch. Biol. 38: 157-165. (In Russian).

Vavilov, N. I. 1922. The law of homologous series in variation. Genetics 12: 47-89.

PRELIMINARY REMARKS ON THE CHARACTERISTICS OF THE ROTIFER FAUNA OF AUSTRALIA (NOTOGAEA)

W. KOSTE* & R. J. SHIEL†

* Ludwig-Brill-Strasse 5, Quakenbrück D-4570, West Germany.
† Zoology Dept., University of Adelaide, Box 498 G.P.O., Adelaide, South Australia

Keywords: Rotifera, Australia, additions to known fauna, giant forms, endemicity

Abstract

Unusually large forms of *Asplanchna sieboldi, Brachionus plicatilis, B. calyciflorus, Filinia pejleri, Trichocerca similis* and *Keratella slacki* were collected from waters of south-eastern Australia. These giant forms are figured, and brief location and ecological data are given. Although the rotifer fauna of the Murray riversystem contains pan-tropical and pan-subtropical species, there exists a greater degree of endemicity than previously considered.

Introduction

With the exception of isolated collections (e.g. Berzins, 1953, 1960, 1963; Russell, 1957, 1961) and single-species descriptions (Sudzuki & Timms, 1977), study of Australia's rotifer fauna has lapsed for more than fifty years. Early workers (Thorpe, 1887, 1891; Shephard, 1896, 1897, 1899) confined their efforts to the population centres of Brisbane, Sydney and Melbourne on the eastern coast. No information has been available from the rest of the continent. Many of the early reports consisted of species lists (e.g. Whitelegge, 1889; Shephard, 1911) with little or no ecological information.

As a result of a survey during 1976-1979 of the zooplankton of the Murray-Darling river system, a drainage basin of more than 1 million km², some 1200 samples were collected from a variety of habitats. These included impoundments, billabongs (= ox-bows or meander lakes) and rivers. Although processing of the material continues, a preliminary report was given by Shiel (1978), and several new taxa *(Brachionus keikoa, Keratella shieli, Lecane ungulata australiensis* and *Filinia pejleri* var. *grandis)* were described by Koste (1979). Synecological information on the lower Murray Rotifera was given by Shiel (1979).

A literature search established 279 valid species records from the continent, and initial Murray samples yielded an additional 52 taxa (Shiel & Koste, 1979). Berzins (pers. comm.) reported three additional species missed in the search, and a further 90 species or sub-species have subsequently been recorded from the Murray-Darling system and from the Alligator River area of the Northern Territory. New species descriptions and ecological information on this material will be published at a later date (Koste, in press; Koste & Shiel, in press, in prep.).

Remarks in this paper are confined to brief notes on the species assemblages, particularly the endemic fauna, and ecological information on several giant forms which occur in the plankton.

Materials and methods

Zooplankton samples were collected in qualitative tows with a 36 µm-mesh cone net, or in quantitative samples by means of a 30 l⁻¹ perspex trap (modified after Schindler, 1969). All samples were preserved in the field in 4% formalin.

Species assemblages

Above the confluence of the two rivers, the rotifer assemblages recorded reflects the distinctive nature of the two systems. The Darling River flows south from around 23°S for 2700 km, carrying water from tropical monsoonal rains in summer and autumn. Waters are characteristically alkaline (pH 8-8.5) and highly turbid (> 120 NTU's). The majority of tributaries are ephemeral, and the Darling has ceased to flow 48 times since 1881 (Woodyer, 1978). There are no significant impoundments, and due to a fall of only 2-3 cm km⁻¹, flow rates are low.

Zooplankton tows usually contain up to thirty rotifer species, including a mixture of pantropical, typical southern hemisphere and endemic Australian forms. Congeneric associations are common, and seasonal variations in species composition occur. Typically, brachionids pre-

Hydrobiologia 73, 221-227 (1980). 0018-8158/80/0733-0221$01.40.
© *Dr. W. Junk b.v. Publishers, The Hague.*

dominate, with *Brachionus angularis, B. calyciflorus, B. diversicornis, B. novaezealandia* and *B. urceolaris* common, *B. budapestinensis, B. falcatus* and *B. keikoa* seasonal, and *B. quadridentatus* and *B. dichotomus* as rare incursions from adjacent billabongs. Usually three or four species of *Keratella* are recorded together: *K. australis, K. cochlearis, K. procurva* and *K. tropica,* with seasonal appearances of *K. quadrata* and *K. slacki.* Three or four species of *Filinia* are also common *(F. longiseta, F. opoliensis, P. pejleri, F. terminalis),* as are several species of *Lecane (L. bulla, L. flexilis, L. hamata, L. luna, L. lunaris, L. signifera, L. stenroosi).* The remaining species are typically from the genera *Cephalodella, Euchlanis, Hexarthra, Polyarthra, Synchaeta, Trichocerca.*

The River Murray, which flows west for some 2500 km from the eastern highlands, carried winter and spring rains and snow melt. Waters are typically less turbid than those of the Darling, but seasonal blooms of *Melosira, Cyclotella* and *Anabaena* often cause increased turbidities. pH range is usually 7.0-7.6. Several large impoundments and a series of locks and weirs regulate the river for irrigation and domestic use, so that it is virtually a series of lacustrine habitats, some of which have extensive macrophyte growth As a result, zooplankton collections from waters of the upper Murray usually contain a mixture of planktonic, littoral and benthic species, with typically 5-14 species occurring in the plankton of river stretches and impoundments, and up to sixty species occurring in single samples taken from billabongs along the river and tributaries.

Common riverine rotifer assemblages are less diverse than those of the Darling, reflecting more uniform conditions and lacustrine sources. Fewer brachionids occur, and typical collections include *B. quadridentatus,* seasonally *B. patulus, K. australis, K. cochlearis, K. procurva, K. slacki,* one or two *Asplanchna (A. brightwelli, A. priodonta, A. sieboldi),* with one or two species from the genera *Conochilus, Euchlanis, Hexarthra, Filinia, Lacinularia, Lecane, Polyarthra, Synchaeta* and *Trichocerca,* with seasonal occurrences of *Lophocharis, Pompholyx, Testudinella* and *Trichotria.*

The rotifer fauna of billabongs will not be discussed here, suffice to note of these habitats that they are typically small, shallow (1-3 m), with abundant macrophyte development and frequent nutrient input from the floodplain. Many are permanent, although decreasingly so following impoundment of the major rivers. Complex communities have developed in most billabongs, so that forty-five to sixty rotifer species are not unusual in single collections.

Below the confluence of the Murray and the Darling a

mixed rotifer assemblage characterizes the potamoplankton, with frequent occurrences of tropical forms brought down the Darling River system. High turbidities result in low population densities–total zooplankton, including Rotifera, Copepoda, Cladocera and Ostracoda rarely exceeds 200 l^{-1} and is more commonly less than 100 l^{-1}. A single species, *Keratella shieli,* has been recorded only from the lower river, and is probably endemic to this habitat.

The material analysed to date from the Murray-Darling system contains endemic species described earlier *(K. australis, K. slacki),* several new species (Koste, 1979; in press), and also several species, including those described here, which differ significantly from the types. It is apparent that a greater degree of endemism exists in the Australian rotifer fauna than has previously been considered.

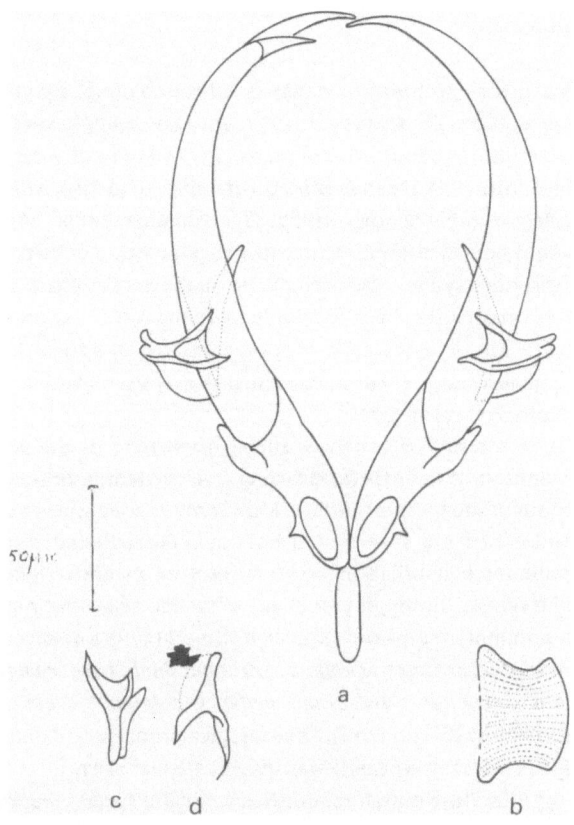

Fig. 1. *Asplanchna sieboldi* (Leydig, 1854), a. trophi ventral, length 340 μm, b. fulcrum, lateral, height 68 μm, length 44 μm, c and d, manubrium from different views, length 40-64 μm.

Remarks on extraordinary species

Asplanchna sieboli (Leydig, 1854)   (Fig. 1 & 2)
In samples from billabongs associated with the River Murray (# 335, 365*) were found females with body lenghts of 1600-2000 μm. They belonged to the 'campanulate morphotype' (Gilbert, 1973: 64; 1976: 3234). Such giant forms ('Riesenform') are cosmopolitan, and known previously from the field (Koste, 1978: 455; Kutikova, 1970: 429; Rylov, 1935; 85) and also reported in culture by Gilbert (op cit.).

In the literature until now, however, trophi measurements for this polymorphic species are between 80-90 μm. In contrast, trophi measurements (seen apically) of specimens from the above Australian samples are as follows:
1600 μm female, trophi of 220 μm (Fig. 2b),
1900 μm female, trophi of 240 μm (Fig. 2a),

---

* Numbers refer to field sampling locations. Collections by one of the authors (RJS) are lodged with the Zoology Department, University of Adelaide.

2000 μm female, trophi of 340 μm (Fig. 1).

These previously unrecorded trophi measurements indicate that the specimens belong to a new ssp. A detailed study is fortcoming (Koste & Shiel, in prep.).

Limited ecological information is available. The specimens came from macrophyte-rich littoral areas of shallow billabongs (less than 1 m). Prey of the *A. sieboldi* females examined belonged to: a species of *Volvox*, *Brachionus falcatus*, *B. novaezealandia*, *Filinia opoliensis*, *F. pejleri grandis* and *Keratella procurva*.

Brachionus calyciflorus Pallas, 1766 (*gigantea* ?n. ssp.)
     (Fig. 3, 4).
In several samples (# 293, 380, 401) polymorphic populations of *B. calyciflorus* were collected (Fig. 4) having resemblances to the f. typ. and also to the *f. anuraeiformis* (Fig. 3). The adult females were distinctive, having an

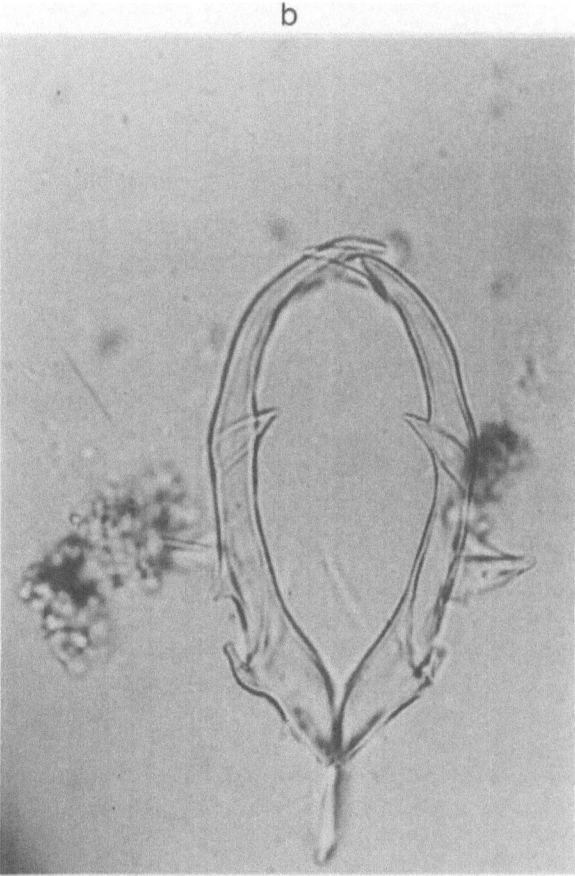

Fig. 2. a. *A. sieboldi*, 240 μm trophi from an animal of 1900 μm body-length. b. 220 μm trophi from an animal of 1600 μm body-length.

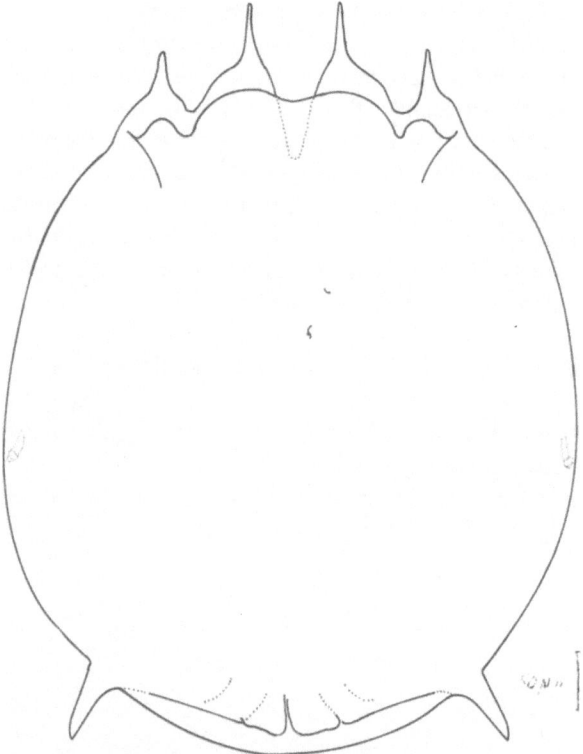

Fig. 3. *Brachionus calyciflorus* Pallas, 1766, ventral lorica. Total length, including spines, 700 μm.

extraordinarily transparent integument and a hitherto unrecorded large lorica size (650-700 μm), A similar but smaller, spined form was described by Apstein (1907) as *f. borgerti* from Formosa and Ceylon (Ahlstrom, 1940: Fig. 20: 7-8). Koste (1978: 89; Fig. 33b: 4) reported on Riesenforms of *f. anuraeiformis* of 600 μm from tropical waters of South America. The latter, however, were characteristically trapezoid with an opaque integument, whereas the Australian animals show a definite circular profile and a transparent integument. The adult females of the new Riesenform had up to five very large subitaneous eggs of up to 180 μm x 140 μm attached to the caudal lorica. Cross-section measurements known until now are only 100 μm x 70 μm.

In view of these features, this is probably a new ssp. The habitat in which it has been collected is invariably highly turbid (> 120 NTU's), alkaline (pH 8.0-8.7) and moderately saline (~750 μS cm⁻¹). The new ssp. has been found only in waters of the Darling River, N.S.W.

Brachionus plicatilis (O. F. M., 1786) (n. ssp ?) (Fig. 5). In a single sample collected by Professor W. D. Williams

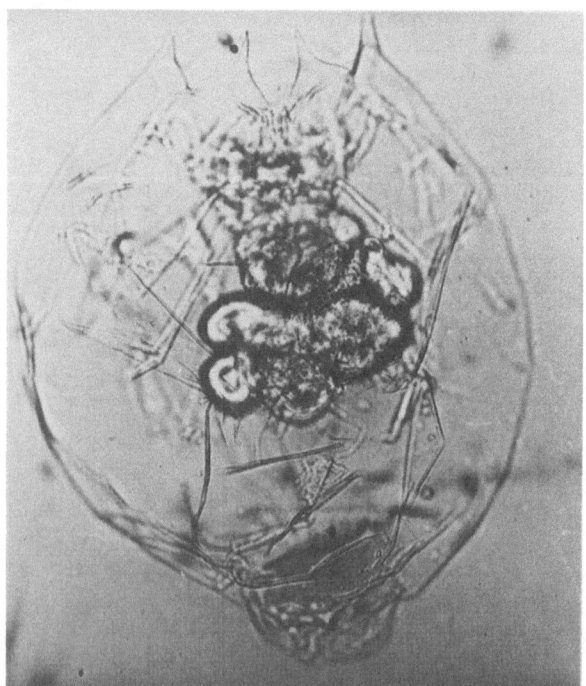

Fig. 4. *B. calyciflorus* Pallas, 1766. Lorica, dorsal view. 690 μm.

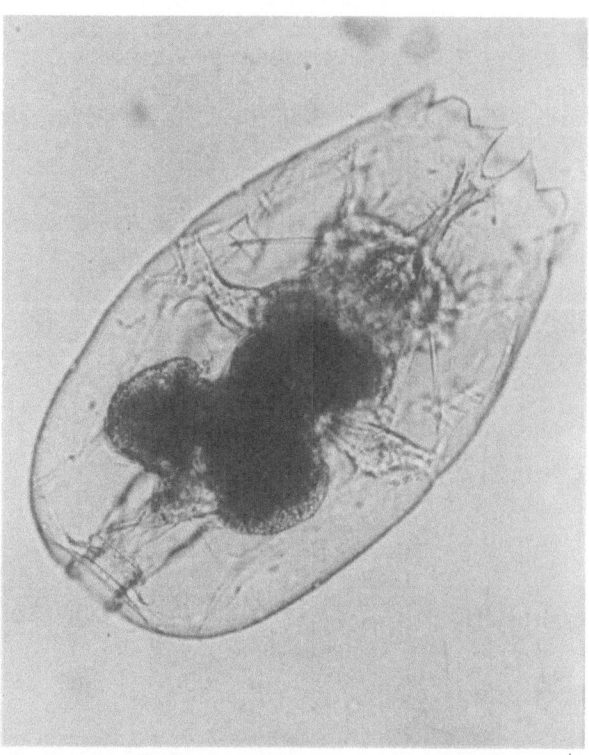

Fig. 5. *Brachionus plicatilis* (O. F. M., 1786). Lorica, ventral. Dorsal lorica length 440 μm, ventral lorica length 400 μm, lorica with 208 μm.

from Lake Colongulac, a saline lake in western Victoria, a *B. plicatilis* population was encountered of up to 440 μm lorica length. Previous lorica measurements for this species range from 125-315 μm (Koste, 1978: 77). Fadeew (1925: 22, Fig. 7) described a similarly alongated but significantly smaller *B. plicatilis* as f. *decemcornis* (Koste, 1978: 77, Fig. 9: 1e). The structure of this new saline-water rotifer will be described elsewhere (Koste, 1980, in press).

Filinia pejleri grandis (Koste, 1979) nov. nom.

(Figs. 6, 7a-c).

Syn: *Filinia pejleri* var. *grandis* Koste, 1979.

Found in many samples from the River Murray, this distinctive *Filinia* form is distinguished from the type (*F. pejleri* Hutchinson, 1964) through its exceptionally large body and bristle measurements. Both representatives of the species occur together in drift samples, but have not

been recorded from the same biotope. *F. pejleri grandis* (previously described by Koste (1979: 251-252, Figs. 25-26) as a variety, but now considered a ssp.) has been recorded sympatrically with an as yet undescribed variety of *F. longiseta*. The almost infinitely variable transition of the caudal bristle from the spindle-shaped body, and the different lengths of the lateral bristles are shown in Figs. 6 & 7.

Keratella slacki (Berzins, 1963)   (Fig. 8, 9).
(Syn: *Keratella valga slacki* Berzins, 1963)

As with the other species groups of the genus, *K. slacki* shows considerable variations with respect to both lorica length and posterior spine development. Size of the subitaneous eggs is relatively constant, so that females of 520 μm and also females of 240 μm carry eggs of 80/84 x 60/64 μm on the posterior border of the dorsal lorica. Ecological information on the billabong habitats from which these

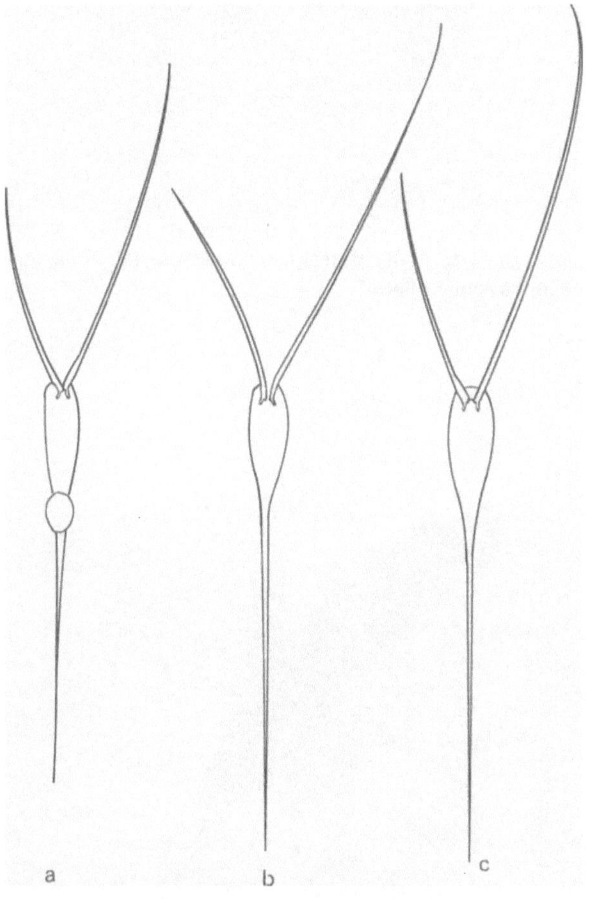

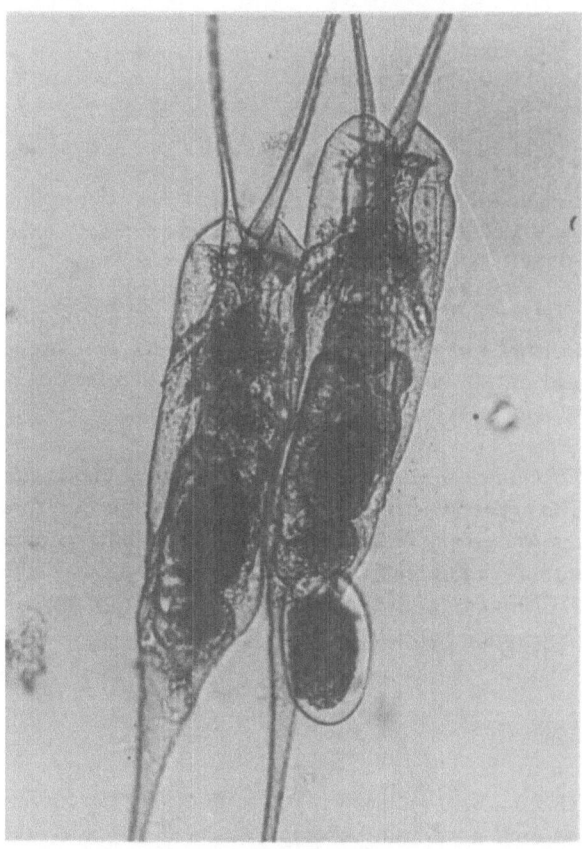

Fig. 6. *Filinia pejleri grandis* (Koste, 1979) a-c different contracted individuals. Lateral bristles 512-550 μm/805-895 μm, caudal bristle 700-715 μm.

Fig. 7. *F. pejleri grandis* (Koste, 1979), contrasted, right individual with subitaneous egg. Body length 330 μm, subitaneous egg 100/65 μm.

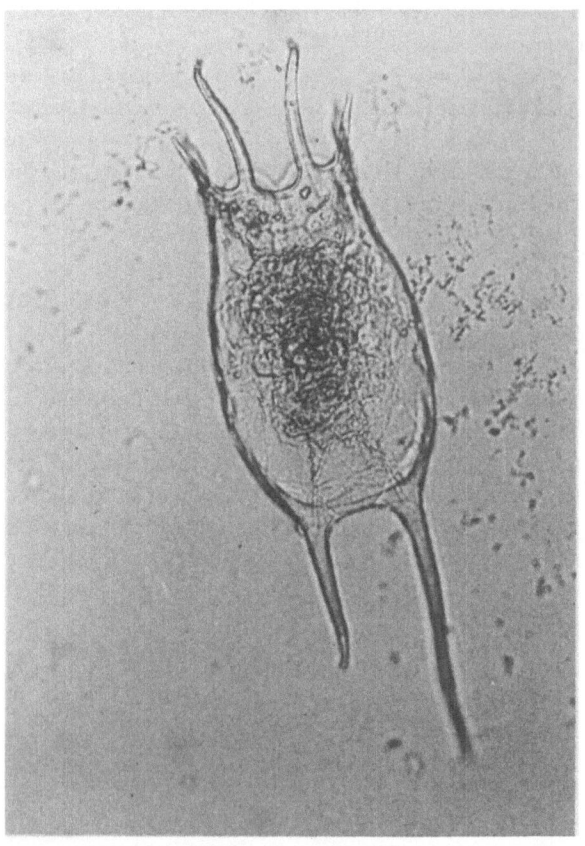

Fig. 8. *Keratella slacki* (Berzins, 1963), giant form, lorica, dorsal. Lorica length 520 μm.

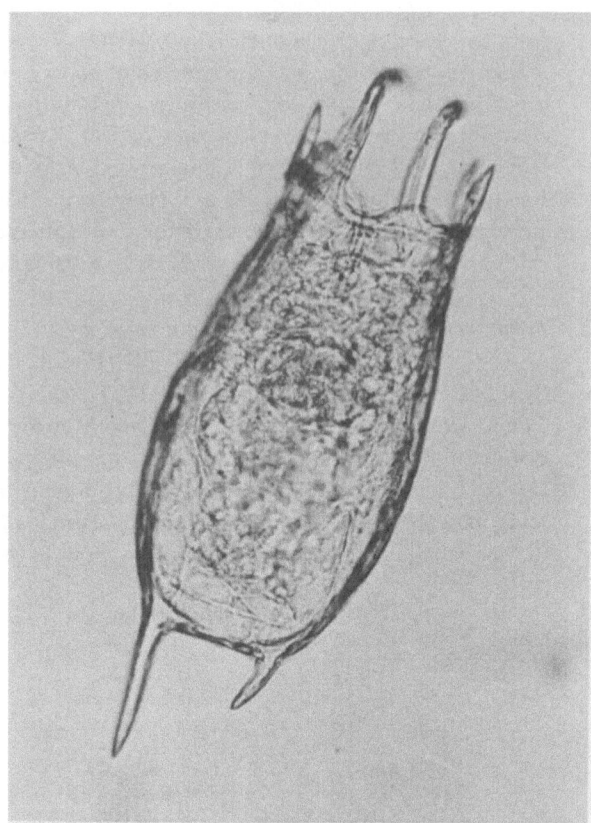

Fig. 9. *K. slacki,* fertile adult female, dwarf form of 240 μm total length including spines.

varieties were collected is sparce, and little can therefore be said on the causes of the extensive variation seen. It is certain, however, that the species is endemic.

Trichocerca similis grandis (Hauer, 1965)   (Fig. 10). This apparently pantropical subspecies was recorded from the Amazon by Hauer (1965), but as yet is not recorded from Africa. A report on the variability in Australia of the *T. similis* group, in conjunction with new descriptions, is in preparation (Koste & Shiel, in prep.).

Summary

This is a brief preliminary report on the state of rotifer-research in Australia. A comprehensive description of the Rotifera of this zoogeographical region will follow in the near future (Shiel, in prep.; Koste & Shiel, in prep.) when analysis of the extensive material is completed. We can

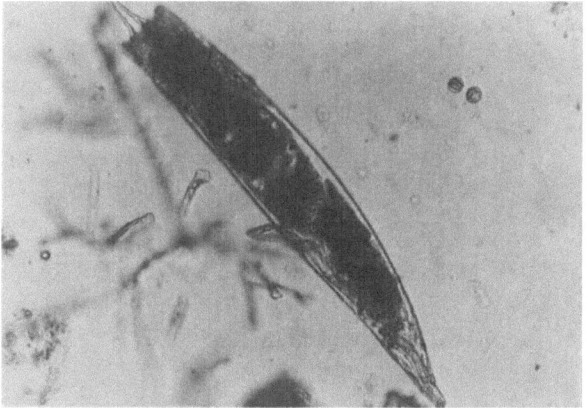

Fig. 10. *Trichocerca (D.) similis grandis* (Hauer, 1965) from northern Australia. Length including toes 450 μm.

already say, however, that the examples given here and recently described, indicate that the Australian region of Notogaea has a species-rich rotifer fauna in which are found several previously unknown endemic species and ssp.

The occurrence of giant forms in several genera is interesting, and worthy of further study. Contributing factors are undoubtedly moderate environmental physico-chemical ranges and the absence of zooplankton predators such as *Leptodora, Diaptomus* and *Mesocyclops edax* from the Australian fauna. The significance of predation by planktivorous macroinvertebrates and fish fry is also unstudied, but the characteristic high turbidities, particularly of the Darling system, would tend to have negative effect on the efficiency of visually orienting predators.

References

Ahlstrom, E. H. 1940. A revision of the Rotatorian genera Brachionus and Platyias with descriptions of three new species and two new varieties. Bull. amer. Mus. Nat. Hist., 80: 411-457.

Apstein, C. 1907. Das Plankton im Colombo-See auf Ceylon. Sammelausbeute von A. Borgert, 1904-1905. Zool. Jb. Syst., 25: 201-244.

Berzins, B. 1953. Zur Kenntnis der Rotatorien aus West-Australien. Lunds Univ. Arsskr. N. F., 249 (8): 12 pp.

Berzins, B. 1960. New Rotatoria (Rotifera) from Victoria. Proc. r. Soc. Vict., 83-86.

Berzins, B. 1963. Two new Keratella, Rotatoria, from Australia. Hydrobiologia, 21/22: 380-383.

Fadeew, N. 1925. Zur Kenntnis der Seen Transkaukasiens. Arb. biol. Stat. N-Kaukasus., 1: 17-26.

Gilbert, J. J. 1973. Induction and Ecological Significance of Gigantism in the Rotifer Asplanchna sieboldi. Science, 181: 63-66.

Gilbert, J. J. 1976. Selective cannibalism in the rotifer Aspanchna sieboldi: Contact recognition of morphotype and clone. Proc. nat. Acad. Sci. USA, 73: 3233-3237.

Koste, W. 1978. Rotatoria. Borntraeger, Stuttgart-Berlin, 2 vols., 673 pp., 234 plates.

Koste, W. 1979. New Rotifera from the river Murray, south-eastern Australia, with a review of the Australian species of Brachionus and Keratella. Aust. J. mar. Freshw. Res., 30: 237-253.

Koste, W. 1980. Lindia deridderi n. sp., ein Rädertier der Familie Lindiidae (Uberordnung Monogononta) aus SE-Australien. Arch. Hydrobiol., 87 (4): 504-511.

Koste, W. 1980. in press. Brachionus plicatilis, ein Salzwasserrädertier. Mikrokosmos, 68.

Koste, W. & Shiel, R. J. 1980. On Brachionus dichotomus Shephard, 1911 (Rotatoria: Brachionidae) from the Australian region, with a description of a new subspecies, Brachionus dichotus reductus. Proc. r. Soc. Vict. 91 (10).

Koste, W. & Shiel, R. J. in prep. New Rotifera from Australia, with descriptions of new taxa.

Kutikova, L. A. 1970. Kolovratki Fauna SSSR. In: Fauna USSR, vol. 104, Akademia Nauk, Leningrad, 744 pp.

Russell, C. R. 1957. Some rotifers from the south Pacific islands and northern Australia. Trans. r. Soc. N. Zeal., 84: 897-902.

Russell, C. R. 1961. The Rotatoria of Queensland, Australia. Trans. r. Soc. N. Zeal., 1: 235-239.

Rylov, W. M. 1935. Das Zooplankton der Binnengewässer. Binnengewässer, 15: 1-272.

Schindler, D. W. 1969. Two useful devices for vertical plankton and water sampling. J. Fish. Res. Bd. Can., 26: 1948-1955.

Shephard, J. 1896. A new rotifer, Lacinularia elongata. Vict. Nat., 13: 22-24.

Shephard, J. 1897. A new rotifer, Lacinularia elliptica. Vict. Nat., 14: 84-86.

Shephard, J. 1899. A new rotifer, Lacinularia striolata. Proc. r. Soc. Vict., 12: 20-35.

Shephard, J. 1911. A list of Victorian rotifers, with descriptions of two new species and the males of two species. Proc. r. Soc. Vict., 24: 46-58.

Shiel, R. J. 1978. Zooplankton of the Murray-Darling system, a preliminary report. Proc. r. Soc. Vict., 90: 193-202.

Shiel, R. J. 1979. Synecology of the rotifera of the river Murray, South Australia, Aust. J. Mar. Freshw. Res., 30: 255-263.

Shiel, R. J. & Koste, W. 1979. Rotifera recorded from Australia. Trans. r. Soc. S. Aust., 103: 57-68.

Sudzuki, M. & Timms, B. V. 1977. A new species of Brachionus (Rotifera) from the Myall lakes, New South Wales. Proc. Linn. Soc. N.S.W., 101: 162.

Thorpe, V. G. 1887. On certain rotifera found in the gardens of the Acclimatisation Society, Brisbane. Proc. r. Soc. Qld, 4: 28-30.

Thorpe, V. G. 1891. New and foreign Rotifera. J. r. microsc. Soc., 301-306.

Whitelegge, T. 1889. List of the marine and freshwater invertebrate fauna of Port Jackson and the neighbourhood. Proc. r. Soc. N.S.W., 23: 163-323.

Woodyer, K. D. 1978. Sediment regime of the Darling River. Proc. r. Soc. Vict., 90: 139-147.

SYNOPSIS OF TAXONOMIC STUDIES ON INDIAN ROTATORIA

B. K. SHARMA & R. George MICHAEL

Department of Zoology, Northeastern Hill University, Sillong - 790014 India

Keywords: taxonomy, India, Rotatoria

Abstract

Taxonomic studies of Indian rotifers have revealed the presence of 241 species which are listed in an appendix. The nature of the Indian rotifer fauna is discussed.

Taxonomic investigations on Indian rotifers were initiated by Anderson (1889) who studied 47 species collected from Calcutta and its environs in West Bengal. Since that time there have been over 70 publications dealing with Indian rotifers. These papers are listed in the references and in Table I they have been grouped according to the state or territory to which they refer. As a result of all these studies and some personal observations of the authors a total of 241 species have so far been recorded from India.

On the whole the rotifer fauna of India seems to be cosmotropical and comparable to the faunas of Indonesia (Hauer, 1937-8) and Sri Lanka (Chengalath *et al.*, 1974). A large number of cosmopolitan species are also represented in the Indian fauna. The Index of similarity of Sorensen (1948) results in a value of 52.6, in a comparison of the rotifer faunas of India and Sri Lanka.

Green (1972) and Chengalath *et al.* (1974) have shown the common occurrence of the genus *Brachionus* and the absence, or near absence, of the boreal genus *Notholca* to be characteristic of many tropical waters. Thus, while the species of *Brachionus* are very common in India, there has been only one record of *Notholca striata* from high altitudes in Kashmir and Ladakh (Edmondson & Hutchinson, 1934).

The rotifer taxa reported from India belong to 21 Eurotatorian families spread over 48 genera. The bulk of these are Monogononta, of which the Order Ploimida is represented by 14 families, while the Order Gnesiotrocha is represented by 6 families. Of the rotifer fauna of this country, the commonest species are *Brachionus angularis, B. bidentata, B. caudatus, B. calyciflorus, B. forficula, B. falcatus, B. quadridentatus, Anuraeopsis fissa, Keratella tropica, K. procurva, Euchlanis dilatata, Mytilina ventralis, Lepadella patella, L. ovalis, Lecane luna, L. crepida, L. papuana, L. bulla, L. closterocerca, Filinia opoliensis. F. longiseta* and *Testudinella patina.* Many among these species are cosmopolitan, but *B. caudatus, B. forficula, B. falcatus, K. tropica, K. procurva* and *L. papuana* are cosmotropical.

In the Lecanidae, the *Lecane* complex *(Lecane, Monostyla* and *Hemimonostyla)* has the maximum number of representatives, with a total of 59 species reported from India. The predominance of the species of this complex may be ascribed to their ability to live in varied habitats. In this group, the validity of number of toes as a character of generic importance has been reexamined and has been accepted only for subgeneric differentiation (Sharma, 1978a). Thus *Lecane (Lecane)* contains 33 species, *Lecane (Monostyla)* is represented by 23 species while *Lecane (Hemimonostyla)* has only 3 representatives.

The Brachionidae come next in order of abundance, with 31 species belonging to 5 genera. The rarity of the genus *Notholca* in India has already been pointed out. Of the other brachionid genera, there are still taxonomical problems in *Brachionus* and *Keratella* (Sudzuki, 1977). The former genus is represented by 19 species and various infraspecific categories (forms and varieties). Species such as *B. calyciflorus, B. caudatus* and *B. quadridentatus* exhibit morphological variations in various regions of this country. Morphometric studies of *B. angularis* have indicated that Indian forms are smaller as compared to their counterparts in other geographical regions. Dhanapathi (1977) studied the distribution of *B. calyciflorus* in India

229

Table 1.

States/Union Terrorities	Number of species recorded	Number of genera recorded	References
Andhra Pradesh	61	21	Naidu, '67; Dhanapathi, '73; '74a, b; '75a, b; '76a, b, c; '77; '78; Rao & Chandra Mohan, '75; '76; '77; Chandra Mohan & Rao, '76a, b.
Andaman & Nicobar Islands	–	–	
Arunachal Pradesh	–	–	
Assam	30	14	Patil, 1978; Sharma, unpublished
Bihar	25	11	Donner, 1949; Nasar, 1973; Laal & Nasar, 1978.
Chandigarh (U.T.)	39	17	Vasisht & Gupta, 1967; Vasisht & Battish, 1971a, b, c, d; Sharma, 1976.
Delhi	–	–	
Gao	–	–	
Gujarat	86	27	Wulfert, 1966.
Haryana	27	12	Sharma, 1976.
Himachal Pradesh	16	9	Sharma, 1976.
Jammu and Kashmir	68	27	Edmondson & Hutchinson, 1934; Das & Akhtar, 1976; Chowdhary et al., '78.
Karnataka	–	–	
Kerala	20	9	Nayar & Nair, 1969; Nair & Nayar, '71.
Madhya Pradesh	–	–	
Maharashtra	13	6	Sharma, 1979g.
Meghalaya	28	16	Patil, 1978; Sharma, unpublished.
Manipur	6	4	Patil, 1978.
Mizoram	–	–	
Nagaland	–	–	
Orissa	36	13	Sharma, 1977; Sharma, 1979h.
Panjab	44	18	Vasisht & Battish, 1971a, c; Sharma, '76.
Rajasthan	36	12	Nayar, 1964; '65a, b; '68.
Tamil Nadu (former Madras state)	37	13	Hauer, '36; '37a, b; Brehm, '51; Pasha, 61; Michael, '66; Wycliffe & Michael, '68; Rajendran, '71; Michael, '73; Santhanam & Krishnamurthy, '75.
Tripura	–	–	
Uttar Pradesh	–	–	
West Bengal	110	48	Anderson, 1889; Sewell, '35; Tiwari & Sharma, '77; Sharma, '78a, b, c; '79a, b, c, d, e, f.

along with the various recorded infraspecific categories. According to him it is probable that in tropical areas the exuberant forms as *amphiceros* arose at high latitudes in summer, whereas *dorcas, borgerti* and *hymani* in lower latitudes had their origin from a typical *pala*-like stem mother.

The brachionid Genus *Keratella* is represented by 7 species. Of these *K. lenzi* is very rare in Indian waters. There are still uncertainties about the taxonomic status of *K. tropica* and *K. valga*. However, presently they are treated as distinct species following Berzins (1955), Arora (1966b) and Pejler (1974). The record of *K. valga* by Arora (1966b), along with *K. tropica* from Nagpur needs further confirmation.

The euchlanids are represented by the genera *Euchlanis, Dipleuchlanis,* and *Tripleuchlanis*. Of these, *T. plicata,* a marine species, has also been reported from freshwaters by Dhanapathi (1977) and Sharma (1979a). Recently,

Dhanapathi (1978) has described a new genus, *Pseudoeuchlanis*, having combined characters of *Euchlanis, Dipleuchlanis* and *Squatinella*.

There is very little information on Bdelloid rotifers from India. Of the five families of Order Bdelloidea, only one (Philodinidae) has been recorded from India. This gap calls for special emphasis to be given in future studies. It is well known that special techniques of collection, preservation and fixation have to be employed in the study of this group.

An overall assessment of the Indian rotifer fauna indicates that this group is quite rich and varied. While many species are ubiquitous, some are restricted to particular habitats. Previous studies have indicated certain lacunae in our present knowledge and leave ample scope for extensive future surveys of other regions in India. Unfortunately, most Indian hydrobiologists rely on foreign monographs for the quick identification of their material. While it is agreed that literature of extra-Indian origin is useful, it must be pointed out that complete dependence on such literature and illustrations alone is liable to be misleading. Hence a detailed monograph of Indian rotifers is much desired. Attempts are made in this department to collect and maintain Indian reference material from various parts of the country.

References

Ahlstrom, E. H. 1940. A revision of the Rotatorian Genera Brachionus and Platyias with descriptions of one new species and two new varieties. Bull. Am. Mus. nat. Hist., 77: 143-184.

Ahlstrom, E. H. 1940. A revision of the Rotatorian Genus Keratella with description of three new species and five new varieties. Bull. Am. Mus. nat. Hist., 80: 411-457.

Anderson, H. H. 1889. Notes on Indian Rotifera. J. Asiatic Soc. Bengal., 58: 345-358.

Arora, H. C. 1962. Studies on Indian Rotifera. Part. I. On a small collection of illoricate Rotifera from Nagpur, India, with notes on their Bionomics. J. zool. Soc. India., 14: 33-44.

Arora, H. C. 1963a. Studies on Indian Rotifera. Part. II. Some species of the Genus Brachionus from Nagpur. J. zool. Soc. India., 15: 112-121.

Arora, H. C. 1963b. Studies on Indian Rotifera. Part. IV. On some species of sessile Rotifera from India (with a description of a new species of Genus Sinantherina). Arch. Hydrobiol., 59: 502-507.

Arora, H. C. 1965. Studies on Indian Rotifera. Part. VI. On a collection of Rotifera from Nagpur, India, with four new species and a new variety. Hydrobiologia, 26: 444-456.

Arora, H. C. 1966a. Studies on Indian Rotifera. Part. III. On Brachionus calyciflorus and some varieties of the species. J. zool. Soc. India., 16: 1-6.

Arora, H. C. 1966b. Studies on Indian Rotifera. Part. V. On spe-cies of some genera of the family Brachionidae, subfamily Brachioninae from India. Arch. Hydrobiol., 61: 482-483.

Arora, H. C. 1966c. Cyclomorphosis (form variations) in some species of Indian planktonic Rotatoria. Int. Rev. ges. Hydrobiol., 51: 623-632.

Berzins, B. 1955. Taxonomic und Verbreitung von Keratella valga und verwandten Formen. Ark. Zool., 8: 549-559.

Brehm, V. 1950. Contributions to the freshwater fauna of India. Part. 2. Rec. Indian Mus., 48: 9-28.

Brehm, V. 1951. Eine neue Brachionus aus Indien (Brachionus donneri) Zool. Anz., 146: 54-55.

Chandra Mohan, P. & Rao, R. K. 1976a. A note on the morphometric studies of the rotifer Brachionus angularis Gosse. Sci. Cult., 42: 287-288.

Chandra Mohan, P. & Rao, R. K. 1976b. Epizoic rotifers observed on Odonata nymphs from Vishakhapatnam. Sci. Cult., 42: 527-528.

Chengalath, R., Fernando, C. H. & Koste, W. 1974. Rotifera from Sri Lanka (Ceylon) 3. New species and records with a list of Rotifera recorded and their distribution in different habitats from Sri Lanka. Bull. Fish. Res. Stn. Sri Lanka (Ceylon), 25: 83-96.

Chowdhary, S. K., Sharma, J. P. & Srivastava, J. B. 1978. On the rotifer fauna of Jammu Ponds. Proc. Indian Sci. Congr., Abstracts.

Das, S. M. & Akhtar, S. 1976. A survey of rotifers of Kashmir with new records of Palaearctic genera and species. Rotifer News, 3: 9-12.

Dhanapathi, M. V. S. S. S. 1973. On the occurrence of the rare rotifer Rotaria neptunia (Ehrenberg) in India. Curr. Sci., 42: 770.

Dhanapathi, M. V. S. S. S. 1974a. A new brachionid rotifer Platyias quadricornis andhraensis subsp. nov. from India. Curr. Sci., 43: 358.

Dhanapathi, M. V. S. S. S. 1974b. Rotifers from Andhra Pradesh, India. I. Hydrobiologia, 45: 357-372.

Dhanapathi, M. V. S. S. S. 1975a. New Record of the rotifer Tripleuchlanis plicata (Levander) from India. Curr. Sci., 44: 130-131.

Dhanapathi, M. V. S. S. S. 1975b. Rotifers from Andhra Pradesh, India. J. Linn. Soc. (Zool)., 57: 85-94.

Dhanapathi, M. V. S. S. S. 1976a. Rotifers from Andhra Pradesh, India. III. Family Lecanidae including two new species. Hydrobiologia, 48: 9-16.

Dhanapathi, M. V. S. S. S. 1976b. A new lecanid rotifer from India. Hydrobiologia, 50: 191-2.

Dhanapathi, M. V. S. S. S. 1976c. Rotifers from Andhra Pradesh, India. II. Euchlanis brahmae sp. nov. with taxonomic notes on Indian species of the genus Euchlanis Ehrenberg. Mem. Soc. Zool. Guntur, 1: 43-48.

Dhanapathi, M. V. S. S. S. 1977. Studies on the distribution of Brachionus calyciflorus in India. Arch. Hydrobiol. Beih., 8: 226-229.

Dhanapathi, M. V. S. S. S. 1978. New species of rotifer from India belonging to the family Brachionidae. J. Linn. Soc. (Zool)., 62: 305-308.

Donner, J. 1949. Horaella brehmi nov. gen. nov. sp. eine neue Radertier aus Indien. Hydrobiologia, 2: 304-328.

Dumont, H. J., Coussement, M. & Anderson, R. S. 1978. An examination of some Hexarthra species (Rotatoria) from western Canada and Nepal. Can. J. Zool., 56: 440-445.

Edmondson, W. T. & Hutchinson, G. E. 1934. Report on Rotatoria. Article IX. Yale North India Expedition. Mem. Conn. Acad. Arts Sci., 10: 153-186.

Green, J. 1972. Latitudinal variation in associations of planktonic Rotifera. J. Zool., 167: 31-39.

Hauer, J. 1936. Neue Rotatorien aus Indien I. Zool. Anz., 116: 77-80.

Hauer, J. 1937a. Neue Rotatorien aus Indien. II. Zool. Anz., 119: 284-288.

Hauer, J. 1937b. Neue Rotatorien aus Indien III. Zool. Anz., 120: 17-19.

Hauer, J. 1938. Die Rotatorien von Sumatra Java and Bali, nach den Ergebnissen der Deutschen Limnologischen Sunda-Expedition. I, II. Arch. Hydrobiol., Suppl., 15: 296-384, 507-602.

Koste, W. 1978. Rotatoria. Die Radertiere Mitteleuropas (Uberordnung Monogononta) Ein Bestimmungswerf, begrundet von Max Voigt. Text-u. Tafelbd. (234 Taf.). Gebrüder Borntraeger. Berlin, Stuttgart.

Laal, A. K. & Nasar, S. A. K. 1977. Rotifer fauna of Bihar, India. Bangladesh J. Zool., 5: 127-8.

Michael, R. G. 1966. A new rotifer, Conochilus madurai sp. nov., from an astatic pool in Madurai, South India. Zool. Anz., 177: 439-441.

Michael, R. G. 1973. A Guide to the study of freshwater organisms. 2. Rotatoria. J. Madurai Univ., Suppl. 1: 23-36.

Murray, J. 1906. Some Rotifera of Sikkim Himalaya. J. R. Microsc. Soc., Lond. 9: 637-644.

Naidu, K. V. 1967. A contribution to the Rotatorian fauna of South India. J. Bombay nat. Hist. Soc., 64: 384-388.

Nair, K. K. N. & Nayar, C. K. G. 1971. A preliminary study of the rotifers of Irinjalakuda and neighbouring places. J. Ker. Acad. Biol., 3: 31-43 (in Malayalam).

Nasar, S. A. K. 1973. The zooplankton fauna of Bhagalpur. Rotifera. Bhagalpur Univ. J., 6: 55-62.

Nayar, C. K. G. 1964. Morphometric studies on the rotifer Brachionus calyciflorus Pallas. Curr. Sci., 33: 469-470.

Nayar, C. K. G. 1965a. Cyclomorphosis of Brachionus calyciflorus Pallas. Hydrobiologia, 25: 538-544.

Nayar, C. K. G. 1965b. Taxonomic notes on the Indian species of Keratella (Rotifera). Hydrobiologia, 26: 457-462.

Nayar, C. K. G. 1968. Rotifer fauna of Rajasthan. Hydrobiologia, 31: 168-185.

Nayar, C. K. G. & Nair, K. K. N. 1969. A collection of brachionid rotifers from Kerala. Proc. Indian Acad. Sci., 69: 223-233.

Pasha, S. M. K. 1961. On a collection of freshwater rotifers from Madras. J. Zool. Soc. India., 13: 50-55.

Patil, S. G. 1978. New records of Rotatoria from Northeast India. Sci. Cult., 44: 279-281.

Pejler, B. 1974. On the rotifer plankton of some East African Lakes. Hydrobiologia, 44: 389-396.

Rajendran, M. 1971. On a new species of Conochilus Rousselet Rotifera: Monogononta, Conochilidae) from Madurai, S. India, with a key for the known species of the genus. Proc. Indian Acad. Sci., 73: 8-14.

Rao, R. K. & Chandra Mohan, P. 1975. Studies on freshwater rotifers from Visakhapatnam. 3rd All. India Congr. Zool., Abstract.

Rao, R. K. & Chandra Mohan, P. 1976. On the occurrence of a rotifer Asplanchnella sieboldii (Leydig) urawaensis (Sudzuki) in Indian Waters. Curr. Sci., 45: 234-235.

Rao, R. K. & Chandra Mohan, P. 1977. Monostyla obtusa Murray (Rotifera, Lecanidae)–a New record from India. Geobios, 4: 118.

Santhanam, R. & Krishnamurthy, K. 1975. On planktonic Rotifera. 3rd All. India Congr. Zool., Abstract.

Sewell, R. B. S. 1935. Studies on the bionomics of freshwater in India. II. On the fauna of the tank in the Indian Museum compound and seasonal changes observed. Int. Rev. ges. Hydrobiol., 31: 203-238.

Sharma, B. K. 1976. Rotifers collected from North-West India. Newsl. Zool. Surv. India, 2: 255-259.

Sharma, B. K. 1977. On a small collection of rotifers (Class: Rotifera) from Orissa. Newsl. Zool. Surv. India, 3: 189-190.

Sharma, B. K. 1978a. Contributions to the rotifer fauna of West Bengal.; Part. I. Family Lecanidae. Hydrobiologia, 57: 143-153.

Sharma, B. K. 1978b. Contributions to the rotifer fauna of West Bengal. II. Genus Lepadella Bory de St. Vincent, 1826. Hydrobiologia, 58: 83-88.

Sharma, B. K. 1978c. Two new lecanid Rotifers from India. Hydrobiologia, 60: 191-192.

Sharma, B. K. 1979a. Record of the rotifer, Tripleuchlanis plicata (Levander) from a freshwater tank in Calcutta, India. Bull. Zool. Surv. India, 2:

Sharma, B. K. 1979b. A note on some epizoic rotifers from West Bengal. Bull. Zool. Surv. India, 2:

Sharma, B. K. 1979c. Rotifers from West Bengal III. Further studies on the Eurotatoria. Hydrobiologia, 64: 239-250.

Sharma, B. K. 1979d. Rotifers from West Bengal. IV. Further contributions to the Eurotatoria. Hydrobiologia, 65: 39-47.

Sharma, B. K. 1979e. Further contributions to the lecanid fauna (Rotifera: Lecanidae) of West Bengal. Acta Hydrobiologica, 20: 53-59.

Sharma, B. K. 1979f. Rotifers from West Bengal. V. General remarks, with key to recorded genera. Bull. C. I. F. E., Bombay (In press).

Sharma, B. K. 1979g. On a small collection of rotifers from Bombay (Maharashtra), India. Bull. C. I. F. E., Bombay (In press).

Sharma, B. K. 1979h. Contribution to the rotifer fauna of Orissa, India. Hydrobiologia (in press).

Sharma, B. K. 1979. Studies on the rotifer fauna of Assam and Meghalaya. (in preparation).

Sorenson, T. 1948. A method of establishing groups of equal amplitude in plant sociology based on similarity of species content and its application to analysis of the vegetation on Danish commons. Biol. Skr., 5: 1-34.

Sudzuki, M. 1977. Some puzzling problems in the taxonomy of Brachionus and Keratella. Arch. Hydrobiol. Beih., 8: 230-231.

Tiwari, K. K. & Sharma, B. K. 1977. Rotifers in the Indian Museum Tank, Calcutta. Sci. Cult., 43: 280-282.

Vasisht, H. S. & Battish, S. K. 1969. The Rotifer fauna of North India. I. Scaridium longicaudum Ehrenberg. Res. Bull. (N.S.) Panjab Univ., 20: 593-594.

Vasisht, H. S. & Battish, S. K. 1970. The rotifer fauna of North India. II. Trichocerca porcellus Res. Bull. (N.S.) Panjab Univ., 21: 515.

Vasisht, H. S. & Battish, S. K. 1971a. The rotifer fauna of North India.: Brachionus Res. Bull. (N.S.) Panjab Univ., 22: 179-188.

Vasisht, H. S. & Battish, S. K. 1971b. The rotifer fauna of North India. Lepadella and Colurella. Res. Bull. (N.S.) Panjab Univ., 22: 189-192.

Vasisht, H. S. & Battish, S. K. 1971c. The rotifer fauna of North India: Platyias, Keratella, Anuraeopsis, Mytilina and Euchlanis. Res. Bull. (N.S.) Panjab Univ., 22: 331-337.

Vasisht, H. S. & Battish, S. K. 1971d. The rotifer fauna of North

India: Lecane and Monostyla Res. Bull. (N.S.) Panjab Univ., 22: 353-358.

Vasisht, H. S. & Dawar, B. L. 1968. The male of the rotifer Cupelopagis vorax (Leidy) Curr. Sci., 37: 466-467.

Vasisht, H. S. & Gupta, C. L. 1967. The rotifer fauna of Chandigarh. Res. Bull. (N.S.) Panjab Univ., 18: 495-496.

Wulfert, K. 1966. Rotatorien aus dem Stausee Azwa und der Trinkwasser Aufbereitung der Stadt Baroda (Indien). Limnologica., 4: 53-93.

Wycliffe, M. J. & Michael, R. G. 1968. Pseudoembata acutipoda gen. et. sp. nov. (Rotifera: Bdeloidea), an epizoic rotifer. Curr. Sci., 37: 106-108.

APPENDIX

* List of Rotifer Taxa recorded from India

Order: PLOIMIDA Delage, 1897

Family EPIPHANIDAE
1. *Epiphanes clavulata* Ehrb., 1832
2. *E. brachionus* (Ehrb., 1837)
3. *E. senta* (Muller, 1773)

Family BRACHIONIDAE
4. *Anuraeopsis coelata* (De Beauchamp, 1932)
5. *A. fissa* Gosse, 1851
6. *Brachionus angularis* Gosse, 1851
7. *B. bidentata* Anderson, 1889;
   f. *adorna* Wulfert, 1966
8. *B. budapestiensis* Daday, 1885;
   var. *punctatus* (Hempel, 1896)
9. *B. caudatus* v. *aculeatus* (Hauer, 1937);
   v. *personatus* Ahlstrom, 1940;
   f. *apsteini* (Fadeev, 1925);
   f. *vulgatus* Ahlstrom, 1940.
10. *B. calyciflorus* v. *dorcas* (Gosse, 1851)
   f. *anuraeiformis* (Brehm, 1909);
   f. *borgerti* (Apstein, 1907);
   v. *hymani* Dhanapathi, 1974;
   f. *amphiceros* Ehrb., 1838
11. *B. dimidiatus* (Bryce, 1931)
12. *B. diversicornis* (Daday, 1883)
13. *B. donneri* Brehm, 1951
14. *B. durgae* Dhanapathi, 1974
15. *B. falcatus* Zacharias, 1898
   v. *lyratus* Lammerman, 1908
16. *B. forficula* Wierzeiski, 1891;
   subsp. *keralensis* Nayar & Nair, 1969;
   f. *minor* (Voronkov, 1913)
17. B. leydigii Cohn, 1862

18. *B. patulus* Muller, 1786
19. *B. plicatilis* Muller, 1786
20. *B. pterodinoides* Rousselet, 1913
21. *B. quadridentatus* Hermann, 1783;
   ssp. *mirabilis* (Daday, 1897);
   v. *cluniorbicularis* (Skorikov, 1897)
22. *B. sessilis* Varga, 1951
23. *B. rubens* Ehrb., 1838
24. *B. urceolaris* Muller, 1773
25. *Keratella cochlearis* Gosse, 1851
26. *K. edmondsoni* Nayar, 1965
27. *K. lenzi* Hauer, 1938
28. *K. procurva* (Thorpe, 1891)
29. *K. quadrata* (Muller, 1786)
30. *K. tropica* (Apstein, 1907)
31. *K. valga* (Ehrb., 1834)
32. *Platyias longispinosus* Arora, 1966
33. *P. quadricornis* Ehrb., 1882
34. *Notholca striata* (Muller, 1786)

Familey EUCHLANIDAE
35. *Euchlanis alata* Voronkov, 1912
36. *E. deflexa* (Gosse, 1851)
37. *E. brahmae* Dhanapathi, 1976
38. *E. dilatata* Ehrb., 1832
39. *E. macrura* Ehrb., 1882
40. *E. menta* Myers, 1930
41. *E. oropha* Gosse, 1887
42. *E. parva* Rousselet, 1892
43. *E. triquetra* Ehrb., 1838
44. *Dipleuchlanis propatula* (Gosse, 1886)
45. *Pseudoeuchlanis longipedis* Dhanapathi, 1978
46. *Tripleuchlanis plicata* (Levander, 1894)

Familey MYTILINIDAE
47. *Lophocharis oxysternon* (Gosse, 1851)
48. *L. salpina* (Ehrb., 1832)
49. *L. naias* Wulfert, 1942
50. *Mytilina acanthophora* Hauer, 1938
51. *M. mucronata* Muller, 1773
52. *M. ventralis* (Ehrb., 1832)

Familey TRICHOTRIDAE
53. *Macrochaetus collinsi* (Gosse, 1867)
54. *M. sericus* (Thorpe, 1893)
55. *M. subquadratus* Perty, 1850
56. *Trichotria pocillum* (Muller, 1776)
57. *T. tetractis* (Ehrb., 1832)

Family COLURELLIDAE
58. *Colurella bicuspidata* Ehrb., 1832
59. *C. caudata* (Ehrb., 1837) = *adriatica* Ehrb., 1837)
60. *C. obtusa* (Gosse, 1886)
61. *C. oxycauda* Carlin, 1939
62. *C. sulcata* (Stenroos, 1898)
63. *Lepadella acuminata* (Ehrb., 1834)
64. *L. aspicora* Myers, 1934
65. *L. aspida* (Harring, 1916)
66. *L. bicornis* Vasisht & Battish, 1971
67. *L. cristata* (Rousselet, 1893)
68. *L. dactliseta* (Stenroos, 1898)
69. *L. ehrenbergi* (Perty, 1850)
70. *L. heterostyla* (Murray, 1917)
71. *Lepadella imbricata* Harring, 1916
72. *L. kostei* Wulfert, 1966
73. *L. ovalis* (Muller, 1786); f. *larga* Sharma, 1978
74. *L. patella* (Muller, 1786)
75. *L. rhomboides* (Gosse, 1886)
76. *L. similis* Lucks, 1912; = *parsimilis*
77. *L. triprojectus* Sharma, 1978
78. *L. triptera* Ehrb., 1830
79. *Squatinella mutica* (Ehrb., 1832)
80. *S. tridentata* (Frøesenius, 1858)

Family LECANIDAE
81. *Lecane (Lecane) acronycha* Harring & Myers, 1926
82. *L. (L.) aculeata* (Jakubski, 1912)
83. *L. (L.) arcuata* (Bryce, 1897)
84. *L. (L.) arcula* Harring, 1914
85. *L. (L.) bifastigata* Hauer, 1938
86. *L. (L.) bidentata* Dhanapathi, 1976
87. *L. (L.) curvicornis* Murray, 1913;
　　　v. *miamiensis* Myers, 1941;
　　　v. *nitida* Hauer, 1938;
　　　v. *padaesparaes* Arora, 1965
88. *L. (L.) crepida* Harring, 1914;
　　　f. *bengalensis* Sharma, 1978
89. *L. (L.) donnerianus* Dhanapathi, 1976
90. *L. (L.) eswari* Dhanapathi, 1976
91. *L. (L.) curvilinealis* Arora, 1965
92. *L. (L.) flexis* (Gosse, 1886)
93. *L. (L.) hastata* (Murray, 1913)
94. *L. (L.) hornemanni* (Ehrb., 1838)
95. *L. (L.) lateralis* Sharma, 1978
96. *L. (L.) lauterborni* Hauer, 1924
97. *L. (L.) leontina* (Turner, 1892)
98. *L. (L.) ludwigi* (Eckstein, 1883);
　　　f. *brevicaudata* Hauer, 1938;

　　　f. *lacinulata* Hauer, 1938;
　　　f. *laticauda* Hauer, 1938
　　L. (L.) ludwigi ercods (Harring, 1914)
99. *L. (L.) luna* (Muller, 1776);
　　　f. *dorsicalis,* Sharma, 1978
100. *L. (L.) methoria* Harring & Myers, 1926
101. *L. (L.) nana* (Murray, 1913)
102. *L. (L.) neali* Wulfert, 1966
103. *L. (L.) nodosa* Hauer, 1938
104. *L. (L.) ohioensis* (Herrick, 1885)
105. *L. (L.) papuana* (Murray, 1913)
106. *L. (L.) ploenensis* (Voigt, 1902)
107. *L. (L.) pusilla* Harring, 1914
108. *L. (L.) schraederi* Wulfert, 1966
109. *L. (L.) sola* Hauer, 1938
110. *L. (L.) tesselata* Arora, 1965
111. *L. (L.) tryphema* Harring & Myers, 1926
112. *L. (L.) ungulata* (Gosse, 1887)
113. *L. (L.) verecunda* Harring & Myers, 1926
114. *L. (Hm.) inopinata* Harring & Myers, 1926
115. *L. (Hm.) sympoda* Hauer, 1929
116. *L. (Hm.) syngenes* (Hauer, 1938)
117. *L. (M.) bulla* (Gosse, 1851)
118. *L. (M.) closterocerca* (Schmarda, 1898)
119. *L. (M.) cornuta* (Muller, 1786)
120. *L. (M.) crenata* (Harring, 1913)
121. *L. (M.) decipiens* (Murray, 1913)
122. *L. (M.) elachis* (Harring & Myers, 1926)
123. *L. (M.) furcata* (Murray, 1913)
124. *L. (M.) galeata* (Bryce, 1892)
125. *L. (M.) hamata* (Stokes, 1896)
126. *L. (M.) harringi* (Ahlstrom, 1934)
127. *L. (M.) lunaris* (Ehrb., 1832)
128. *L. (M.) obtusa* (Murray, 1913)
129. *L. (M.) pawlowski* Wulfert, 1966
130. *L. (M.) perplexa* (Ahlstrom, 1938)
131. *L. (M.) punctata* (Murray, 1913)
132. *L. (M.) pyriformis* (Daday, 1905)
133. *L. (M.) quadridentata* (Ehrb., 1832)
134. *L. (M.) sinuata* (Hauer, 1938)
135. *L. (M.) stenroosi* (Meissner, 1908);
　　　f. *lineata* Wulfert, 1966
136. *L. (M.) styrax* (Harring & Myers, 1929)
137. *L. (M.) tethis* (Harring & Myers, 1926)
138. *L. (M.) thienemanni* (Hauer, 1938)
139. *L. (M.) unguitata* (Fadeev, 1925)

Family PROALIDAE
140. *Proales indirae* Wulfert, 1966

141. *P. decipiens* (Ehrb., 1831)

Family NOTOMMATIDAE

142. *Cephalodella catellina* (Muller, 1786)
143. *C. exigua* (Gosse, 1886)
144. *C. forficula* (Ehrb., 1832)
145. *C. gibba* (Ehrb., 1832)
146. *C. hiulca* Myers, 1924
147. *C. panarista* Myers, 1924
148. *C. megalocephala* v. *rotunda* Donner, 1950
149. *C. misgurnus* Wulfert, 1937
150. *C. mucronata* Harring & Myers, 1924
151. *C. wiszniewskii* Edmondson & Hutchinson, 1934
152. *Eosphora anthadis* Harring & Myers, 1921
153. *E. najas* Ehrb., 1830
154. *Itura aurita* (Ehrb., 1830)
155. *Notommata copeus* Ehrb., 1834
156. *N. epaxia* Harring & Myers, 1924
157. *N. glyphura* Wulfert, 1935
158. *N. pseudocerberus* De Beauchamp, 1908
159. *N. tripus* Ehrb., 1838
160. *Scaridium longicaudum* (Muller, 1786)

Family GASTROPODIDAE

161. *Ascomorpha ecaudis* Perty, 1850
162. *A. saltans* Bartsch, 1870;
       v. *indica* Wulfert, 1966
163. *Gastropus hyptopus* (Ehrb., 1838)

Family TRICHOCERCIDAE

164. *Trichocerca brazieliensis* Murray, 1913
165. *T. brachyura* (Gosse, 1886)
166. *T. cavia* (Gosse, 1886)
167. *T. cylindrica* (Imhof, 1891)
168. *T. flagellata* Hauer, 1937
169. *T. iernis* (Gosse, 1887)
170. *T. longiseta* (Schrank, 1802)
171. *T. myersi* (Hauer, 1931)
172. *T. nitida* Harring, 1914
173. *T. porcellus* (Gosse, 1886)
174. *T. rattus* (Muller, 1776)
175. *T. ruttneri* Donner, 1953
176. *T. scipio* (Gosse, 1886)
177. *T. similis* (Wierzejski, 1893)
178. *T. stylata* (Gosse, 1851)
179. *T. tigris* (Muller, 1786)
180. *T. tropis* Hauer, 1937
181. *T. weberi* (Jennings, 1903)

Family ASPLANCHNIDAE

182. *Asplanchna brightwelli* Gosse, 1850
183. *A. intermedia* Hudson, 1886
184. *A. priodonta* Gosse, 1850
185. *A. sieboldi urwaensis* (Sudzuki, 1956)
186. *Asplanchnopus bimaveraensis* Dhanapathi, 1975
187. *A. hyalinus* Harring, 1917
188. *A. multiceps* (Schrank, 1793)

Family SYNCHAETIDAE

189. *Synchaeta littoralis* Rousselet, 1902
190. *S. oblonga* Ehrb., 1832
191. *S. pectinata* Ehrb., 1832
192. *S. stylata* Wierzejski, 1893
193. *S. tremula* (Muller, 1786)
194. *Polyarthra euryptera* Wierzejski,
195. *P. longiremis* Carlin, 1943
196. *P. multiappendiculata* Arora, 1962
197. *P. trigla* Ehrb., 1834

Family DICRANOPHORIDAE

198. *Dicranophorus dolerus* Harring & Myers, 1924
199. *D. forcipatus* (Muller, 1786)
200. *D. epicharis* Harring & Myers, 1928
201. *D. myriophylli* (Harring, 0000)
202. *D. tegillus* Harring & Myers, 1927
203. *Encentrum longipes* Wulfert, 1966

Order: GNESIOTROCHA

Family FLOSCULARIDAE

204. *Floscularia conifera* Hudson, 1886
205. *F. ringens* (Linnaeus, 1758)
206. *Ptygura furcillata* (Kellicott, 1889)
207. *Lacinularia flosculosa* (Muller, 1773)
208. *Limnias ceratophylli* (Schrank, 1803)
209. *Beauchampia crucigera* (Dutrochet, 1812)
210. *Limnia melicerta* Weisse, 1848
211. *Ptygura stephanion* (Anderson, 1889)
212. *Sinantherina spinosa* (Thorpe, 1893)
213. *S. socialis* (Linnaeus, 1758)
214. *S. triglandularis* Arora, 1963

Family CONOCHILIDAE

215. *Conochilus madurai* Michael, 1966
216. *C. arboreus* Rajendran, 1971
217. *C. hippocrepis* (Schrank, 1803)
218. *C. dossuarius* (Hudson, 1886)
        v. *asetosus* Arora, 1962

Family HEXARTHRIDAE
219. *Hexarthra bulgarica* (Hudson, 1871)
      ssp. *nepalensis* Dumont *et al.,* 1978
220. *H. fennica* v. *oxyuris* (Sernov, 1903)
221. *H. intermedia* (Wizniewski, 1929)
222. *H. mira* (Hudson, 1871)

Family FILINIIDAE
223. *Filinia opoliensis* Zach. 1898
224. *F. pejleri* Hutchinson, 1964
225. *F. longiseta* (Ehrb., 1834)
226. *F. terminalis* (Plate, 1886)

Family TESTUDINELLIDAE
227. *Testudinella patina* (Hermann, 1783)
228. *T. incisa* (Ternetz, 1892)
229. *T. mucronata* (Gosse, 1886)
230. *Pompholyx complanata* (Gosse, 1851)
231. *P. sulcata* Hudson, 1885

Family TROCHOSPHAERIDAE
232. *Horaella brehmi* Donner, 1949

Order BDELLOIDEA

Family PHILODINIDAE
233. *Philodina citrina* (Ehrb., 1832)
234. *Rotaria macroceros* (Gosse, 1851)
235. *R. mento* (Anderson, 1889)
236. *R. rotatoria* (Pallas, 1766)
237. *R. vulgaris* (Schrank, 0000)
238. *R. ovata* (Anderson, 1889)
239. *R. neptunia* (Ehrb., 1832)
240. *R. neptunis* (Milne, 1886)
241. *Pseudoembata acutipoda* Wycliffe & Michael, 1968

* The system of classification followed is after Koste (1978).

AN HISTORICAL SURVEY OF THE COLLECTION AND STUDY OF ROTIFERS IN BRITAIN

Charles G. HUSSEY

British Museum (Natural History), London SW7 5BD

Keywords: Rotifera, British Isles, history

Abstract

With reference to the author's compilation of British rotifer records, the main contributions to our knowledge of rotifers in the British Isles are outlined and the impetus provided by local natural history societies is commented upon. The usefulness and limitations of early records are examined and the scope offered to future collectors is discussed.

The arrival of the microscope in the seventeenth century made possible the discovery of rotifers. The first record of these animals in Britain appears, according to Hudson (Hudson & Gosse, 1886), to have been a paper contributed by John Harris, Rector of Winchelsea in Sussex (Harris, 1696) to the Philosophical Transactions of the Royal Society, when he wrote: 'I saw here also an animal like a large maggot which would contract itself up into a spherical figure and then stretch itself out again; the end of its tail appeared with a forceps, like that of an Ear-wig ...and they seemed to be busy with their mouths as if in feeding.' Letters by Leuwenhoek (1703) to the same journal on animalcules found in Holland further alerted the British to their presence and provided the first Figures. Books by Sir John Hill (1752) and Henry Baker (1753) also make mention of rotifers, the latter work being particularly detailed so that several modern species may be identified. The first male rotifer was found and described by Thomas Brightwell (1848) when he discovered a new species of *Notommata* in a pond in a brick-ground near Norwich. This species has subsequently been identified as *Asplanchna brightwelli* Gosse.

Certain conditions were required before a group such as the Rotifera could be adequately studied. First it was necessary to know their habit and how to obtain speci- mens, and one reason why rotifers became popular amongst amateur naturalists was their ease of collection. Adequate microscopes were required and so also was a comprehensive guide for their identification. Finally there had to be a body of people possessing sufficient time and patience, and an enthusiasm towards the study of the natural world. These conditions were fulfilled in Britain in the latter half of the eighteenth century when local natural history and microscopical societies and field clubs began to spring up in nearly every major town. A fascinating account of the phenomenon of the naturalist in Britain is given by Allen (1976). Pritchard (1842) became the standard work of reference with about one sixth devoted to the Rotifera. It was based largely on Ehrenberg's 'Infusions thierchen' (1838) and covered all species known at that time. It went through four editions, more than doubling in size, and remained in use until replaced by Hudson & Gosse's classic monograph (1886, 1889), which is still referred to today.

By 1886 over 230 species had been described, a major contribution having been a catalogue of Rotifera produced by Gosse (1851) listing 108 species. Further notable surveys were produced by Hood (1891), Murray (1906), Stevens (1912), Harris (1936), Galliford (1946, 1954), Kemp (1956), Wright (1957) and finally, Berzins (1967, 1978).

The author has been engaged in the preparation of a checklist of British rotifer records which has been compiled from all available published records and from data relating to specimens in the collections of the British Museum (Natural History). Figures (1, 2 and 3) are taken from a preliminary analysis of these data.

The first figure shows the number of references to rotifers found in the published literature and includes excursion reports and notes of specimens exhibited at

237

Hydrobiologia 73, 237-240 (1980). 0018-8158/80/0733-0237$00.80.
© *Dr. W. Junk b.v. Publishers, The Hague.*

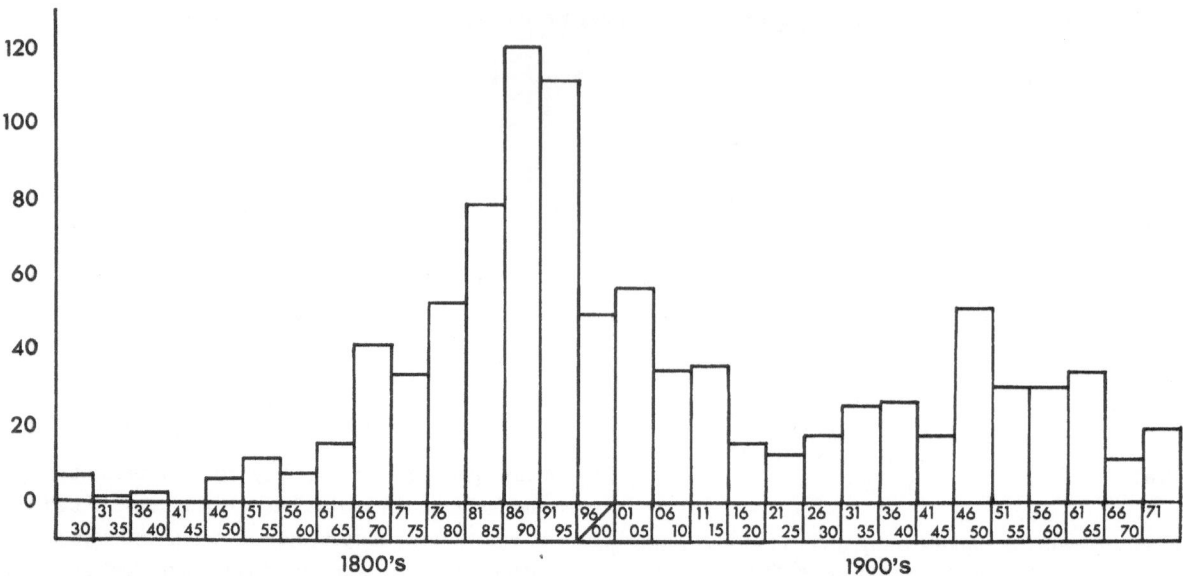

Fig. 1. Histogram by five-year periods of the number of references to rotifers in the British Isles and Ireland.

meetings of societies as well as original articles. The spurt of interest in rotifers that occurred at the end of the last century, particularly once Hudson and Gosse's monograph had appeared, can be clearly seen. The continuing production of papers through to the present day should also be noted; much of the more recent work being contributed by Wright and Galliford.

The second figure displays the number of names and combination of names used in print for species found in Britain. This is compared with the actual number of valid species which has been obtained by combining all synonymous species and discarding all those that are now considered unrecognisable. The contribution made by Gosse's catalogue of 1851 may be clearly seen. The proliferation of names after 1880 contrasts with the fairly steady rise in the number of new records of valid species. Further increases in the number of names appear in 1930-1945 as Harring's (1913) revision of the nomenclature began to be adopted. The most recent upsurge in names is mainly due to re-arrangement of species as subspecies or varieties.

The third figure, showing the number of new records for Britain of monogonont and bdelloid rotifers, grouped by five-year periods, also shows up Gosse's catalogue and the work of Hudson & Gosse (1886). The recent increase of 1965-70 is due to apparently new British records in Berzins (1967) and the 1970's records are mainly from Martin (1976, 1977), collecting at Thursley Common in

Surrey, and Maitland (1977). The work on the bdelloid rotifers is almost wholly contributed by Bryce and Murray in the 1890's and first fifteen years of this century. Few new bdelloids have been added since their time.

Although the rotifers are easy to collect, they have proved difficult to preserve satisfactorily and so we are unable to refer to type specimens and must rely instead on original descriptions and figures when working on their taxonomy and systematics. The location and interpretation of the earlier literature is complicated by the confused nomenclature and obscurity of some of the journals. Many early descriptions are too casual to be of any use and lack measurements or even adequate Figures. However, much fine work has been contributed by men such as Rousselet, Murray, Bryce, Hudson and Gosse, and their descriptions are worth examining even today.

Some of the value of the more thorough, earlier reports lies in their records of the associations of rotifers, and extended series of excursion reports can contribute to our knowledge of the seasonal occurrence of some species. It would also be of interest to re-visit the sites of early surveys and record the species found today.

Much of the work on the rotifers in Britain has been supplied by amateurs and this is still true today. I believe that this should be taken into account when producing identification guides, particularly as regards to price and availability. Pontin (1978) achieves this well, but care should be taken (as in Ruttner-Kolisko, 1974) to appraise

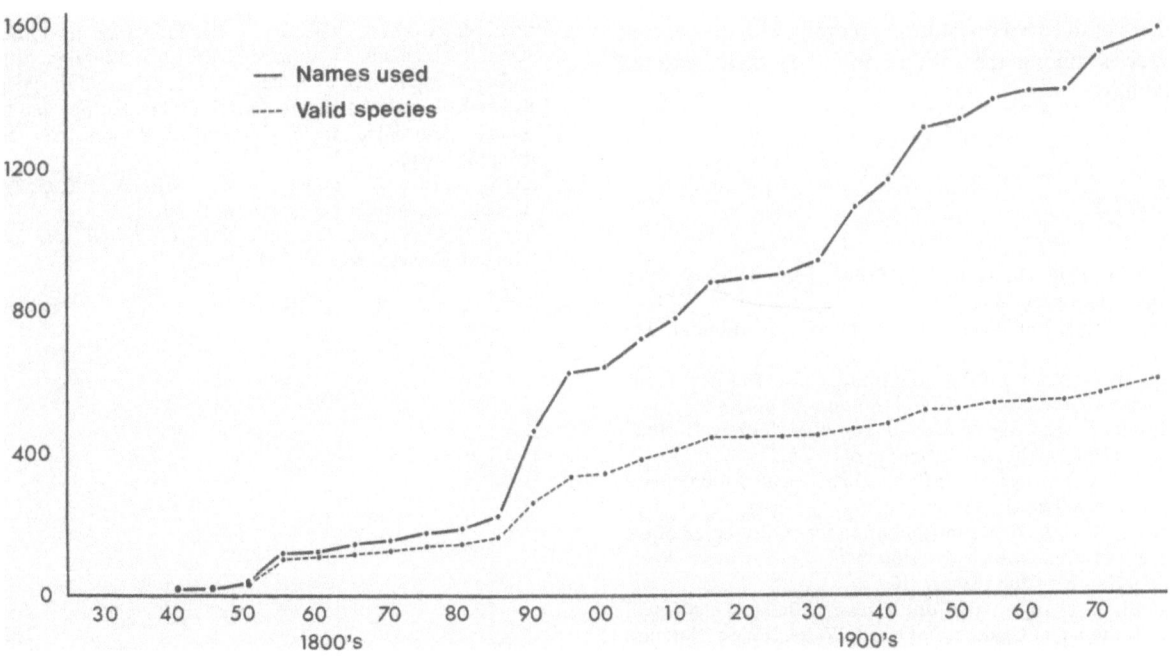

Fig. 2. Names of rotifers and combinations of names appearing in British literature, compared to actual number of recognised species.

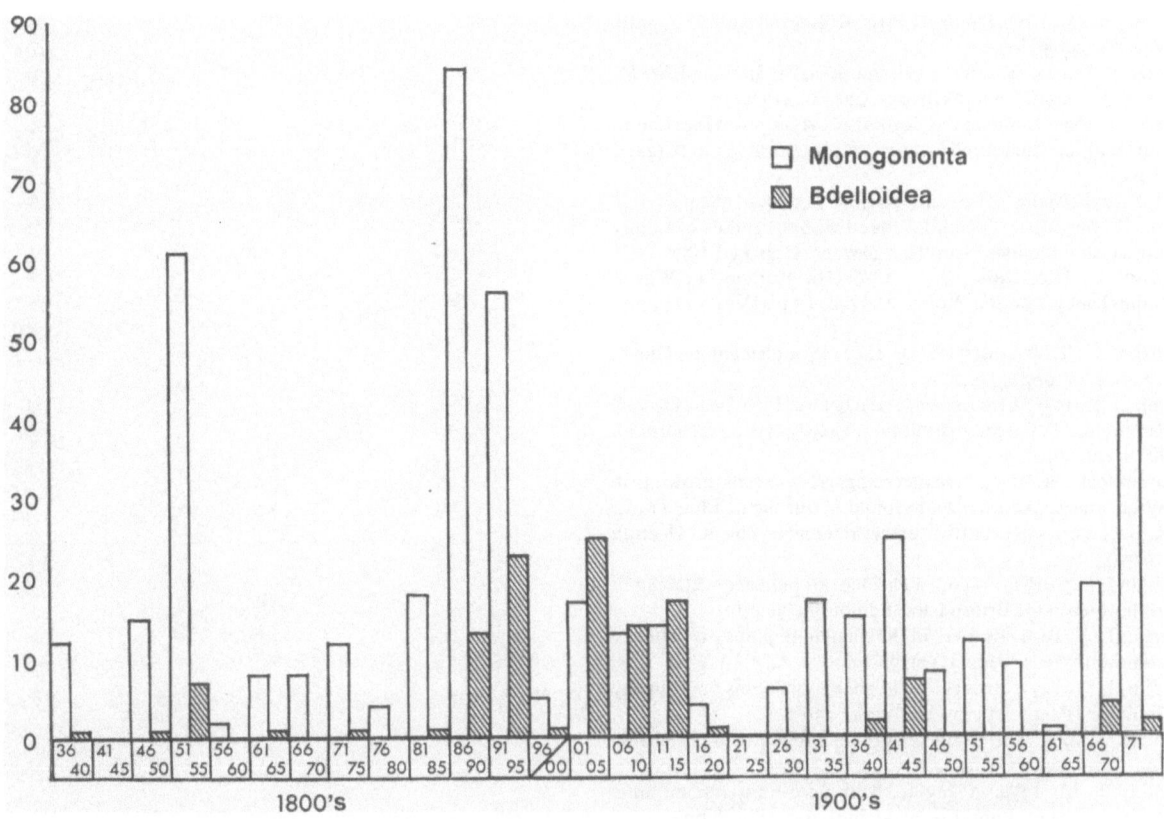

Fig. 3. Histogram by five-year periods of new records for Britain and Ireland of monogonont and bdelloid rotifers.

workers of the need for caution in coping with the concept of species among the rotifers and with their seasonal variability.

References

Allen, D. E. 1976. The naturalist in Britain, a social history. Allen Lane, London, 292 pp.

Baker, H. 1753. Employment for the microscope. London, 452 pp.

Berzins, B. V. A. 1967. Rotatoria. In: Illies, J. (ed.) Limnofauna Europaea. Stuttgart, 35-68 [2nd edition 1978 pp. 54-91].

Brightwell, T. 1848. Sketch of a fauna infusoria for East Norfolk. (published by author), Norwich, 40 pp.

Ehrenberg, C. G. 1838. Die Infusionsthierchen als vollkommene Organismen. Leipzig, 547 pp.

Galliford, A. L. 1946. A contribution to the rotifer fauna of the Liverpool area, with a description of a new species. Proc. Lpool Nat. Fld Club 1945: 10-16.

Galliford, A. L. 1954. Notes on the distribution and ecology of the Rotifera and Cladocera of North Wales. NWest. Nat. new ser. 1: 513-529.

Gosse, P. H. 1851. A catalogue of Rotifera found in Britain; with descriptions of five new genera and thirty-two new species. Ann. Mag. nat. Hist. ser. 2, 8: 197-203.

Harring, H. K. 1913. Synopsis of the Rotatoria. Bull. U.S. natn. Mus. 81: 226 pp.

Harris, A. E. 1936. Rotifera of Glamorgan. In: Tattersall, W. M. (ed) Glamorgan County History, Cardiff, 413-417.

Harris, J. 1696. Some microscopical observations of vast numbers of animalcula seen in water. Phil. Trans. R. Soc. 19 (220): 254-259.

Hill, J. 1752. An history of animals. [&c.] London, 584 pp.

Hood, J. 1891. List of Rotifera found within a radius of twenty miles round Dundee. Scot. Nat. new ser. 5: 20-25, 71-80.

Hudson, C. T. & Gosse, P. H. 1886. The Rotifera: or Wheel-Animalcules. London, Vol. 1, 128 pp., 15 pls; Vol. 2, 144 pp., 15 pls.

Hudson, C. T. & Gosse, P. H. 1889. [Supplement to above] London, 64 pp., 4 pls.

Kemp, J. R. 1956. Freshwater invertebrates. In: Walsh, G. B. & Rimington, F. C. (eds.) The natural history of the Scarborough district. 2: 78-90.

Leuwenhoek, A. 1703. ...concerning green weeds growing in water, and some animalcula found about them. Phil. Trans. R. Soc. 23: 1304-1311 [for further references, consult Harring (1913)].

Maitland, P. S. 1977. A coded checklist of animals occurring in fresh water in the British Isles. Edinburgh, 76 pp.

Martin, L. V. 1976. Rotifers in the sphagnum pools on Thursley Common. Microscopy 33 (2): 90-93.

Martin, L. V. 1977. Rotifers in the sphagnum pools on Thursley Common. Part 2. Microscopy 33 (4): 236-241.

Murray, J. 1906. The Rotifera of the Scottish Lochs. Trans. R. Soc. Edinb. 45: 151-191.

Pontin, R. M. 1978. A key to British freshwater planktonic Rotifera. Scient. Publs Freshwat. biol. Ass. 38: 178 pp.

Pritchard, A. 1842. A history of the Infusoria... London, 439 pp.

Pritchard, A. 1861. A history of the Infusoria; including the Desmidiaceae and Diatomaceae, British and foreign. (4th ed. enlarged and revised by Arlidge, J. T. et al.) London, 968 pp.

Ruttner-Kolisko, A. 1974. Plankton rotifers: biology and taxonomy. [translated from German] Binnengewässer 26, 1 (Suppl) 146 pp.

Stevens, J. 1912. Some of the Rotifera of Devon. Rep. Trans. Devon. Ass. Advmt Sci. 44 (ser. 3, 4): 681-691.

Wright, H. G. S. 1957. The rotifer fauna of East Norfolk. Trans. Norfolk Norwich Nat. Soc. 18 (5): 1-23.

ASYMMETRY AND VARIATION IN KERATELLA TROPICA

J. GREEN

Zoology Dept., Westfield College, Hampstead, London NW3 7ST, U.K.

Keywords: tropical, rotifer, variation, spines, Calanoida

Abstract

Keratella tropica usually has two posterior spines. Both may vary in length, but the left is always shorter than the right. Sometimes the left spine is absent. Spine length varies with the length of the lorica, but there are other, external, influences. Three separate lines of evidence indicate that factors which promote spine length are related in some way to the presence of calanoid copepods.

Introduction

Keratella tropica is widespread in tropical freshwaters, and extends into sub-tropical areas as a summer form. The typical form bears two posterior spines on the lorica, with the left spine shorter than the right. Sometimes the left spine is absent. This type of asymmetry is known in other species of *Keratella* such as *K. valga* and *K. procurva*. As far as I am aware there has never been a detailed investigation or satisfactory explanation of this peculiar phenomenon. The purpose of the present paper is to provide data from a range of tropical lakes which may contribute to at least a partial understanding of the nature of this variation and the conditions under which it occurs.

The material for this study has been collected over the last twenty years, and has involved the use of tow nets in order to obtain a sufficient number of individuals to give a statistically valid sample. The general aim was to measure 50 individuals from each sample, but as will be seen in Table 1, this ideal was not always achieved. Three measurements of each individual were made, as shown in Fig. 1. Means and standard deviations were calculated for each sample. The mean values are given in Table 1. The standard deviations have been omitted; in general they were low, about 3 or 4 μm, so that the standard error of

the mean for a sample of 50 would be about 0.5 μm for most of the means. Some examples are given in Table 2 for the Kigezi lakes.

The relationship between the length of the left posterior spine and the length of the lorica is shown in Fig. 2. The largest spines are found only on the largest individuals. In the lower part of the size range there is great variation, so that at a length of 102 μm it is possible to find samples in which all the individuals lack left spines and other samples in which the mean length of the left spine is about 50 μm.

The upper limit of left spine length in relation to lorica length seems to fall along a fairly smooth curve, but many populations fall short of the upper limits.

Fig. 3 shows the relationship between the right posterior spine and the length of the lorica. Again there is a general tendency for the length of the spine to increase with the length of the lorica, but there is considerable variation. The mean length of the right spine in these samples never fell below 36 μm.

Figs. 2 and 3 show the range of variation found in samples taken over a wide geographical area. If attention is focussed on a single locality or a single sample some additional features are found. In both Figs. 2 and 3 the data for Lake Albert are closely grouped, and show relatively little variation, with both spines being relatively short. Fig. 4 shows the data for Lake Albert plotted on a seasonal basis. There is remarkably little variation and the population seems to be morphometrically stable in this large stable lake. Fig. 5 shows that it is the spine lengths which have become most stable, both right and left spines fluctuate together, and neither shows any significant trend with the mean lorica length.

Within a single sample there is often a strong relationship between the length of the right spine and the length

Hydrobiologia 73, 241-248 (1980).   0018-8158/80/0733-0241$01.60.
© Dr. W. Junk b.v. Publishers, The Hague.

Table 1. *Keratella tropica*, ranked by lorica length.

Locality	Date	N	Mean lorica length μm	Mean length left spine μm	Mean length right spine μm	Diaptomids
Kafue,Zambia	Jun 65	10	122.5	77.6	79.5	++
" "	Jan 65	50	117.3	73.3	78.8	++
Bahr el Jebel,Sudan	Dec 76	50	112.3	66.9	127.8	++
L.Chad-Yobe Mouth	Jul 67	50	107.8	59.9	62.8	++
L.Mutanda,Uganda	Oct 62	6	105.7	59.3	69.0	++
L.Albert,(Buhuka)	Aug 62	50	105.5	18.2	40.9	−
L.No,Sudan	Dec 76	21	104.9	52.3	97.1	++
L.Victoria,Uganda	Oct 62	10	102.5	51.1	65.7	++
L.Albert (Mid)	May 62	25	102.4	18.6	41.1	−
L.Kundi,Sudan	Jan 76	50	102.4	0.0	79.6	++
L.Edward,Uganda	Jul 62	10	101.5	28.2	49.1	−
L.Bunyonyi,Uganda(S)	Dec 64	20	99.8	26.9	54.8	−
L.Mulehe,Uganda	Oct 62	6	99.0	34.0	60.8	−
L.Edward,Uganda	Nov 62	10	98.9	20.9	42.3	−
L.Bunyonyi	Jul 62	50	98.9	31.5	55.6	−
" (D)	Dec 64	10	98.5	22.1	52.4	−
L.Bur Akok,Sudan	Apr 78	50	97.9	24.5	62.2	++
Ranu Pakis,Java	Aug 74	50	97.8	12.1	55.0	−
Ranu Lading,Java	Aug 74	20	97.6	14.8	55.5	−
Ranu Bedali,Java	Jul 74	50	97.4	16.0	55.3	−
L.Bunyonyi,Uganda	Oct 62	50	96.9	32.6	54.9	−
Telaga Pasir,Java	Aug 74	50	96.8	12.5	49.2	−
L.George,Uganda	Nov 62	20	96.8	14.9	38.6	−
L.Kyoga,Uganda	Nov 62	30	96.5	36.8	55.7	+
L.Mohasi,Ruanda	Aug 75	10	92.8	18.1	41.6	−
L.Nkugute,Uganda	Jan 64	6	90.3	21.2	36.3	−
Mboandong,Cameroon	Apr 70	10	86.0	5.3	58.4	−

of the lorica. Fig. 6 shows the relationship for three of the samples with the longest right spines. The levelling of the relationship in the sample from the Bahr el Jebel appears to be real, and may imply that there is an upper limit to the length of the right spine irrespective of any further increase in lorica length. The points shown on this Figure are means; some individuals had right spines up to 146 μm in length.

In Table 2 the data for the lakes in the Kigezi district of Uganda are presented in more detail. The most striking feature is the long spines of *K. tropica* in Lake Mutanda in 1962. There have been marked changes in the zooplankton of L. Mutanda between 1962 and 1975 (Green, 1976). These changes include the disappearance of three species

of *Daphnia* and the copepod *Metadiaptomus aethiopicus*. A similar change occurred in Lake Bunyonyi sometime between 1930 and 1962. These changes in the composition of the zooplankton are accompanied by changes in the posterior spines of *K. tropica*, so that the specimens from L. Mutanda are now very similar to those from L. Bunyonyi.

As the change in L. Mutanda was accompanied by the disappearance of a diaptomid copepod it was decided to see if any similar relationship could be found elsewhere. Examination of Table 1 shows that the larger specimens of *K. tropica* with larger spines are found in lakes where diaptomids are abundant. There is a significant clumping of diaptomid localities at the top of the Table. Exceptions

Table 2. *Keratella tropica* in the Kigezi lakes, Uganda.

Lake	Date	Lorica mean length	left spine mean length	right spine mean length	N	Diaptomids
Mutanda	Oct 62	105.7 ±2.9[+]	59.3 ±4.0	69.0 ±3.3	6	++
Mutanda	Aug 75	93.7 ±3.3	26.1 ±3.9	52.4 ±6.6	35	−
Bunyonyi	Jul 62	98.3 ±2.4	31.5 ±2.5	55.6 ±2.9	50	−
Bunyonyi	Oct 62	96.9 ±2.6	32.6 ±3.7	54.9 ±4.5	50	−
Bunyonyi (D)	Dec 64	98.5 ±1.5	22.1 ±3.9	52.4 ±3.5	10	−
Bunyonyi (S)	Dec 64	99.8 ±2.3	26.9 ±4.4	54.8 ±5.2	20	−
Bunyonyi	Aug 75	95.7 ±1.9	29.0 ±2.3	53.8 ±2.8	14	−
Mulehe	Oct 62	99.0 ±2.8	34.0 ±2.2	60.8 ±4.1	6	−
Mulehe	Aug 75	93.4 ±2.7	22.0 ±5.0	51.6 ±3.4	12	−

[+]-standard deviation
(D)- deep station
(S)- shallow station.

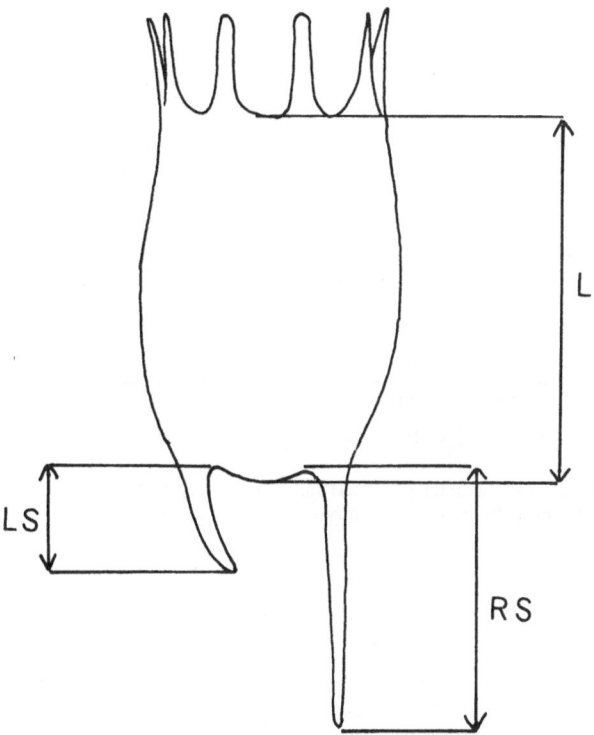

Fig. 1. Outline of the lorica of *Keratella tropica* to show the measurements made. L – lorica; LS – left spine; RS – right spine.

are found. The two Lake Albert samples have relatively large loricas, but their spines are shorter than in other samples with the same lorica length. The samples from Lake Albert were lacking in diaptomids. This supports the idea that conditions which support diaptomids are favourable for the development of the posterior spines of *K. tropica,* and when diaptomids are absent the spines tend to be shorter. There are still some anomalies in the Table, such as Lake Bur Akok and Lake Kyoga, and these will require further investigation.

Further support for the general idea comes from my data on the River Sokoto (Green, 1960, 1962). Fig. 7 shows data for the left spine for two seasons. In 1956 this spine was longer than in 1957, and the diaptomid *Tropodiaptomus laurentii* was much more abundant. However it should be noted that the left spine of *K. tropica* went through a cycle of increase and decrease in length even though *Tropodiaptomus* was sparse throughout the 1957 season. This indicates that any relationship between abundance of diaptomids and spine length in *K. tropica* is only a part of the total complex determining spine length.

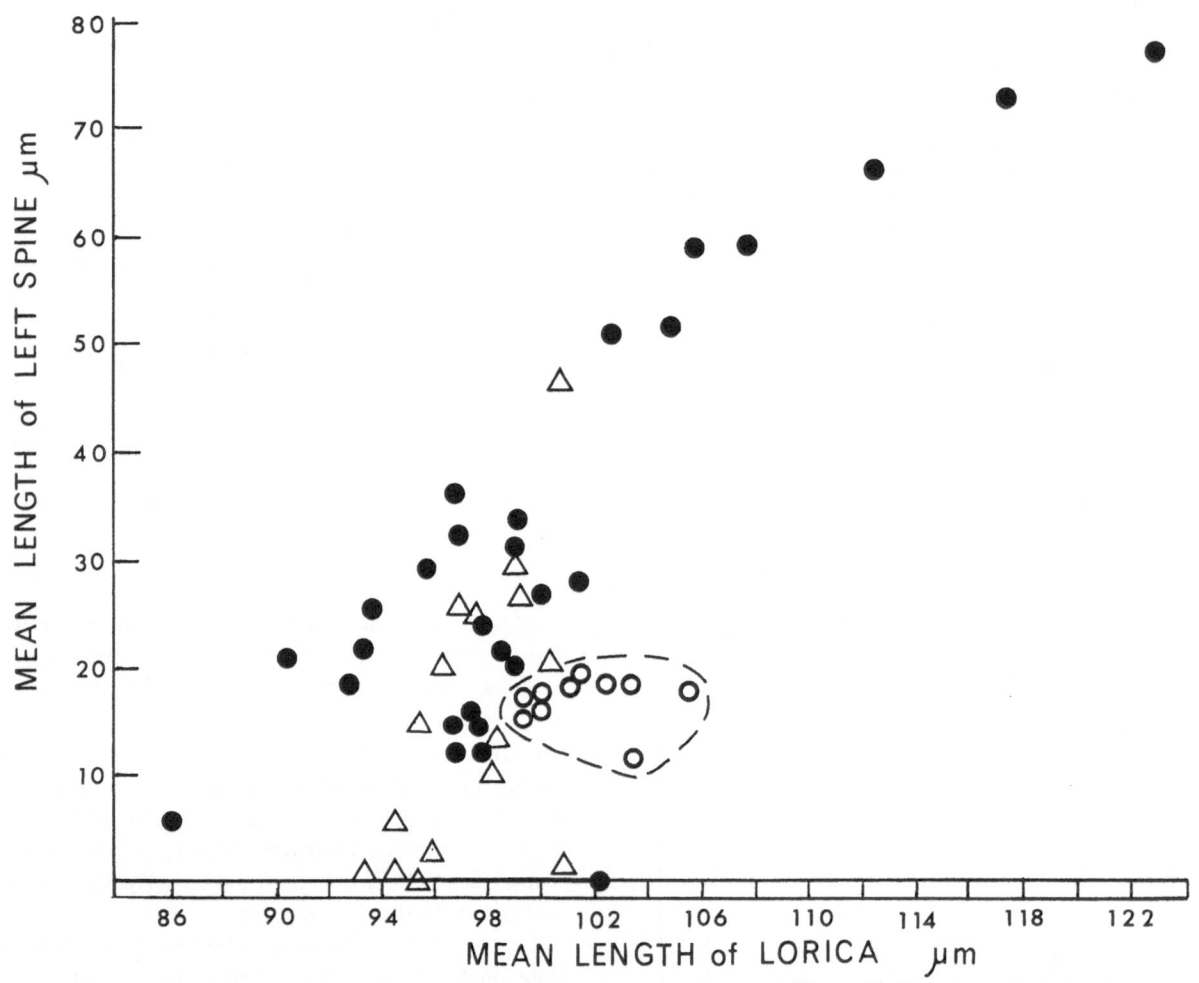

Fig. 2. *Keratella tropica:* relationship between mean length of lorica and mean length of left spine in 54 samples. Open circles – Lake Albert, Uganda. Open triangles – River Sokoto, Nigeria. Black dots – all other localities – see Table 1.

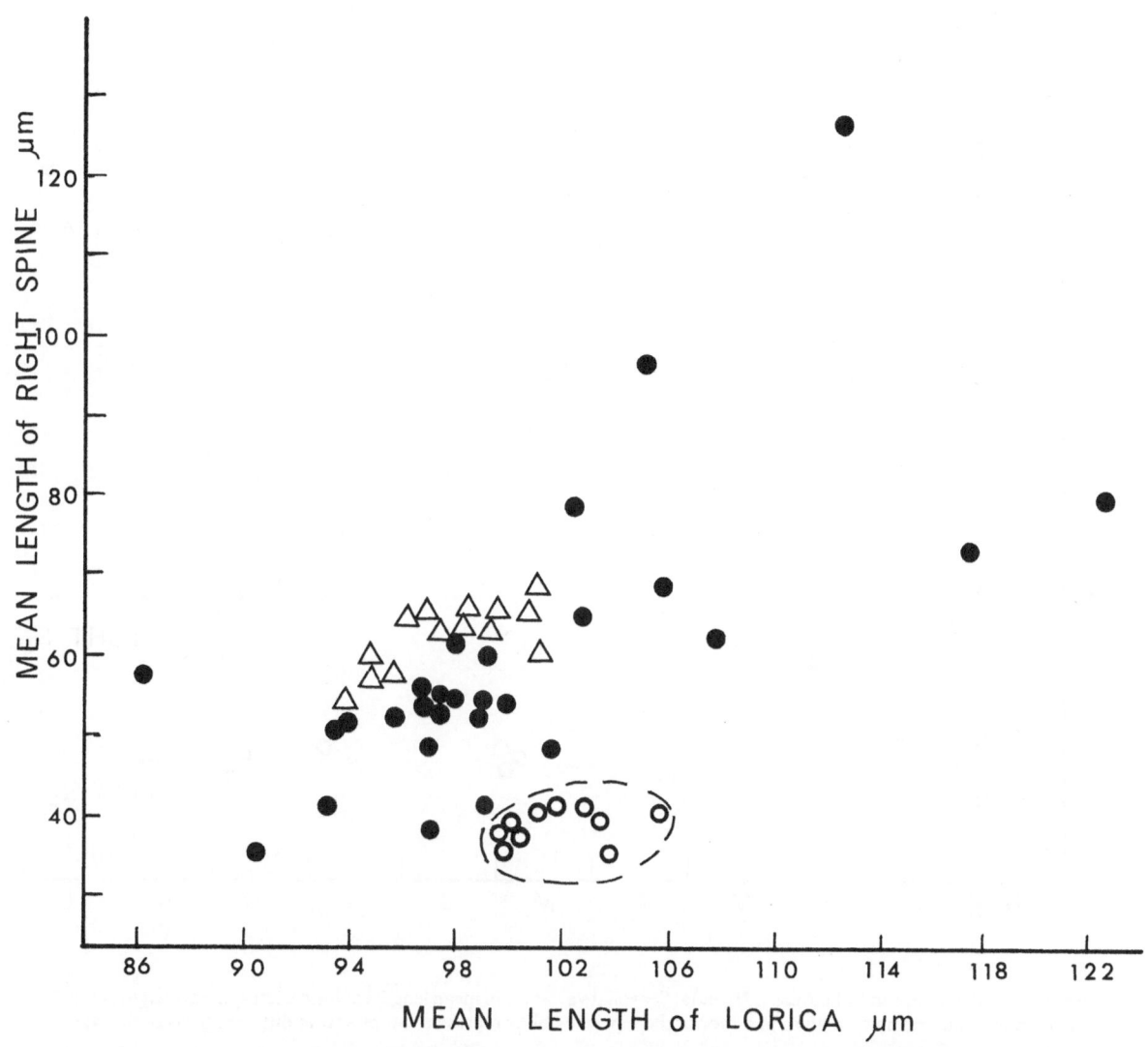

Fig. 3. *Keratella tropica:* relationship between mean length of lorica and mean length of right spine in 54 samples. Open circles – Lake Albert, Uganda. Open triangles – River Sokoto, Nigeria. Black dots – all other localities – see Table 1.

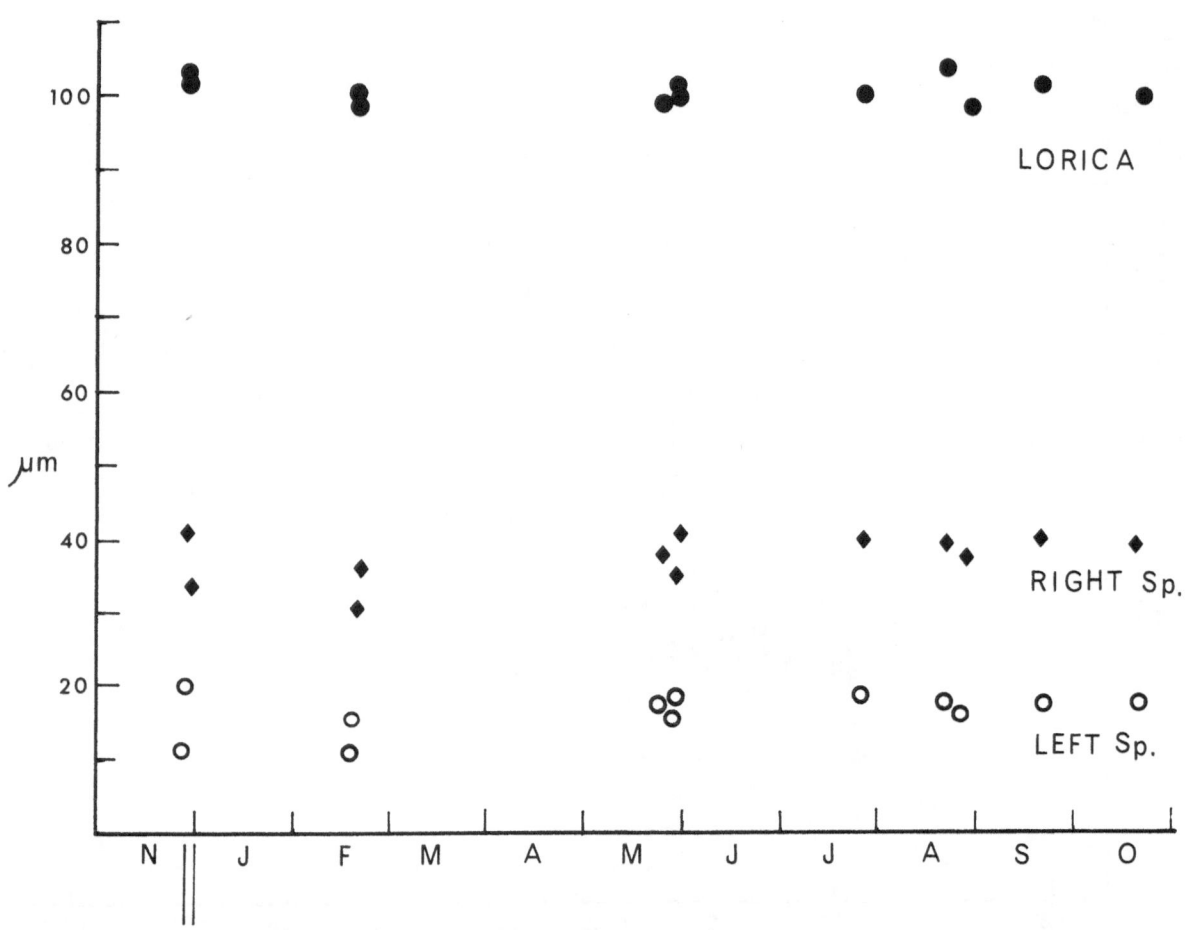

Fig. 4. *Keratella tropica* in Lake Albert, Uganda. Seasonal variation in mean lengths of lorica, left spine and right spine. Where more than one point is given in a month the separate points represent samples from different parts of the lake.

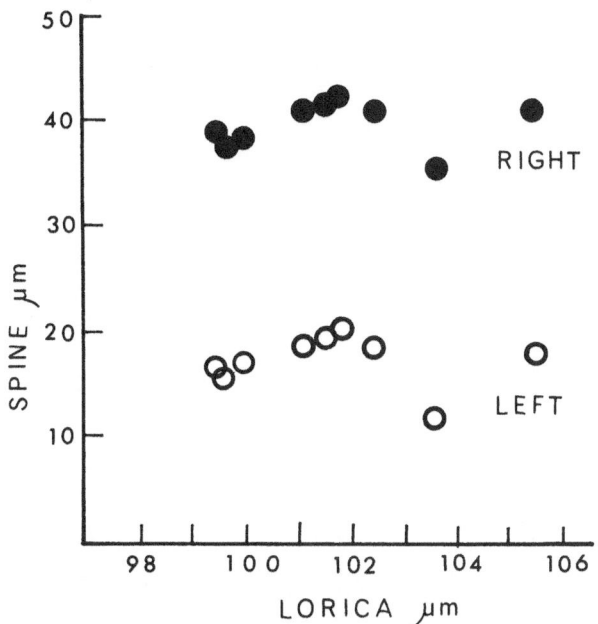

Fig. 5. *Keratella tropica* in Lake Albert, Uganda. Detailed plot of mean lengths of left and right spines in relation to mean length of lorica.

The data presented here indicate that variation in the length of the spines of *K. tropica* is a complex phenomenon. Part can be explained by the relationship with lorica size, but there is also considerable variation not related to lorica size. Part of this variation may be associated with the presence or absence of diaptomid copepods. Any suggestion of a causal relationship has been carefully avoided, but the data from three separate sources: 1) the Kigezi lakes, 2) the ranked list, and 3) the River Sokoto, suggest that the possibility of a common factor is worthy of investigation.

The data presented here have not helped towards an understanding of the asymmetrical aspect of the variation. Perhaps the right spine is part of a stabilising system for swimming and the left spine increases under more biological influences, and decreases when these influences are absent.

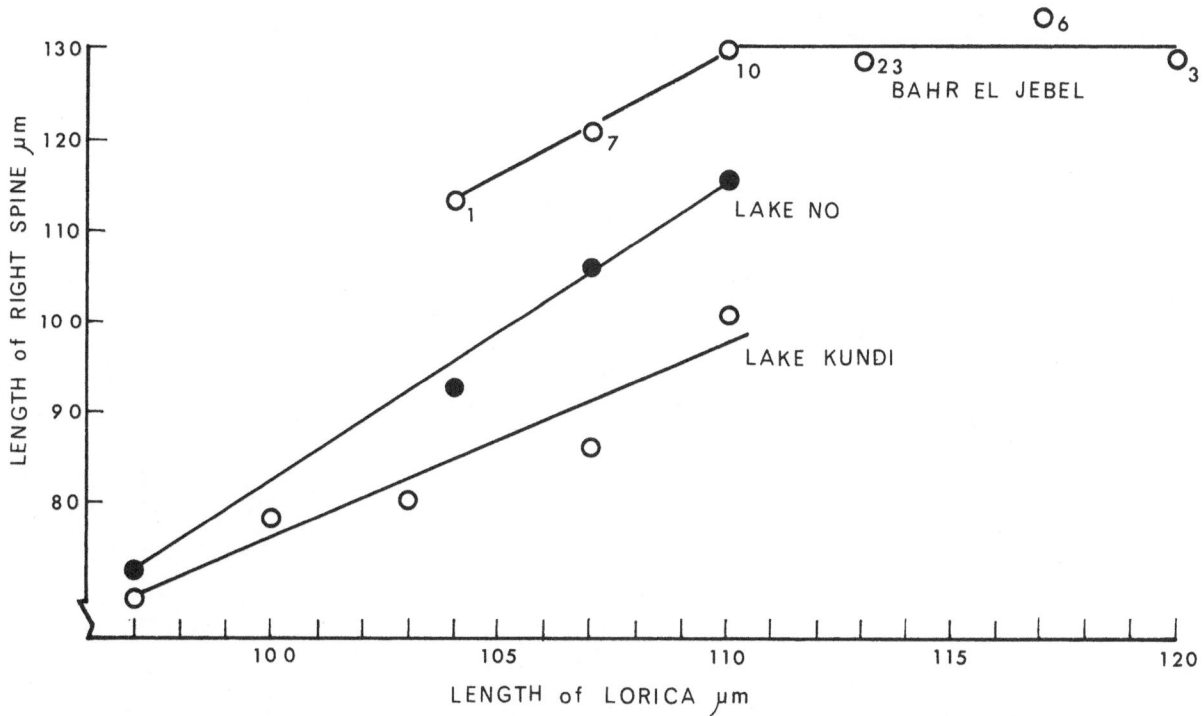

Fig. 6. *Keratella tropica:* relationship between length of lorica and the lenght of right spines in samples with long spines. Each point is based on a variable number of individuals; details are given for the Bahr el Jebel.

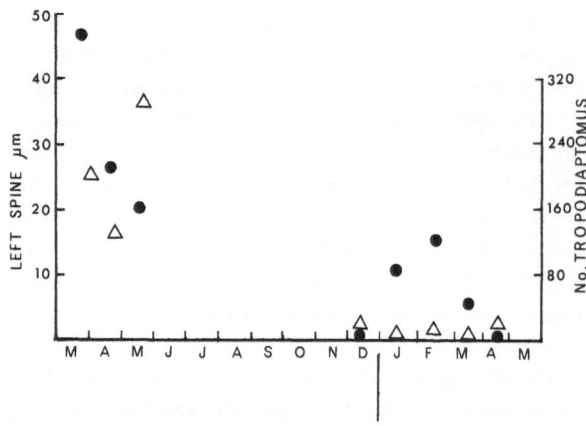

Fig. 7. *Keratella tropica* in the River Sokoto, Nigeria. Seasonal variation in length of the left posterior spine in relation to the abundance of *Tropodiaptomus laurentii*. Black dots – length of left spine. Open triangles – numbers of Tropodiaptomus per sample of $\frac{1}{3}$ m^3.

References

Green, J. 1960. Zooplankton of the River Sokoto. The Rotifera. Proc. zool. Soc. Lond. 135: 491-523.

Green, J. 1962. Zooplankton of the River Sokoto. The Crustacea. Proc. zool. Soc. Lond. 138: 415-453.

Green, J. 1976. Changes in the zooplankton of Lakes Mutanda, Bunyonyi and Mulehe (Uganda). Freshwater Biology 6: 433-6.

SOME PECULIAR ELEMENTS IN THE ROTIFER FAUNA OF THE ATLANTIC SAHARA AND OF THE ATLAS MOUNTAINS*

Marc COUSSEMENT & Henri J. DUMONT

Zoological Institute, The State University of Gent, Belgium

* Contribution n° 28 from project 'Limnology of the Sahara, under contract n° 2.0009/75 with the Fonds voor Kollektief Fundamenteel Onderzoek, Belgium

Keywords: Rotifers, atlantic Sahara, Atlas mountains, Biogeography

Abstract

In addition to ubiquitous forms, the Adrar of Mauretania has a number of rare tropicopolitan species, which occur here in relative abundance. In the Moroccon Atlas mountains, West-African forms meet boreo-alpine forms.

A population of *Hexarthra* from Atar, Mauretania, was intermediate between the *fennica*-group and the *jenkinae*-group.

Introduction

It has long been known that coastal deserts or coastal zones of deserts which continental dimensions stand, climatologically speaking, rather apart. Thus, the atlantic or western Sahara is characterized by relatively low temperatures, a high nebulosity, a relatively high atmospheric humidity, and a relatively high rainfall that may occur in all seasons (Dubief, 1971). It is, moreover, a windy area, especially in the south (Mauretania, Rio de Oro). This zone is sometimes called the 'western rain-bridge', and is supposed to have been an important pathway for both cultural and faunal exchange between the mediterranean shorelands and equatorial Africa until the recent past, and even until today. A branch of the bird migrating pathways also follows the atlantic coast-line. Finally, it is interesting to note that a mountain range pertaining to the alpine folding, the Atlas, runs almost parallel to the coast. It is only weakly glaciated today, and only in its highest parts around the Jebel Toubkal massif, while its southernmost chain, the Anti-Atlas, projects deep down into desert country.

Very little is known about the rotifer fauna of this area, which comprises most of Morocco, Rio de Oro (the former spanish Sahara, presently under political dispute), and western Mauretania.

A report on a small rotifer collection from the north of Rio de Oro (the Saguiat el Hamra) was published by Dumont & Coussement (1976) (12 species identified). However, in the framework of the same research project on the aquatic biota of the Sahara, numerous plankton samples were also collected in Morocco and Mauretania. In all, about 70 rotifer species have now been found in this material, 25 of which belong to the sole genus *Lecane*. A few species also appear to be undescribed, but will not be dealt with in this paper, which will be concerned with some morphologically and chorologically interesting species from this list.

Results

1. *Hexarthra* cfr. *fennica* (Figs. 13-15) and *H. fennica* (Levander) (Table 1).

In June 1975, a *Hexarthra* population was sampled in a concrete, almost completely covered drinking water reservoir in the town of Atar, Adrar of Mauretania. The animals were very small (body length 140-150 µm), and a remarkably high proportion of the population was producing resting eggs.

249

Hydrobiologia 73, 249-254 (1980). 0018-8158/80/0733-0249$01.20.
© *Dr. W. Junk b.v. Publishers, The Hague.*

Table 1.

	fennica	jenkinae	libica	cfr. fennica
–filaments on dorso-lateral arm	7	5-8	7	7
on ventro-lateral arm	8-9	6-8	9	6
on dorsal arm	8	4-5	5	5
–ventral arm				
filaments	8-9	6	8	6
spinules	+		+	
naked		+		+
spines	4/4	2/2, 3/2, 4/3, 4/4	2/3	4/4
–mastax: teeth on unci	7-8	9-10	9	7
–length ventral arm/body length	> 1	≤ 1	≤ 1	0.5 ≤ x ≤ 1

Upon examination of diagnostic characters, an amalgam of features appeared, summarized in Table 1, where they are compared with known species likely to occur in the area.

It appears that our material is very close to *H. jenkinae* (Beauchamps), and to *H. libica* (Manfredi) which is now widely considered to be conspecific with the former, but the low teeth index on the unci is that of *H. fennica* (Levander).

Ruttner-Kolisko (1974) suggests that the number of teeth on the unci increases with increasing salinity, and that the ratio length ventral arm/body length decreases in warmer climates, both for unknown reasons. The Atar population confirms this suggestion, and raises questions about the specific difference between *H. fennica* and *H. jenkinae*.

The mere discovery of the Atar population is, however, no sufficient evidence for synonymizing both taxa. Indeed, some 50 km SE. of Atar, a typical population of *H. fennica* was found in a saline rockpool at Agmeimime, and further records for typical animals are from Semara, Rio de Oro (Dumont & Coussement, 1976), from Dayat Srij, a shallow saline pond in the moroccon presahara (East of the Atlas chain), but, surprisingly, also from Lake Ifni (alt. 2400 m) in the high Atlas, the only genuine high-mountain lake of Morocco, and a typical freshwater as well.

2. *Filinia saltator* (Gosse): Fig. 9.

A little known species, reported only from South America and from the Antilles (Pourriot, 1975). Two populations were found in the Adrar of Mauretania:
The individuals were rather small. Koste (1978) gives 120-135 μm for body length and 255-320 μm for the length of the two appendages. Our specimens had a body length of 100-110 μm, and extremely short appendages, not longer than 165 μm. We do not, at present, consider this difference sufficient to warrant the creation of a new taxon.

3. *Pseudoploesoma greeni* Koste: Fig. 10-12.

Representatives of the genus Pseudoploesoma were to date known from North America and West Africa only, the african species having been found only once, in the River Sokoto, Nigeria (Green, 1960; Wulfert, 1965; Koste, 1978). It therefore came as a surprise to us to find a rich collection of specimens in the litoral of small, circular lake Iffer in the middle Atlas, Morocco, July 1971. Diagnostic characters are the structure of the asymmetrical mastax (Fig. 12), which is figured here in greater detail than previously, and further the presence of a cephalic palp and of a tubercle on the toes.

The occurrence of this rare species in Morocco confirms that faunal exchange between the Maghreb and West Africa can take place in rotifers, but also that suitable niches are available to it in an environment so different from the Sokoto area as the Middle Atlas.

Moroccon specimens are somewhat larger than West-African ones. Koste (1978) gives 216-290 μm for body length and 170 μm for height of the lorica. Our specimens have body length 260-360 μm and lorica height 170-240 μm.

4. *Keratella procurva* (Thorpe): Fig. 6.

A rare species, although apparently widespread in the tropics and subtropics of Asia, Australia, and Africa. In the latter continent, it was found on Madagascar, the Sudan Republic, and lake Victoria in Uganda (Berzins, 1955).

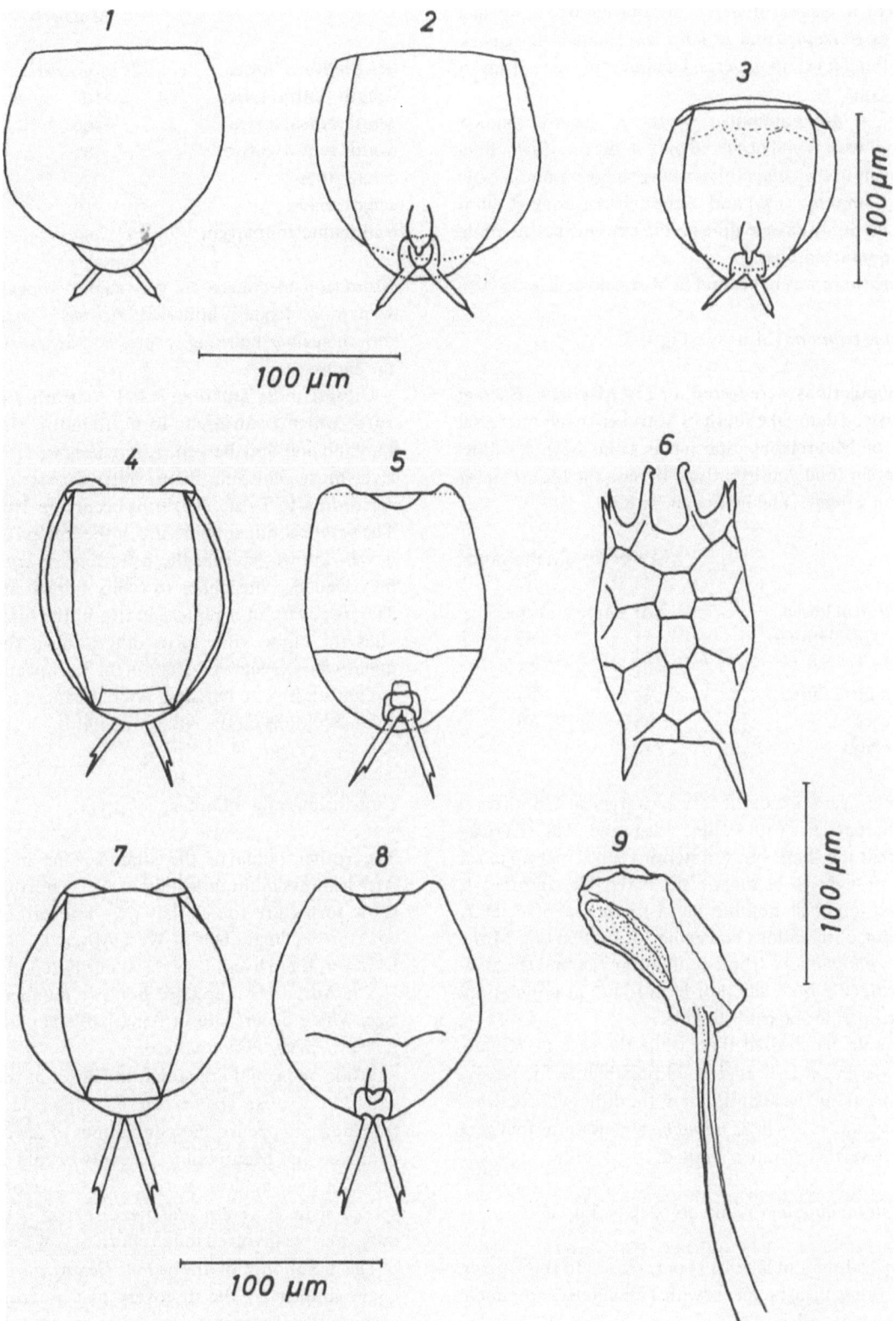

Where it occurs, it is usually outnumbered by such forms as *K. tropica* and *K. lenzi*. We found it, in remarkable abundance, in several localities of the Adrar of Mauretania.

In fact, it did outnumber by far *K. tropica* that co-occurred with it and occurred only in the f. *reducta*. Such a reversals in the competitive situation are common in the desert (Dumont, 1979) and, although not understood in detail, certainly have to do with the extreme nature of the local aquatic biotopes.

K. procurva was not found in Morocco or Rio de Oro.

5. *Lecane papuana* (Murray): Fig. 4-8.

Two populations were found, one in Morocco (Barrage Cavagnac, a dam-lake south of Marakech) and one in the Adrar of Mauretania. Specimens from both localities were rather (and consistently) different, the Mauretanian ones being bigger, and before all, wider.

	Morocco	Mauretania
length dorsal lorica	101 μm	115 μm
length ventral lorica	111	120
width dorsal lorica	88	95
width ventral lorica	85	88
length toes	46	46
length claws	10	9

The anterior margin of the lorica was typical in both cases and co-occurrence with *L. luna* was found. Russell (1956) states that the shape of the anterior ventral margin of the lorica, thought to be diagnostic, is flexible, and that *L. papuana* might be nothing but a fixation artifact of *L. luna*. Our observations contradict this statement. Moreover, *L. papuana* is a tropicopolitan species and the 'fixation artifact' has to date not been found in populations from temperate and cold climates.

It should be stressed that, while the case of *Pseudoploesoma greeni* indicated good possibilities for passive transport along the atlantic coast, the differences between the two populations of *L. papuana* tend to show that gene flow is nevertheless rather limited.

6. *Lecane* cf. *latissima* Yamamoto, 1955: Fig. 1-3.

Specimens found in lake Ifni come closest to this species. Comparative measurements with Yamamoto's specimens are given hereunder.

	Morocco	Japan
length dorsal lorica	99 μm	90 μm
length ventral lorica	108	103
width dorsal lorica	106	112
width ventral lorica	82	85
length toes	30	33
length claws	10	6
width anterior margin	99	84

Moroccon specimens are very slightly longer, and somewhat more slenderly built than Japanese. An irregular pattern of surface markings is present but was not drawn on the figures.

Closest to *L. latissima* stands *L. rotundata* Olofsson, 1918, which is an arctic form, found in Nova Zembla, Lapland and Spitsbergen. *L. latissima* was found near sea level on a peninsula in the extreme north of Honshu. According to Koste (1978), it occurs on Hokkaido too. The severe continental nature of the climate of these areas is well known, so that the *L. rotundata-latissima* group may well be considered to typify boreal environments. The discovery of a representative in the high Atlas now adds the alpine zone to its range, and it should consequently be expected to occur in the European alps as well.

Doubtlessly, *L. latissima* is a remnant of former extensive glaciations on the Atlas mountains.

Conclusions and Summary

The rotifer fauna of the atlantic zone of North-West Africa shows multiple affinities. As far north as Morocco some forms are found that were known only from the equatorial climate belt of West-Africa; in the same area, however, there lives also at least one boreo-alpine species.

The Adrar of Mauretania houses a number of rare species, which occur here in remarkable abundance: *Keratella procurva*, *Filinia saltator*.

While one could argue that the range of such forms as *Pseudoploesoma greeni* can be explained by passive transport, it appears that populations of *L. papuana* from Morocco and Mauretania can easily be told apart on evidence of their habitus. Yet, *L. papuana* is far more widespread than *P. greeni*. Passive dispersal is thus not the only variable involved in the chorology of rotifers.

The taxonomy of the genus *Hexarthra* does not become simpler by the discovery of a puzzling, possibly somewhat degenerated population in Atar, Mauretania.

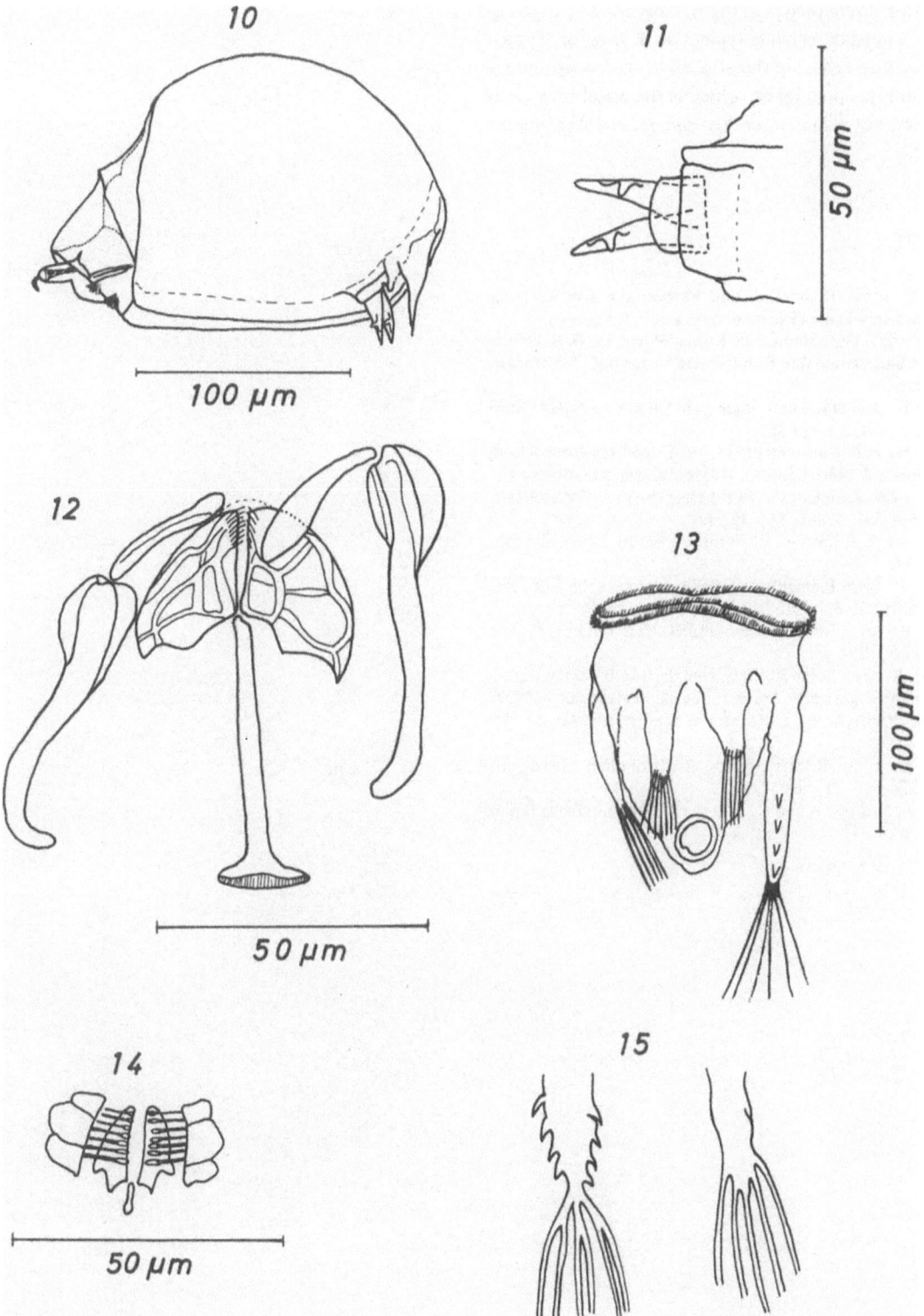

It shows all the characters of the *H. jenkinae-libica*-group, but with a mastax which is typical of *H. fennica*. Typical *H. fennica* also occurs in the area. In Morocco, specimens were found in saline lakes, which is the usual biotope of the species, but also in a weakly mineralized high-mountain lake.

References

Berzins, B. 1955. Taxonomie und Verbreitung von Keratella valga und verwandten Formen. Ark. Zool., 8: 549-559.

Dubief, J. 1971. Die Sahara, eine Klima-Wüste. In: H. Schiffers, ed., Die Sahara und ihre Randgebiete, 1: 227-348. Weltforum, München.

Dumont, H. J. 1979. Limnologie van Sahara en Sahel. State University of Gent. 557 pp.

Dumont, H. J. & Coussement, M. 1976. Rotifers from Rio de Oro (North-Western Sahara). Hydrobiologia, 51: 109-112.

Green, J. 1960. Zooplankton of the river Socoto. The Rotifera. Proc. zool. Soc. Lond., 144: 383-402.

Koste, W. 1978. Rotatoria. Borntraeger, Berlin. 2 vols: 673 pp., 234 plates.

Manfredi, P. 1939. Plancton delle acque interne della Tripolitania. Atti Soc. it. Sci. nat., Mus. civ. Milano, 78: 99-107.

Pourriot, R. 1975. Rotifères des Antilles. Cah. ORSTOM, Sér. Hydrobiol., 9: 81-90.

Russell, C. R. 1957. Some Rotifers from the South Pacific Islands and Northern Australia. Trans. r. Soc. N. Zealand, 84: 897-902.

Ruttner-Kolisko, A. 1972. Rotatoria. Binnengewässer, 26: 99-234.

Wulfert, K. 1965. Rädertiere aus Afrikanischen Gewassern. Limnologica, 3: 347-365.

Yamamoto, K. 1955. A new Rotifer (Order Ploima) from Japan. Annot. zool. Jap., 28: 33-34.

ON MORPHOLOGICAL VARIATION IN KERATELLA COCHLEARIS POPULATIONS FROM HOLSTEIN LAKES (NORTHERN GERMANY)

W. HOFMANN

Max-Planck-Institut für Limnologie, Abt. Allgemeine Limnologie, Plön, BRD

Keywords: *Keratella cochlearis* (Rotatoria), morphological variation, taxonomy, Lauterborn cycle

Abstract

Keratella cochlearis occurs in many Holstein lakes (northern Germany) as three well defined and separated forms: *'cochlearis'*, *'hispida'*, and *'tecta'*, each showing very little variation between the lakes. The present data show that the *'tecta'* form did not originate from a Lauterborn cycle.

Introduction

The morphological variation in the widely distributed rotifer *Keratella cochlearis* (Gosse) has resulted in extensive studies on its taxonomic and ecological implications. However, as Koste (1978) mentioned in his recent revision of the European rotifers, many problems still exist.

Very little is known about the *K. cochlearis* populations of northern Germany lakes. Voigt (1903, 1905) and Naber (1933) hinted that besides the typical *cochlearis* form, other forms like *hispida* and *tecta* may occur in Holstein lakes.

During summer 1975 a limnological investigation of 13 lakes in the vicinity of Plön was carried out. The wide range of ecological conditions, as for example mean depth, conductivity, nutrient content, and productivity, represented by the lakes under study led to an examination of the distribution of the various *K. cochlearis* forms, especially because in this species a close correlation between morphological characters and ecological conditions is generally assumed (Ruttner-Kolisko, 1972).

These lakes may roughly be divided into three groups with respect to mean depth (2 to 18 m) and mean secchi depth during summer (0.2 to 4.0 m) (Fig. 1). Likewise, there were tremendous differences in PO_4-P content of the upper water layer ranging from 0 to 953 μg P/l.

With the exception of a brackish water lake and a shallow water, *K. cochlearis* was found in high numbers in all of the lakes and was dominant among the rotifers (relative abundance > 10%) in ten of them.

Examining the samples, three forms could easily be distinguished regarding their morphological characters. They are called here *'cochlearis'*, *'tecta'*, and *'hispida'*, and refer to *K. cochlearis cochlearis* (Gosse), *K. cochlearis* var. *tecta* f. *typica* (Lauterborn), and *K. cochlearis* var. *hispida* f. *typica* (Lauterborn) (Koste, 1978).

Table 1 lists the lakes in the decreasing order of secchi depths: *hispida* was not found in the sites with maximum algal biomass, whereas *tecta* was absent in the samples from the least productive lake. In ten lakes all three forms were co-occurring and these populations were subjected to biometric analysis.

In Fig. 2 the mean values of lorica length and the length of the caudal spine of five syntopic populations are compared. In each case, the same relationship in body length between the three forms is seen, *cochlearis* being the smallest one, *hispida* the largest one and *tecta* in beween. As shown by the standard deviations, the differences between these forms are valid in most cases. As to the length of the caudal spine, *hispida* generally surpassed *cochlearis*. There was, however, great variation in this respect.

The impression of a constant pattern concerning the relations of dimensions of the forms from different lakes is confirmed when the individual values are considered and body length is plotted against spine length. In this way for each lake emerged an almost identical point cluster. This pattern is also discernable when the values from several lakes are plotted together (Fig. 3): The three forms are

Hydrobiologia 73, 255-258 (1980).   0018-8158/80/0733-0255$00.80.

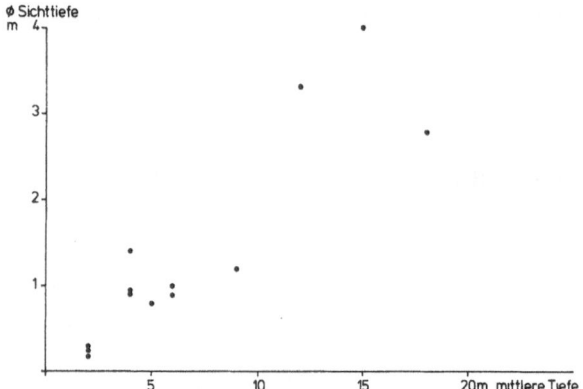

Fig. 1. Mean secchi depths of 13 Holstein lakes during summer 1975 plotted against mean depths.

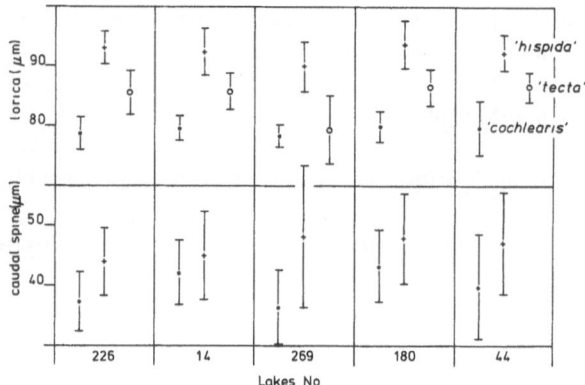

Fig. 2. Mean lengths of lorica and of caudal spine (and standard deviations) of syntopic 'populations' of *cochlearis*, *hispida* and *tecta* from five lakes.

each represented by a clearly separated point cluster.

As mentioned above on the basis of the mean values, body length increases from the *cochlearis* form over *tecta* to the *hispida* form. Moreover, the individual values show that there is hardly any overlapping between *cochlearis* and *hispida* even if populations from different lakes are compared. This pattern of size relations was not only found in the five lakes considered in Fig. 3 but in all lakes inhabited by these *K. cochlearis* forms.

However, in the first sample series (July) from the deep and less productive lakes (Fig. 1) an additional form was found which had no particular morphological characteristics but is clearly separated from the above mentioned *cochlearis* form by its larger size of both lorica and caudal spine (Fig. 4). So far it resembles *K. cochlearis* f. *macracantha* (Lauterborn). This form disappeared during summer.

Comparing the body lengths in *cochlearis* and *hispida* at the two sampling dates in July and October, it is obvious

that both forms increased in size in nearly the same proportion with decreasing water temperature. The point clusters remained in the same position relative to each other and were quite well separated in both cases. This separation is more distinct if syntopic forms from the same lake are compared than if population of different lakes are involved (Figs. 3, 4).

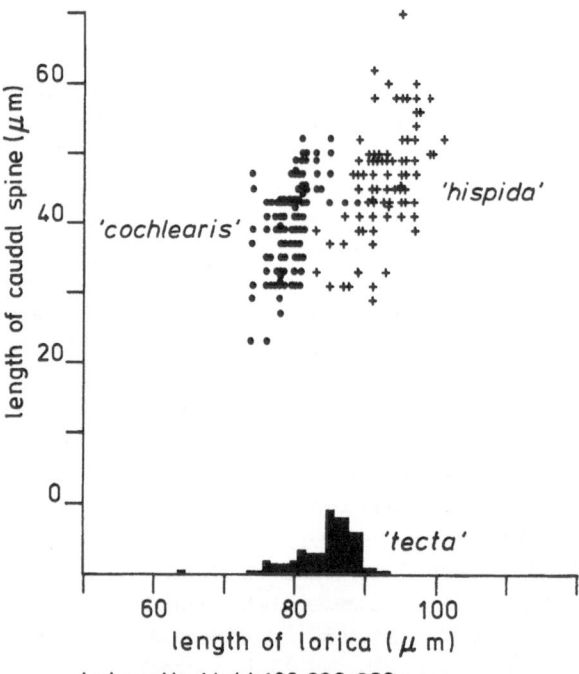

Fig. 3. Individual values of lorica length and spine length of syntopic 'populations' of *cochlearis*, *hispida*, and *tecta* from five lakes (same lakes as in Fig. 2).

Table 1. Occurrence of *Keratella cochlearis* forms in 11 Holstein lakes arranged in decreasing order of Secchi depth.

	'cochlearis'	'hispida'	'tecta'
Selenter See (250)	x	x	
Stocksee (255)	x	x	x
Schluensee (229)	x	x	x
Belauer See (14)	x	x	x
Tresdorfer See (269)	x	x	x
Dobersdorfer See (44)	x	x	x
Passader See (180)	x	x	x
Schierensee (226)	x	x	x
Stolper See (256)	x		x
Postsee (198)	x		x
Rottensee (210)	x		x

256

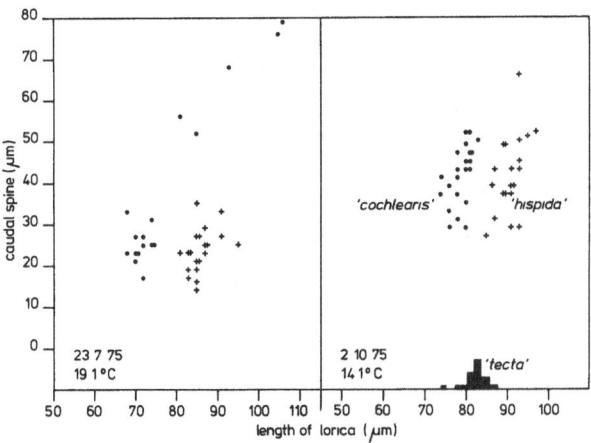

Fig. 4. Individual values of lorica length and spine length of syntopic 'populations' of *cochlearis, hispida,* and *tecta* in one lake in July and October.

Discussion

Recapitulating the situation, in the lakes in question distinct forms of *Keratella cochlearis* co-occur, which are clearly separated by morphological characteristics both qualitative and quantitative. If it were not *K. cochlearis* these results would lead to the assumption of three distinct species. However, Lauterborn (1900) found the above mentioned forms *cochlearis, hispida,* and *tecta* developing from one morphologically uniform type occurring through the winter. Up to now the taxonomy of *K. cochlearis* is based on such transitional series in the sense of Lauterborn (Koste, 1978).

In the case of *tecta* continuous transitions from long spined *macracantha* forms to the spineless *tecta* proved this view (Lauterborn, 1900; Pejler, 1962).

The variation in spine length was explained by allometric growth: When growth is accelerated this will result in small specimens with short spines or without spines. Conversely, retarded growth leads to large specimens with relative longer spines (Buchner *et al.,* 1957; Pejler, 1962; Lindström & Pejler, 1975).

On the other hand, cases are known where the spineless *tecta* form did obviously not derive from a typical *cochlearis* form (Ahlstrom, 1943; Pejler, 1957; Nauwerck, 1978). This holds also true for the *tecta* forms under discussion: Their descent from the syntopic *cochlearis* form is unlikely because (1) no transitional forms were found, because (2) they were significantly larger than *cochlearis,* and (3) because *tecta* forms were very abundant in late summer and autumn when spine length in *cochlearis*

was increasing (Fig. 4). Hence, the view mentioned by Pejler (1957) about polyphyletic origin of the spineless forms called *tecta* in the literature is supported. It seems as if there are forms which derive from *micracantha* forms and others which show no relation to spined *cochlearis* forms. Therefore, *tecta* is obviously no taxonomic unit.

It is interesting from the competition point of view that in the syntopic populations under discussion *tecta* was intermediate in size relative to *cochlearis* and *hispida,* which might show a mechanism of reducing niche overlap.

Unfortunately, no samples were taken in spring and autumn. So there is no information on the origin of the *hispida* form. Thus, it can be stated only that during summer *hispida* appeared as a distinct and well separated form, which was rather uniform when populations form different lakes were compared.

A discontinuous morphological variation in *cochlearis* forms, as shown in Fig. 4, was also found for instance by Pejler (1957, 1962) in Swedish lakes and by Nauwerck (1978) in Lake Ontario. Such discontinuities question the validity of the theory that the different *cochlearis* forms in general originate from Lauterborn cycles. But even if the theory holds, the taxonomic ranks of these forms remain doubtful: Lauterborn (1900), faced with the fact that the different forms may develop in the same biotope under identical ecological conditions, presumed that the homogenous phenotype of the winter form masks different genotypes. Hence, the crucial point is to decide if those genotypes are genetically isolated, e.g. if these forms are morphs of one polymorphic population or if they represent different species. This is known to be a puzzling problem in the case of *K. cochlearis,* because in lake populations bisexual periods are often absent. It is this reproductive isolation which could preserve discontinuities, as mentioned by Pejler (1957). This would imply that the splitting up into different forms would have evolved endemically in each of the lakes, which seems inconsistent with the homogeneity of the particular forms from different lakes.

Further studies are required, which in the sense of the biological species concept (Mayr, 1968; Pejler, 1977) concentrate on syntopic populations (or assemblages of forms) in order to look for the existence of Lauterborn cycles and their relation to bisexual periods in *K. cochlearis.* In addition, as Nauwerck (1978) claimed, experimental studies are needed in this respect.

References

Ahlstrom, E. H. 1943. A revision of the rotatorian genus Keratella with descriptions of three new species and five new varieties. Bull. am. Mus. nat. Hist., 80: 411-469.

Buchner, H., Mulzer, F. & Rauh, F. 1957. Untersuchungen über die Variabilität der Rädertiere. I. Problemstellung und vorläufige Mitteilung über die Ergebnisse. Biol. Zbl., 76: 289-315.

Koste, W. 1978. Rotatoria. 2 Vols. Borntraeger, Berlin. 637 pp., 234 Plates.

Lauterborn, R. 1900. Der Formenkreis von Anurea cochlearis. Ein Beitrag zur Variabilität bei Rotatorien. I. Morphologische Gliederung des Formenkreises. Verh. Nat. hist.-Medizin. Ver. Heidelberg N. F., 6: 412-448.

Lindström, K. & Pejler, B. 1975. Experimental studies on the seasonal variation of the rotifer Keratella cochlearis (Gosse). Hydrobiologia, 46: 191-197.

Mayr, E. 1967. Artbegriff und Evolution. Parey-Hamburg-Berlin.

Naber, H. 1933. Die Schichtung des Zooplanktons in holsteinischen Seen und der Einfluß des Zooplanktons auf den Sauerstoffgehalt der bewohnten Schichten. Arch. Hydrobiol. 25: 81-132.

Nauwerck, A. 1978. Notes on the planktonic rotifers of Lake Ontario. Arch. Hydrobiol., 84: 269-301.

Pejler, B. 1957. On variation and evolution in planktonic Rotatoria. Zool. Bidr. Uppsala, 32: 1-66.

Pejler, B. 1962. On the variation of the rotifer Keratella cochlearis (Gosse). Zool. Bidr. Uppsala, 35: 1-17.

Pejler, B. 1977. General problems on rotifer taxonomy and global distribution. Arch. Hydrobiol. Beih. 8: 212-220.

Ruttner-Kolisko, A. 1972. Rotatoria. Binnengewässer, 26: 99-234.

Voigt, M. 1903. Das Zooplankton des kleinen Uklei- und Plus-Sees bei Plön. Forsch. Ber. Biol. Stat. Plön, 10: 1-11.

Voigt, M. 1905. Die vertikale Verteilung des Planktons im Großen Plöner See und ihre Beziehungen zum Gasgehalt dieses Gewässers. Forsch. Ber. Biol. Stat. Plön, 12: 115-144.

NOTE ON SOME BRACHIONIDAE (ROTIFERS) FROM THE NETHERLANDS

P. LEENTVAAR

Research Institute for Nature Management, Leersum, The Netherlands

Keywords: Rotifers, biogeography, biological indication, thermal pollution

Abstract

Some Brachionidae indicative of polluted water are normally found in brackish but unpolluted waters in The Netherlands. The circumtropical *K. tropica* is now regularly recorded in the Hollands Diep, which is slightly thermally polluted, but it also seems to occur, at times, in waters that receive no heated water effluents.

For biological examination of surface waters in The Netherlands, a general plankton survey is made, at the occasion of which Rotifers are recorded among other animal groups. As it is not possible in the context of this work to make a special study of Rotifers, attention is paid to observations of rare species only or species which are found in water where, ecologically, they should not occur according to the literature. This aspect is of special importance to us as we use Rotifers in the assessment of water quality. Conclusions must be sound and not dependent on anomalies. An example is the occurrence of *Brachionus urceolaris* and *Brachionus calyciflorus* in unpolluted brackish water in our country, while according to the saprobic systems, these species are indicative of polluted water. This disaccordance of our observations with the saprobic system might be due to the fact that the saprobic system is deduced from fresh and running water-environments, while in our country most water is brackish and often stagnant. Probably, for these two species the brackish character is physiologically comparable to fresh water loaded with organic or inorganic matter, and therefore they cannot be used as indicator organisms for pollution in our waters.

In the genus *Keratella,* special attention was given to the occurrence of *Keratella valga,* as this species appears to be indicative of mesotrophic environments. It was found in several pools on the higher sandy soils which were meso-oligotrophic in character. In pools which are very acid-oligotrophic only *Keratella serrulata* was found, which is known as typical for this environment. Pejler (1977) states that *Keratella valga* is distributed in northern regions, which may be related to the lower temperatures in these regions. In a letter, I drew his attention to the fact that the distribution of *Keratella valga* is also governed by water quality. As in northern regions meso-oligotrophic water is common, *Keratella valga* can be expected to occur there more often than in regions where eutrophic water is dominant. I am eager to know if *Keratella valga* is also found in eutrophic waters of northern regions.

I paid special attention to *Keratella valga* in our samples as an indicator for meso-oligotrophic water. When examining in 1959 samples of the fresh-water tidal zone of the river Rhine (Biesbosch), I was struck by the occurrence in this eutrophic, polluted, tidal water of a form resembling *Keratella tropica,* in August and September (Leentvaar, 1961). Berzins verified the identification and supposed that the species had been brought in by ships. Recently, my assistant J. Sinkeldam collected samples in the adjacent region of the Hollands Diep in the period August-November 1977, and now this species was found in every sample (Table 1). Since 1970, when the Haringvlietdam was closed, the water in Biesbosch-Hollands Diep became nearly stagnant. This may have favoured *Keratella tropica,* e.g. by the changing of food conditions and the interruption of the water flow. On the other hand, the water is warmed up by the discharge of cooling water from power plants, and this too may have favoured *Keratella tropica,* as its range lies mainly in the southern subtropical and tropical regions (Berzins, 1956).

Temperature measurements in the Hollands Diep were carried out in 1977. Only at the point of discharge of cooling water a rise in temperature was found. This certainly is not enough to explain the regular occurrence of *Keratella tropica* here, but it is noteworthy that since 1956 the water of the lower Rhine probably warmed up slightly by discharge of cooling waters in Germany, as there is now no more ice in winter. However, in 1976 a specimen of *K.*

259

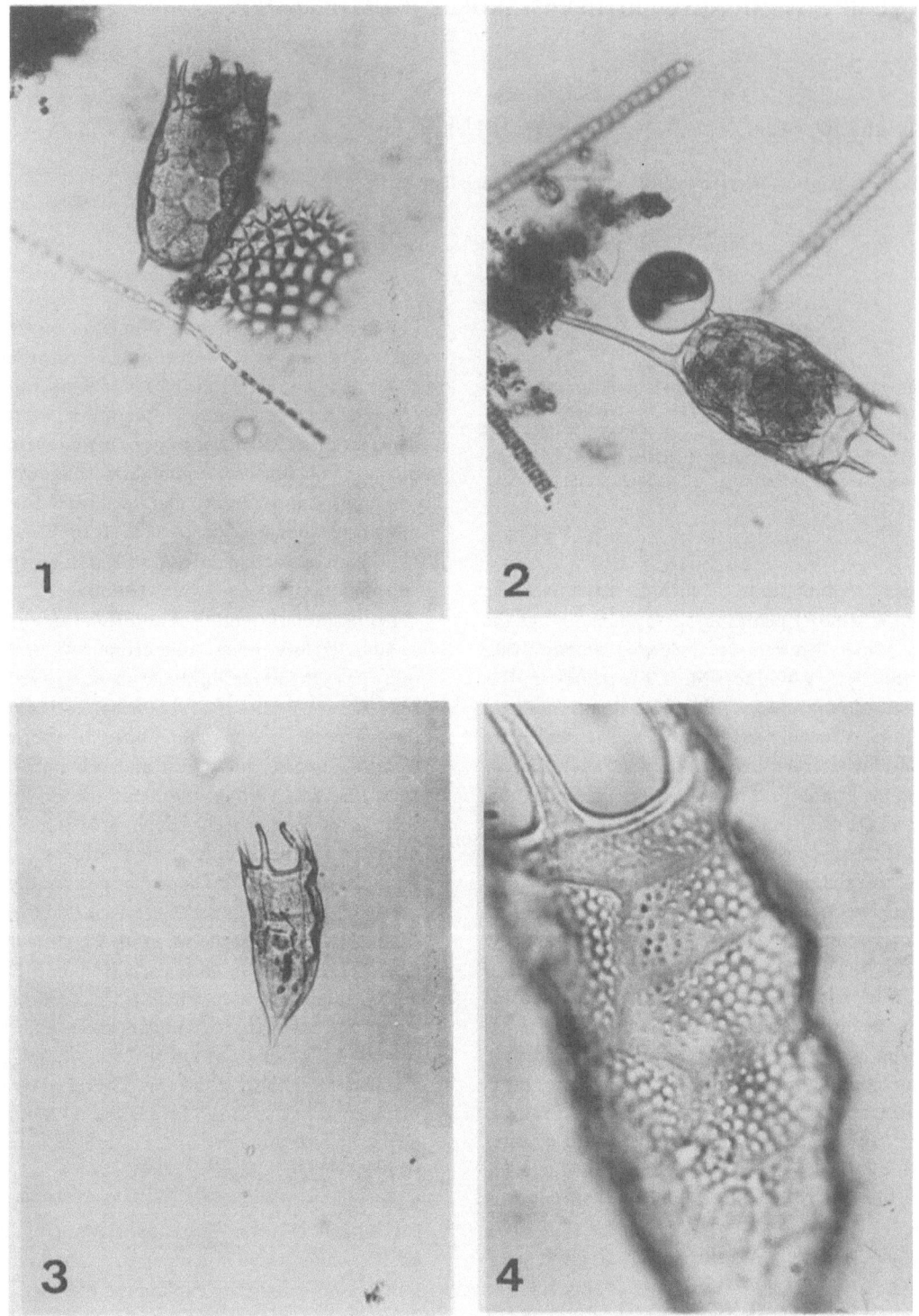

Fig. 1. Keratella tropica; magn. 240x; Biesbosch, 12-8-1971.   Fig. 2. Keratella tropica; magn. 240x; Hollands Diep, 1-8-1977.   Fig. 3. Keratella paludosa; magn. 230x; Ven, Zandenbos, Nunspeet, 24-6-1978.   Fig. 4. Keratella paludosa; magn. 1120x; Ven, Zandenbos, Nunspeet, 24-6-1978.

Table 1. Zooplankton from Shell-Moerdijk-Hollands Diep 1977.

Method / Sampling date	H 1–8	V 26–9	H 26–9	V 24–10	H 24–10	V 23–11	H 23–11
Crustacea:							
Daphnia longispina	1			1			1
Bosmina longirostris	1	1	1	1	1	1	1
Cyclops sp.	1					1	1
naupii:	2	2	2	1	1	1	1
Alona rectangula						1	
Rotatoria:							
Keratella cochlearis	1	2	2	2	2	1	1
K. cochl. f. tecta	2	1	1	1	1	2	1
K. tropica	2	1	1	1	1	1	
K. quadrata		1	1	2	2		1
Brachionus angularis	1	1	1	1	1	1	
B. calyciflorus	2			1	1	1	
B. calyc. f. amphiceros	2	1	1	2	1		
B. urceolaris					1	1	
Polyarthra cf. vulgaris	2		1	1	1	1	1
Trichocerca pusilla	1						
T. sp.		1			1		
Filinia terminalis		1	1				1
F. brachiata			1				
Anuraeopsis fissa	1						
Asplanchna sp.				1			
Notholca acuminata						1	
Lecane sp.				1		1	
Kellicottia longispina							1
Protozoa:							
Tintinnopsis lacustris	1	1	1	1	1	1	1
Tintinnidium sp.	1	1	1	2	2	1	1
Vorticella sp.	1		1	1	1	1	1
Arcella sp.	1					1	1

H = horizontal haule
V = vertical haule

1 = few specimens
2 = also few, but more regularly present
3 = moderate numbers
4 = many specimens
5 = numerous

tropica was found in a sample from the Reeuwijkse Plassen, which is a broad in the midwest of Holland not warmed up by discharges, while it has also been observed in 1974 in Lake Donk, Belgium (Coussement, 1977), and a strong pulse of the species occurred in the Watersportbaan, Ghent (Belgium) in 1974 (Coussement, pers. comm.). So, it still remains an open question whether *Keratella tropica* has widened its range to the north. Perhaps nobody in the past took notice of the presence of this species in The Nether-lands. This may also be the case with the first record of *Keratella paludosa* in our country. The sample was collected in an oligotrophic pool on sandy soil in the centre of The Netherlands, the Zandenbos near Nunspeet, on 24 June 1978, and analyzed by T. Dresscher.

References

Berzins, B. 1956. Taxonomie und Verbreitung von Keratella valga und verwandten Formen. Ark. Zool. 8: 549-559.

Coussement, M. 1977. Nieuwe gegevens omtrent de Rotatoria-fauna van het Donkmeer in Oost-Vlaanderen. Natuurwet. Tijdschr. 58: 138-146.

Leentvaar, P. 1961. Quelques rotateurs rares observés en Hollande. Hydrobiologia 18: 245-251.

Pejler, B. 1977. On the global distribution of the family Brachionidae (Rotatoria). Arch. Hydrobiol. suppl. 53: 255-306.

INDEX AUCTORUM

CHARRS
Salmonid Fishes of the Genus Salvelinus
edited by Eugene K. Balon

Volume 1 of the new book series: PERSPECTIVES IN VERTEBRATE SCIENCE

CHARRS: SALMONID FISHES OF THE GENUS *SALVELINUS* is an anthology of reviews and original studies by the world's leading experts of current knowledge about charrs. The evolution of charrs in many aspects parallels the evolution of man. The charrs evolved from more specialized, planktonophagous ancestors of a warm climate, into paedomorphic, less specialized, eternal juveniles, as did man from their pithecine ancestors into neotenic hominids. The neotenic charrs were able to invade the most inhospitable habitats; so were neotenic men.

Salmonid fishes are the subject of more published work than any other fishes. This is influenced by the fact that salmonids share their environment with a group of hominids that up to now have produced most of the world literature. It is certainly not a reflection of the food or sport value of salmonids for the whole of mankind. But the significance of charrs, as of many other organisms, may be in more than just their usefulness to man. This book attempts to demonstrate this at least in part.

CONTENTS

1980. IX + 919 pp., 353 figs., 126 tables and 16 color plates. Cloth. ISBN 90 6193 701 9
Dutch Guilders 400,--/US $ 210.60

DR. W. JUNK BV PUBLISHERS, P.O.BOX 13713, 2501 ES THE HAGUE, THE NETHERLANDS